P9-CRK-014

Rational Rhetoric

Rational Rhetoric

The Role of Science in Popular Discourse

David J. Tietge

Parlor Press
West Lafayette, Indiana
www.parlorpress.com

Parlor Press LLC, West Lafayette, Indiana 47906

Printed in the United States of America

S A N: 2 5 4 - 8 8 7 9

Library of Congress Cataloging-in-Publication Data

Tietge, David J., 1966-

Rational rhetoric : the role of science in popular discourse / David J. Tietge.

p. cm.

Includes bibliographical references and index.

ISBN 978-1-60235-069-4 (pbk. : alk. paper) -- ISBN 978-1-60235-070-0 (hardcover : alk. paper) -- ISBN 978-1-60235-071-7 (adobe ebook)

1. Science--Social aspects. 2. Science--Philosophy. I. Title.

Q175.5.T547 2008

306.4'5--dc22

2008024748

Cover design by David Blakesley.
Printed on acid-free paper.

Parlor Press, LLC is an independent publisher of scholarly and trade titles in print and multimedia formats. This book is available in paper, cloth and Adobe eBook formats from Parlor Press on the World Wide Web at http://www.parlorpress.com or through online and brick-and-mortar bookstores. For submission information or to find out about Parlor Press publications, write to Parlor Press, 816 Robinson St., West Lafayette, Indiana, 47906, or e-mail editor@parlorpress.com.

Contents

Foreword and a Note on Methodology

While the body of work available in the area of rhetoric of science is fairly expansive, few outside specialized academic programs in rhetoric have any knowledge of it. A peer of mine declared in a review once that "there is a lot out there on the rhetoric of science." When I tell this to people outside of the sub-field of rhetoric of science, even to career academicians, they frequently reply, "what *is* the 'rhetoric of science'? I never thought that these two words could go together." The paradox of a large body of scholarship that is virtually unknown to anyone not interested in this area of research is that, while there may in fact be "a lot out there," no one except experts in the field is reading it, and this leads to a self-referentiality and academic inbreeding that guarantees the formation of barriers for anyone not thoroughly steeped in the restricted discourse of this specialized academic community. However, this is not unusual in the world of academe. One would not expect an accountant to read about neurobiology any more than one would expect a chemist to read a literary analysis of *Mrs. Dalloway.* Yet, there is a growing need for people of all economic and educational levels to be informed about science, and just as importantly, about the language that science uses to achieve its information. In my earlier book, *Flash Effect,* I admittedly underplayed the depth of research available on the topic of the rhetorical use of scientific ideas because it was an interdisciplinary work that I wanted to make accessible to a broader range of people interested in history, rhetoric, cultural studies, and political discourse; had I chosen the traditional scholarly route for that book, it would have made it unreadable in any practical way for even the most well-rounded layperson, and it is the intent of this current project to make the material accessible not only to scholars in other fields, but also above all to the average reader.

One of the working premises in this text is that scientific ideas have *not* been made accessible in any truly critical way to the average person whose interest in science should be encouraged. By "truly critical" I mean in a way that instructs readers in the skills of humanistic "critical thinking," while not merely "dumbing down" (a favorite and somewhat tiresome phrase of the conservative scientific elite) scientific material in the process. It is from the vantage point of rhetorical analysis that we can draw a basic conclusion about academic publishing: Far too many aloof academics are preoccupied with guarding their own sphere of interest at the expense of sharing knowledge with the rest of the reading and thinking public, and to lament that the reading and thinking public no longer exists only further emphasizes this point. If we are honest with ourselves, then we might ask how complicit we academics are in keeping the public ignorant. The success of the *For Dummies* and *Idiot's Guide* book series shows how "experts" often patronize the very people who should be thinking about the implications of ideas central to modern life (although *Fondue for Dummies* could probably be sacrificed without a corresponding blow to civilization), a process that not only alienates the reading public and helps keep them shielded from matters that impact them directly, but also undermines the very democratic process that we assume is part of a broad humanistic and liberal arts education. Without education (not the *indoctrination* that seems to dominate public and private education, or, worse, mere job training designed to keep people busy with the technology but not with the more difficult—and dangerous—task of assessing what they do in the workplace) there is nothing on which to base a truly enlightened understanding of scientific enterprises. Such enterprises, in fact, cannot be enlightened without the commencement of public interest and knowledge. Otherwise, we blow limply with whatever direction experts and "authorities" tell us, and it only further compounds the confusion of the general public when they see that the arguments made by the experts on which they rely so heavily are as diverse as the fields they represent. Some may like the idea that the American public seems easily swindled, that they can be manipulated with little or no real evidence or suasive acumen, but I not only firmly believe that this is a caricature of the American "Everyman" (and woman), but also feel that if it *is* true, it is another symptom that our great civilization is in its death throes. If this is true, humanism truly is the hardest faith to keep.

Education, so long held as the panacea for all social ills and as the scapegoat for all social failures, is in dire straights indeed; we invoke its power as a word while rarely cultivating its real worth and potential as a common condition for every person. It is up to the academic, more than anyone else, to lead the way. School boards have failed, as have administrations on all levels, from the local to the federal. Why? Because those who are often in the position to make the important and difficult political decisions regarding education are too often influenced by people who have little experience in, knowledge about, or genuine concern for, the state of education. (Teachers, by the way, fail only when they are set up to fail by their bosses, who are often lame duck administrators performing the will of *their* superiors. Those at the highest levels of educational administration are very frequently nothing more than political automatons, and they give far too much credence to the reactionary whims of their constituency. It's time we stopped blaming teachers for our own shortcomings.) Academics, as long as the tenure system remains intact (which we should in no way assume) have the happy luxury of "Academic Freedom," which, at least ideally, means they can speak honestly and candidly without fear of reprisal. For those who would deny us this last bastion of intellectual liberty—those who think tenure is a haven for the lazy, the kooky, and the ideologically subversive—please consider the alternative, a world with no dissenting voice and a populous beaten into submissive silence, unaware of its own state and unequipped to foster change. This is not the image most of us have of a functional and vibrant democracy. We will no longer have any means for exploring new visions or producing a more enlightened public because universities will have adopted a business model that allows for the termination of its employees without substantial cause or justification—in many cases, for simply discussing an unpopular idea. We will produce, hire, and reward professors who toe the party line, mere puppets of the status quo with the same intellectual relevance to the advancement of our knowledge and learning as the twenty-dollar-an-hour unionized assembly line worker. The specializations that Ivory Tower academics covet so carefully as a venue for bantering among themselves the minutiae of sub-sub-sub-specialties while watching the rest of society crumble from educational decay will certainly no longer be needed nor desired. We ignore the common person—the one who possesses an innate intellectual curiosity and a natural human intelligence but no means (nor incentive) to express it—at our peril. The real point here is simply this: This book strives to reach the American (and anyone else, but Americans especially, since we have "fallen behind" the "rest of the world" in so many areas, another specious cry for economic strength through occupational training) who has a healthy interest in the world around her, who wants to know what is happening in the hallowed halls of the scientific community, who wants to be able to decipher the scientific code that has such a huge bearing on her life. This book is for the common reader, the one with the purely human sense of inquiry and curiosity. This book is for the reader who has had his natural inquisitiveness stripped from him by years of misguided education, by the tedious and overwhelming demands of modern life on his intellectual energy, by the white noise that distracts his attention away from ideas of consequence and meaning, by a reward system that encourages a shallow solipsism over philosophical contemplation. This book is for the garden-variety thinking human being.

For the record, then, I wish to assure my peers and the reader that I am neither ignorant of nor bound by the literature that has been made available within the academy on topics similar to the one dealt with here, for while such a corpus of material is essential reading for those interested in the rhetoric of science as a professional subdivision of rhetoric studies, as a whole it does not accomplish what I hope to accomplish. Many academics dismiss interdisciplinary work because they feel that no one person can master all the fields necessary to say anything meaningful about a subject that has many different topical perspectives, and many different intellectual access points, to illuminate. Such objections cloak a disciplinary insecurity and are, I think, ultimately anti-intellectual. The world, contrary to positivistic wishful thinking, is not partitioned into neat and discrete categories for our convenient cataloging and recall. It is much more random, much more untidy than all that, and this apparent chaos should not be shunned or feared; we should acknowledge it willingly, as we so often do when we invoke our natural skills with symbolism. A symbol, after all, is a shorthand, a way of packaging the messiness and arbitrarily associative properties of experience that routinely govern our thinking and acting. That we think and communicate in symbols naturally should be a clue to us that this capacity is one of our best access points into understanding the world and the people, animals, objects, and ideas within it. This study is not the purview of any one discipline or any myriad disciplines acting independently; it is, rather, a humanistic enterprise, and one that should draw on everything it can to make sense of ourselves and our environment while at the same time realizing that no such study can or should attempt to be comprehensive.

I am not aware of a book that attempts to describe and explain scientific rhetorical discourse holistically (that is, as an entire mode of discourse, not just as a specialized vocabulary in a specialized field) nor one that further attempts to straddle the arenas of public consciousness and academic intellectualism in a way that will be broad, and therefore, helpful to the reader wishing to interpret scientific discourse for his or her own educational and informational purposes. There is *much* fine work done in the academic sub-specialty of rhetoric of science, but nearly all of it pin-points specific loci and provides a close reading of a single discursive phenomenon, material that is often only of interest to specialists in the fields of rhetoric or communication studies. There are exceptions, but they are rare and almost always tightly focused on a single discursive moment. I see these works as further reading for those who would like a more detailed account of specific areas of scientific discourse, and I encourage anyone interested in the rhetoric of science to refer to the bibliography, where she will find a vast array of perspectives on this growing and important field. As for this book, it was written with the realization that little has been done to provide an accessible, broadly-based and *practical* discussion of the history, the spokespersons, and a method of analysis for how to decode the scientific language that defines and directs a very large division of our ideological make-up. This book, I hope, helps fill a void.

As I have implied already, an unfortunate trend in academe (and in American society generally) is the tendency to overspecialize. Gone, it seems, are the humanistic studies that, while based in a scholarly specialization, also demonstrated a broad base of knowledge outside of a "mother" discipline so that many people could come away with an increased understanding of the issue under study. The trend to be bound inexorably to a discipline is part of the economics of our culture, an economics with which science is intimately tied. As technology does for us what we once had to do for ourselves, financial survival depends on one's ability to avoid expendability and redundancy. "Labor" has changed in a way that necessitates knowledge of the products of technology and the bureaucracies that manage those technologies. It is the so-called "information age," and one irony of our postmodern condition is that our information is generated and dispersed at the expense of a broader wisdom required to make it meaningful. In other words, we amass "facts" at the expense of "knowledge," and since the sheer bulk of facts available make managing it all impossible, we must rely on people whose expertise is in sifting and sorting minute portions of information. This situation sees an analogy in humanistic disciplines that tend to be saturated with scholars who feel the pressures from the schools that employ them to get published in areas that are becoming increasingly inaccessible to anyone outside of their own narrow areas of interest and expertise. More broadly, our universities cater to the student/consumer and generate curricula designed with a professional goal solely in mind, and we increasingly see attacks against the so-called "general education" or "core" curriculum because such courses of study are, sadly, viewed as superfluous. On the other hand, the professors who teach in these institutions are often asked to teach almost exclusively general education courses, with the opportunity to only occasionally teach a course in the professor's area of expertise, a situation which further widens the gap between the "expert" and the "layperson." In other words, we demand highly specialized professors while expecting them to teach only the most general of courses. We produce diploma mills and educations of convenience, as if picking up a degree should be as effortless and accommodating a process as picking up one's dry-cleaning, but we are so preoccupied with credentialing that we overlook the need for high-quality, general educators. Put another way, students aren't the only ones asking, "when in 'life' will I ever 'use' this?" and the question reflects a woefully limited perspective, one that may be the direct result of the shift in attitude from "education" to "training." It equates "life" with "professional identity," and it assumes that the only goal of an education is to forge this one-dimensional aspect of one's distinctiveness.

When I expressed a similar concern with the "what will I do with an English degree?" question early on in my collegiate career, my father, a very wise man in many respects, responded, "Get an education first; then you'll have what you need to decide what to do with your life." I was fortunate to be the recipient of such sage advice, but most other young people embarking on a college journey are not so well-guided. As a humanist, I frequently see the need for students to become generally educated—learning I would inadequately define as an "intellectual sampling" which, if it does nothing else, creates a habit of thinking that allows one to see the intersections between many disciplines and to hold in one's mind contrasting, at times even contradictory, ideas without becoming neurotic or, worse, apathetic. Without this ability, I fear future generations may have an especially rough time reconciling some of the problems they will inevitably inherit and even more problems that, through sanctioned ignorance and shortsightedness, they will create for themselves and for future generations.

In the area of rhetoric studies, however, there seems to be a push to resist the educational "means to an end" movement that I have just described. And this is fortunate, because it means that we in rhetoric studies have a certain measure of liberty to pursue topics with a scope that can incorporate the advantages of a number of disciplines. To be responsible for everything ever published that touches an exploration such as the one contained within these pages is not only impossible; it's undesirable. The method, then, is a combination of close reading, historical analysis, textual criticism, hermeneutic interpretation, and cultural cross-referencing. It is *not* "objective" in the usual sense of that term, nor will it attempt to impose the quasi-scientific methods often employed by sociology because, as I'm sure many hard scientists would agree, such methods are inappropriate and inadequate when studying the ambiguous disarray of language and symbol systems as people tend to use them. It will not invoke "quantitative" or even "qualitative" data, because I will not pretend that by marching dutifully through such procedures alone that I have accomplished anything meaningful or established anything conclusive. In fact, this study is in many ways designed to challenge the notion that science and its self-professed rigor can be called upon to solve all problems of the "human condition," a pattern of research that may be as much the cause of our most controversial issues as it is the solution because it sees science as the only valid epistemology. I am thinking here particularly of psychotherapy, sociology, and public education, areas that have adopted an ostensibly scientific methodology much to the detriment of the human "subjects" in their charge. I realize, too, that by making such a claim, I will have alienated a certain segment of my potential reading audience, but to such readers I can only ask that they please read on: It is not my intention to oversimplify the very complex issues debated in scientific circles and reduce them to some rudimentary straw man. I do, however, point out many instances where this is done for public comprehension, often by scientists themselves. I would echo a sentiment that Stephen

Jay Gould once articulated when he was writing about something that he anticipated would be seen as beyond his professional "authority": "Broad generalizations always include exception and nuanced regions of 'however' at their borders—without invalidating, or even injuring, the cogency of the major point" (*Rocks of Ages* 56). If the reader will indulge my major point, he or she might see that it is the nature of human inquiry that gives us a meaningful starting point for addressing particulars and solving the problems that each of us, individually, find most important.

Rational Rhetoric

Introduction: A Case For Rhetorical Studies

Since science has a well-documented scholarly history as a communicative, discursive phenomenon, this book's primary task is to discuss how and for what purposes science communicates to a large general audience. The range of possible applications for rhetorical studies, even within just the specialized sub-field of rhetoric of science, has been impressive. Such applications have been especially useful in decoding the discourse of popular culture, providing a vocabulary for understanding complex social phenomena like the proliferation of pseudo-science and the appeal of technological modes of discursive delivery. James A. Berlin, in his landmark book, *Rhetoric and Reality,* identified three main areas of discourse (or "rhetorics") that he viewed as the main contributors to the ways in which we ideologically frame the world (Berlin is talking about education, specifically writing instruction, but his observations are pertinent on a much broader scale): cognitive, expressionist, and social-epistemic.[1] Of the three, epistemic rhetoric is for Berlin the most serviceable, because of what it reveals about the nature of language and knowledge:

> From the epistemic perspective, knowledge is not a static entity located in the external world, or in subjective states, or even in correspondence between external and internal structures. Knowledge is dialectical, the result of a relationship involving the interaction of opposing elements. These elements in turn are the very ones that make up the communication process: interlocutor, audience, reality, language. The way they interact to constitute knowledge is not a matter of preexistent relationships waiting to be discovered. The way they interact with each other in forming knowledge

> emerges instead in acts of communication. Communication is at the center of epistemic rhetoric because knowledge is always knowledge for someone standing in relation to others in a linguistically circumscribed situation. [. . .] Language forms our conceptions of our selves, our audiences, and the very reality in which we exist. (166)

Berlin favored epistemic rhetoric over cognitive and expressionist ways of knowing because he felt it most accurately modeled the way that we actually gain knowledge, how knowledge is not a static and stable "thing" waiting to be uncovered, but rather is formed at the intersection between the communicative elements that affect us all. He was a proponent of the "cultural formation" school of knowledge, which held that the impact of cultural forces, interchanges, and structural mechanisms had as much to do with how our reality is shaped as any accumulation of "facts." He saw the study of rhetoric as a study of our reality because it is through the examination of how language is really used that we can more fully understand what we think and why we think it. Beyond that, an understanding of our communicative realities also lends us insight into understanding *why we act* based on our understanding of the reality that we have constructed.

This is not nearly as radical as it seems. It is only through the Western positivist tradition (and bias) that we would view the idea of multiple realities with suspicion. Yet we give lip service to its principles all the time. When we embrace tolerance for other cultures or try to understand differing belief systems, we are acknowledging that reality for another group may be significantly different from reality as we see it. Yet, the fact that, in the United States, any discussion of the "real world" involves a capitulation to the economic forces that are the staple of our political and social system should tell us that our understanding of "reality" is thin indeed. We are in need of a little rhetorical analysis, if only to expand our notion of what is "real." Rhetoric, as a heuristic for describing different realities, touches everything human, which for scholars like Berlin means that it touches everything that is subject to the domain of language (including, significantly, rhetoric itself). That encompasses a lot. Berlin puts it in perspective when he says, "in studying the way people communicate—rhetoric—we are studying the ways in which language is involved in shaping all the features of our experience. The study of rhetoric is necessary, then, in order that we may intentionally direct this process rather than be unconsciously controlled by it" (166).

Berlin's statement is one of the best rationales for the rhetorical study of communicative phenomena available, but it is not without precedent. The importance of recapturing the centrality of rhetoric as a critical mechanism goes back to the 1930's and 40's when I.A. Richards and Kenneth Burke begin to argue for a new articulation of the rhetorical project; the notion that an understanding of the richness of language is an important part of any general education goes back at least as far as 1958, when Harold Martin published his seminal statement on the goals of general education at Harvard, long viewed as the flagship institution for new pedagogical approaches. He argued, quite convincingly from today's vantage point, that first-year writing study must be "concerned with language not simply as a medium by which a transmission of knowledge takes place, but as a phenomenon of particular interest itself" (87). What is perhaps more radical is the implication that language, as a focus of study in and of itself had, up until this point, been more or less neglected as an important component of college-level core courses. By 1970, Young, Becker, and Pike were making similar statements about the ideological function of rhetoric in their book, *Rhetoric: Discovery and Change,* suggesting that a study of the nature of communication can not only reveal the bases of conflicting ideologies, but also offer remedies through constructing a rhetorical system that acknowledges "that the 'truths' we live by are tentative and subject to change, that we must be discoverers of new truths as well as preservers and transmitters of old, and that enlightened cooperation is the preeminent ethical goal of communication" (9). By the 1980s, the renaissance of rhetoric studies is in full bloom, taking cues from rhetoricians like Kenneth Burke, Chaim Perelman, and Wayne C. Booth.

By 1987, rhetorician Walter H. Beale was making the argument in his *A Pragmatic Theory of Rhetoric* that a study of rhetoric was the study of symbolic intersections, that all aspects of daily life that had at their root a linguistic mechanism driving our actions: "All human making," he said, "all human systems, all human institutions involve convergences, which we are hard pressed to understand, of mind and matter, intelligence and action, knowing and doing—*understandings of reality* and *participation of reality*" (55). He further notes, in the next sentence, that the mixture of institutional convergence is not only necessary but also fundamental to grasping the significance of our utterances in relation to our actions. "Without this mixing," Beale notes, "intelligence on the one hand and action on the other are lifeless and meaningless" (55). Here, finally, is acknowledgement that the simple dichotomy often forwarded by literalist America that "there are talkers, and there are doers" was a naïve capitulation to some very basic, and erroneous, assumptions about the nature of words. It is also one of the earlier attempts to recapture the philosophical essence of rhetorical studies, to note that any examination of language is at its base a philosophical question. Rhetoric was beginning to make philosophy practical.

Because of thinkers like these, suddenly there was an explosion of interest in the pragmatic study of rhetorical operations and the philosophical (and, in particular, the epistemological) implications of language in education, in institutions, in society, and in life. While rhetoric had been around for a long time (going back at least as far as Aristotle's attempt to systematize a primer for rhetorical instruction), it had always been limited by the narrow applications for which it seemed especially suited. Because of scholars like I.A. Richards and Kenneth Burke, the rhetorical canon had been both challenged and added to—Plato, Aristotle, Cicero, and Quintilian were no longer the last word on rhetoric, but the first word in a whole new conversation about the ubiquity of rhetorical performances. With the publication of Chaim Perelman's *The New Rhetoric,* serious study in the field was validated. Even scholars like George A. Kennedy, long seen as the definitive translator of Aristotle's *On Rhetoric,* for example, pushed the limits of rhetorical relevance to include such recent contributions as *A New History of Classical Rhetoric* and *Comparative Rhetoric: A Historical and Cross-Cultural Introduction.* By 1990, Richard Cherwitz had compiled a reader called simply, *Rhetoric and Philosophy,* containing work that, in his words, "explores alternative ways in which the attempt has been made to find a philosophical grounding for rhetoric" (xv). Likewise, a little-known Italian Vico scholar named Ernesto Grassi was attempting to find that same philosophical grounding in a humanist tradition that was already hundreds of years old in his 1980 book, *Rhetoric as Philosophy: The Humanist Tradition,* a book that predated Cherwitz but was not made available to the American scholar until the mid-1990s. Many works about rhetoric that had festered in obscurity were beginning to come to the fore, and new sub-specialties of rhetoric were being formed. Suddenly, the Rhetoric of Science had its own cadre of scholars, organizations, and journals featuring names like Trevor Melia, Alan Gross, John Angus Campbell, John Lyne, and Lawrence Prelli.

As one example of the range of inquiry that can be pointed to to illustrate this bulging sub-field, Todd S. Frobish demonstrates an important overlap between technology and the acceptance of fringe religions that, using traditional recruiting methods, have experienced little widespread support but, through the use of technological interfaces like the World Wide Web, have gained significantly in popularity. Such a topic demonstrates the pliability of rhetoric for a large array of issues—issues that display the overlap between such seemingly disparate and otherwise sequestered areas of inquiry such as religion, pseudo-science, speech communication, and technology. Beyond rhetoric, there exist virtually no other academic fields that are equipped to deal discursively and critically with an issue that touches so many different spheres of human communication so comprehensively. In his 2000 *American Communication Journal* article entitled "Alter Rhetoric and Online Performance: Scientology, Ethos, and the World Wide Web," Frobish asks a two-prong question that only rhetoric can help answer: How can religions with credibility problems in other, more traditional venues create a degree of credibility online that it lacks offline; and in what way does the World Wide Web function as an ethical strategy for these religions (1). He concludes, through rhetorical analysis of religious and technologically-based ethos and an examination of the history of Scientology, "that religious organizations can use the WWW to establish personas in ways that are much more effective online than offline" (1).

I mention this article as a representative example of rhetoric's versatility and to show that there are many interesting social and cultural phenomena that would go academically unexamined if not for the methods that rhetoric is able to bring to bear on them. The current social situation also speaks to the chasm that separates formal academic study and so-called "popular culture." Increasingly, however, the need to analyze popular and public discourse is coming to the forefront as these myriad discursive communities reflect issues that require a more sophisticated understanding than they have been afforded in the past. Many academics decry the decline in educational standards, but as they point the finger at popular media, crumbling public education, and the corporate entertainment industry that takes away from more lofty intellectual pursuits, the three fingers pointing back at them represent an inability to adjust to the educational needs of a new generation of students, a willingness to sacrifice their own standards because they assume that students in their very own classes are not capable of dealing with more ostensibly refined material, and an impulse to endorse a cultivated division between academia and the overall population. It is little wonder that the public has lost touch with what happens in our nation's universities; as universities continue to recast themselves in a homogenous corporate image, public relations becomes a higher priority than public information. The result of the public relations emphasis in higher ed is a stylistically packaged image of a school that has little bearing on the reality of the learning it offers, an administrative orientation that could itself benefit from rhetorical analysis. College advertising is deemed a necessary component when the bottom line rules and requires an increased commitment to recruit the most and, it is hoped, the best students and faculty. The PBS special documentary, "Declining By Degrees," (based on the book edited by Richard H. Hersh and John Merrow) does a good job of illustrating this corporate "reality" by showing that the preoccupation of college presidents—even whole administrative departments—is to put a positive spin on a school to make it an attractive option for prospective students. Telemarketing, alumni fund raising, public relations directed by specialists, student "satisfaction questionnaires," and "selling credits" all illustrate that college is a commodity, a monetary investment, and a pathway to economic advancement, not an opportunity for an idealistic education of personal fulfillment and enhancement nor even, for that matter, to educate generalists to participate intelligently in an increasingly complex society.

This situation reflects a failure to inform the general public in an honest way. This, in turn, further dilutes the understanding of what an education is all about. The situation demands a readjustment in our educational objectives—a forthright statement of purpose that has less to do with corporate competition and more to do with admitting that certain schools and certain disciplines do certain things better than others instead of insisting on the corporatized model of homogenous consumer offerings in higher ed. But as state and federal money gets routinely yanked from college subsidies, most colleges and universities find themselves having to sink or swim just as any other business, and the resulting short-sightedness and rush to compete with other universities for student enrollment becomes more commonplace. Should colleges be dictated by "market forces" in the same way as, say, the household product industry is? Rhetoric and its companion disciplines may be best equipped to supply the critical penetration necessary to examine the situation and offer possible solutions. Certainly economics is a factor, but not in the same way as it is a factor in determining whether a new line of dog food will make a profit. For some fields, data-based outcomes assessment is appropriate, but the problem is that a one-size-fits-all approach to assessment has taken hold in nearly all accreditation processes, internal and external to the school for which they are being applied. It is not possible to determine, for example, whether someone's writing has improved in the same way that we can determine whether or not certain math skills were acquired by evaluating them through an objective test. Yet, we try to do it because we want to be able to show with "hard data" that a school is achieving a certain level of success in educating its students, and numbers seem to speak louder than more "subjective" (a dirty word just like "liberal") criteria that the general public and administrations can't understand without rhetorical training. Moreover, the field of rhetoric not only helps identify areas that need improvement, but also helps make the argument for reform. It can enlist students into the inner sanctum of academic, political, and social activity by making language relevant in all majors and disciplines. It can break down the class barriers between the Ivory Tower institution and the Person on the Street. Rhetoric is the great equalizer, a kind of blue-collar intellectual discipline, and this is one reason that it has seen significant success in all of its scholarly applications.

If more fields were willing to bridge the gap between "high-brow" intellectualism and "low-brow" public interest, we might go far in easing the so-called educational crisis we are confronted with today. The public views scholars and professors with suspicion because they don't know what we do. They assume that we all have cushy teaching posts, the duties for which are to meet with two classes a week, spout some nonsense, and collect an exorbitant salary. Those of us in academia know how misinformed this is, but we have made virtually no efforts to correct the error or inform the public of the current state of the Academy.[2] In an on-line search, I found many examples of public outreach regarding higher education, but all in highly specialized areas: "Public Attitudes Toward Agricultural Biotechnology in South Africa"; "Public Preferences for Informed Choice Under Conditions of Risk Uncertainty"; "Public Perception in the Advancement of the Chemical Sciences"; etc. There were, however, a couple of interesting exceptions. One of these, "Accountability and External Ethical Constraints in Academia," written by H. Paul LeBlanc III of Southern Illinois University at Carbondale, takes a decidedly rhetorical look at the social issues that require an external ethical accountability of educational institutions of higher learning over traditional self-regulation. His two main areas of concern are tenure and graduate education (Leblanc). Of these two areas, tenure is by far the most contentious in the public mind, and this is due to a fundamental misunderstanding about what tenure is and why it is in place. Part of the problem here is that the public tends to conflate public primary and secondary school tenure with tenure in higher education, a decidedly different status with a decidedly different criteria of achievement. For the majority of the uninitiated public, tenure is seen simply as an opportunity to rest on one's laurels, to receive a pay raise, and to cease being productive. It also seems to many to be anti-capitalistic; since the incentive to produce is ostensibly removed when tenure is granted, many feel that professors in such a position simply don't care any longer. Such a generalization reveals how little the public knows about the operation and politics of universities. For the public, as for LeBlanc, tenure is an issue of accountability; for professors, it is an issue of academic and political freedom, an opportunity to construct a needed culture of dissension in an otherwise dangerously indistinct political and cultural landscape.

A slight diversion is necessary here in order to underscore the often combative relationship between the members of higher educational institutions and the public which helps fill its coffers, if only because the issue demonstrates a serious rupture between these two entities—a rupture so severe that it can only be damaging to the integrity of the university system and the public it serves. The fundamental misunderstandings that drive attitudes about college life promote a strictly utilitarian and vocational role for universities in the public eye, which in turn creates external pressures for universities to function like corporations and, ultimately, adds to the anti-intellectualism that informs public attitudes. Such attitudes generate a bottom-line mentality that chips away at more idealistic educational goals, creating a society of technicians rather than thinkers, a community of people driven only by market competition and not by more thoughtful democratic objectives. Tenure is perhaps the most misunderstood of all, and as LeBlanc points out, a number of universities (the University of North Dakota and proposals before the Texas state university system, for example) have capitulated to public pressure to reform tenure policies because their objections are so vocal and because it seems impossible to efficiently reconcile a case for tenure in the face of rising tuition costs and the apparent necessity to provide every single person in the country with a college education. Tenure at such schools means little; the tenured faculty are still required to submit annual performance reports, student evaluations, etc. in order to retain their jobs (Leblanc). Like the apparently acceptable policy of the corporate world, where a worker can be fired without cause or justification, universities may be quickly following suit, and very few people seem to be seriously challenging this trend.

The public in general sees the university, then, not as a place where ideas are exchanged freely in the ongoing pursuit of new knowledge, but, at best, as a very expensive meal-ticket, and, at worst, as a financial con. This generalization is based on years of dealing with parents and students who have the apparent endorsement of the administrations that capitulate to this attitude, and to mass media representations of higher education ("Animal House" and its vast array of imitations has done far more to damage the credibility of higher education than any of the meek efforts to inform the public about its reality has done to improve it). Whereas certain concerns about the expense of education may be warranted, the blame has been squarely misplaced. Schools that once relied on state and federal subsidies are finding the need to supplement that income to create realistic operating budgets as allowances are slashed in favor of other governmental expenses (most notably, a long and protracted war that is bleeding this country dry in ways more than financial).[3] If a school cannot find enough private investors and benefactors, then the slack must naturally be taken up with student tuition and increased fees. Lack of state and federal funding also reduces the amount of financial aid that will be subsidized by the government, leaving student loans in the hands of private banks that will charge higher interest rates to compensate for the increased risk of non-guaranteed student loans while at the same time reducing the overall aid awards. These problems are hardly caused by the university system at large, though, of course, mismanagement of funds may happen in certain educational institutions just as it does elsewhere. The larger influence is simple inflation; it costs more to get an education today than it did twenty years ago because it costs more to build new facilities, to buy equipment, to supply departments and offices, and to pay faculty and staff. LeBlanc sees the blame placed on universities themselves as a public preoccupation with accountability, but the difficulty with accountability a la the corporate model is that, while it is relatively easy to track the performance of a stock or the quality of a product, one cannot as clearly evaluate the outcomes of an education. One must consider whether it is possible to judge what a student knows except to use dubious standardized tests or the even more dubious gauge of job placement after graduation. Do these measures really show us what knowledge students leave an institution with? Can these questions be objectively and accurately quantified, and can the results be verified? Will the arbitrary release of faculty make a bad situation—if it exists in the ways that we assume—better?

Public perception is certainly part of the problem here, and the media only confounds already silly caricatures of the academic "lifestyle." Deborah Churchman, in her essay "Voices of the Academy: Academic's Responses to a Corporatized University" captures the problem this way:

> Academia is presented in the media in a certain way. Academics are cited as experts in support of media items and often filmed in book-lined rooms. The public perception of the absent-minded professor dedicated to the pursuit of knowledge is alive and well and is reinforced in films and literature. These images bear little relevance to the entrepreneurial skills which are touted by government as essential for institutions to prosper in the current climate. They do, however, provide a convenient backdrop against which staff who wish to perpetuate traditional academia can operate. (11)

Many people have little or no further exposure to the realities of academia beyond this, either because of the remote temporal distance of their experience, reifying an outmoded picture of what university life is like, or because they have never been to college in the first place. The perception of the university and its members is in stark contrast to the actual conditions, as an interviewee of Churchman's makes clear:

> I think we have a perception that academia is like the Oxbridge model, sitting at the feet of greatness of a Professor and learning from that experience, being involved in challenge and debate. The reality is today's climate is that that model cannot exist anymore because you have a larger number of students, do more with less support, do it faster, etc. I am not saying there is no quality there, it's different—the model is different. It's if I have 200 first year students in a particular course and work has to be assessed and we do it in the most time efficient way, and that is not engaging students in those sort of out of hours sessions. (11)

The university as a consumer-driven business becomes, under conditions such as these, a thoroughly anti-educational enterprise that morphs into a series of requirements fulfilled for the purpose of credentialing, credentialing that is becoming increasingly meaningless as the requirements become mere tasks to complete as opposed to knowledge to be gained or wisdom to be acquired. The corporate model is one based on data analysis, market forces, macro and micro economics, cost/benefit ratios, and a for-profit mentality. It is, moreover, a quasi-positivistic social and economic *science* that engages the economic world in barren, statistically probable terms and measures outcomes by a single standard: the amount of money that is made for providing a product or service. An education based on this blunt property alone guarantees the production of unquestionably one-dimensional citizens.

The inappropriateness of such a principle for educational purposes should be clear and obvious. Education, even by the most generous accounts, cannot be considered a "service" like a dry cleaning business or an airline because it deals with offering people an inroad into knowing themselves and their world through the act of intellectual contribution—to the cumulative knowledge of faculty with highly varied experiences and expertise. Knowledge, even in the most material or practical sense, most certainly cannot be equated with a product, since its acquisition is an ongoing, life-long effort to improve one's intellectual knowledge base. Yet colleges and universities across the nation are pressured to formulate just these sorts of market equations and apply them to both the operations of the institution and the recipients of an education. It is an attempt to make an economic science out of a wide-ranging landscape of needs and solutions for educational goals and turn this diversity into a commodifiable jello-mold that shapes academic success. So inundated is this country with the two-prong *Weltanschauung* of science and capitalism that it is assumed that market-forces, if properly assessed and analyzed, will always give us the ready-made answers we seek. In large part it is because education has incrementally eroded in substance over the years that we suffer from an inability to see other possibilities, a condition that reflects our worsening state of imagination. It is impossible for us to consider educational options that do not a) slavishly follow a corporate paradigm, and b) use a social scientific methodology to assess the quality and outcomes of our efforts in education.

Rhetorical studies, while certainly not a magic bullet for this large and unwieldy problem, can provide remedies for those who want something out of their educational experience besides a passport to the nine-to-five business world. It provides an alternative vocabulary to the scientistic mode of thinking that dominates most areas of inquiry, and it allows for a reifying analysis of the discourse that contributes to our culture and society by allowing students access to the language that shapes their attitudes and identities on a meta-discursive level. ("Scientism," I should note, is a term used throughout this book to denote not merely the methodology, technique, and outcomes of science, but the further internalization of science as a mode of conduct and an epistemological belief system—the term implies, that is, the *ideological* dimensions of science.) The increasingly indistinctive undergraduate education that exists in today's universities and colleges provides little in way of understanding the language students are required to assimilate; students are expected to absorb discipline-specific discourse in order to be familiar with it and even use it, but they are rarely asked to assess the properties of meaning behind the discourse itself. Educators are increasingly strong-armed into complying with a standardized "content" orientation for teaching courses with little or nothing contributing to students' understanding of what they are learning or why it is important. Still rarer is any critical scrutiny of the discourse students are being taught. While professors often have students review and comment on relevant literature, the activity is usually meant to assess its content and pertinence to a particular question. It is unusual for anyone outside of rhetorical studies, philosophy, literary criticism, or communication theory to question how it is that the language students encounter has *meaning*. On the undergraduate level, the search for meaning is rarely addressed, even in above-cited areas of inquiry. Graduate students in these fields are the only ones likely to ever encounter questions about issues like authorial intent and authority, the rhetorical efficaciousness of language acts, or how a text generates meaning for one audience but not another. Rhetorical studies sees these questions as fundamental; it is impossible to be thoroughly educated without at least a working understanding of how language means. Otherwise, education devolves into training, and training is a poor substitute for the fuller, richer educational experience that—all unsubstantiated nostalgia aside—once was the standard in a humanistic higher education. Rhetoric offers students an opportunity to delve deeply into our most important human resource, language, in order to get a better idea of how and why it is that humans respond to the institutional symbolism that defines who we are and what we think.

Outside of linguistics, there is no science for such a project. Even linguistics, while useful for understanding the structure and technical dimensions of language, is not systematically equipped to deal with the more philosophical questions that give language meaning. Rhetoric, however, is a discipline specifically suited for this purpose. Through it, students can learn not only *what* something can possibly mean, but *how* it achieves that meaning. Rhetoric becomes one of the most serviceable of the meta-disciplines because it must generate a way of talking about language by using language. It differs from a more scientific approach, however, in that it doesn't assume to be above the very rhetorical operations it purports to describe. It is, in this sense, aware of its own rhetorical and ideological faculty, a feature James A. Berlin pointed out in *Rhetoric and Reality* over twenty years ago. Rhetoric is perhaps the most salient of all hermeneutics because its practitioners are aware of its own rhetorical thrust, and those who do practice rhetoric are comfortable with this realization. Unlike the more positivistic philosophers and scientists who futilely attempt to discover a "pure" language that is unfettered by the uses and abuses of our myriad discursive acts, rhetoricians not only acknowledge, but embrace, the need for ambiguity and the importance of depth in words as they contribute to the richness of language. As John S. Nelson et al. point out in "Rhetoric of Inquiry," even Ludwig Wittgenstein saw that his initial judgment about the nature of language, the positivistic preoccupation that defined the *Tractatus,* was in fact in error, and he reformed his ideas in the posthumously published *Philosophical Investigations:*

> Ludwig Wittgenstein was another who wanted to free scholarship from the philosophers, or at least from philosophers who wanted to separate it from practical life. From the start of Wittgenstein's sparring with philosophy, he manifested a preoccupation with the corruption of ordinary language. To be sure, he tried first to assimilate language to the alleged certainties of logic and mathematics. But he later renounced this craving for certainty in favor of a practical and rhetorical emphasis on human languages as games among speakers, listeners, and actors. The metaphor of the game encourages attention to the back and forth, the give and take, of real argument. (Nelson et al. 8–9)

Wittgenstein, like Nietzsche, saw the philosophy that was being practiced at the time as so remote from ordinary experience that it had little bearing on practical wisdom or life. It was philosophy for the sake of doing philosophy, and what Wittgenstein was more interested in was determining the relationship between words and ideas (as well as words and other words, sentences and other sentences) so that they can convey an "accurate symbolism" (see Betrand Russell's introduction to *Tractatus* 7). Wittgenstein was primarily preoccupied with what conditions would be necessary for a logically perfect language, while realizing that such a language could not in reality ever be produced. What he concluded in *Tractatus* was interesting, for it reveals a fundamental limitation of language as it is used by philosophers and laypersons alike: "That which has to be in common between the sentence and the fact cannot . . . be itself in turn *said* in language. It can, in this phraseology, only be *shown,* not said, for whatever we may say will still need to have the same structure" (8). What Wittgenstein seems to be saying is that there must be an identical structure between a "fact" (or an idea, or an abstraction) and the manner in which it is articulated which, given the syntactic and practical limitations of language, can never really exist.

If linguistics and formal logic are incapable of determining meaning in any absolute sense, rhetoric can help fill the logistic and theoretical gaps by acknowledging that language is, by both its nature and usage, an imprecise operation. In fact, most modern theories of rhetoric contend that a skillful user of rhetoric will rely on the ambiguity and vagueness of language in order to produce a desired result. Our reflexive reaction to this is to see rhetorical practitioners as verbal magicians who manipulate language for devious and deceptive motives, but beyond the blunt misinformation driving this attitude, it becomes a case of the pot calling the kettle black; we all innately understand the advantages of linguistic imprecision and have learned not only to accept it, but also to use it ourselves. One of my favorite examples of this is an incident that involved my daughter, who at the time was only about three years old. She came into my study one day and announced, "Daddy, the lamp got broke." Already, at the tender age of three, having spoken for barely more than a year, she understood the tremendous opportunities that rhetoric afford for telling the truth while at the same time hiding the agent of responsibility by simply invoking the passive voice. To embrace this advantage and to understand it opens up the sheer richness of human language; a language that said exactly what it meant at all times with no room for interpretation, ambiguity, or outright confusion would be a thin language indeed. While we might actually solve some of the philosophical problems that interested Wittgenstein, one might ask what we would have left to do or how this knowledge would ultimately benefit us. It is only vanity that pushes us to solve problems that inherently have no intrinsic or necessary solution; human beings thrive on problems, for without them, our existence seems pointless. Rhetoric helps capture the ways in which we wrestle with language—how it hides our motives, diverts our attention, smirks at multiple meanings, and generates puzzlement. Rhetoric isn't devious and untrustworthy; that feature is reserved for language itself. Rhetoric simply recognizes this fact and gives us a way to talk about why and how we use language in the many ways that we do.

Even some psychoanalysts recognize the depth of language and the human need to exploit it and the folly of pursuing a "pure" language unencumbered by the natural traits many see as undesirable. Ernest G. Schachtel, for instance, has observed that anyone serious about using language must contend with what he terms the "temptation of language" that seduces us into settling on the familiar and convenient words (Carlston 146). He adds that the quest for a language that offers a direct correlation between the thing described and the symbols used to describe it is an illusion. Once we put something into words, the words appropriate the meaning and significance of the object or subject it replaces:

> The danger of the schemata of language, and especially of the worn currency of conventional language in vogue at the moment when the attempt is made to understand and describe an experience, is that the person making the attempt will overlook the discrepancy between experience and language cliché or that he will not be persistent enough in his attempt to eliminate this discrepancy. Once the conventional schema has replaced the experience in his mind, the significant quality of the experience is condemned to oblivion. (qtd. in Carlston 147)

An admission of this sort—one where it is understood in so-called "postmodern" terms that language literalizes an experience or idea and touches off a symbolic transformation that substitutes the words for the signified object of those words—should tell us that no language is bereft of this phenomenon. Science, too, is subject to this transformation, perhaps more so than other forms of inquiry since it deals so often in things that are unfamiliar, attempts to articulate phenomena that have no precedent in experience, such that the very act of naming corrals a scientific discovery and envelopes it in a linguistic holding pen (see Grassi, *Rhetoric as Philosophy: The Humanistic Tradition*). Without a methodology for interpreting scientific discourse, the language of science can become a symbolic puzzle that both resists our attempts to understand it and changes the structure of the theories we are able to understand by framing the meaning in terms that are accessible but distorted. Add to this the tendency of other agents to seize the discourse of science for journalistic, commercial, or political ends, and the result is a far from objective picture of "reality" in any pure sense of the word; scientific language becomes, rather, a discursive apparatus like any other, wielded for purposes that only loosely represent the activities of science or their unique perspective on the world.

Randy Allen Harris has made the observation that the rhetoric of science has seen slow acceptance by the academic community in general and the scientific community in particular because the disciplinary ethos that these respective areas of knowledge occupy appear so radically counter to one another; I think it can be safely said that the perpetuation of this disciplinary essentialism in the academy is in fact a myth. As many of the rhetoricians, psychologists, sociologists, historians, philosophers, communication scholars, and even scientists themselves (those who consider the linguistic layers of scientific discourse) have pointed out, science is never done in an intellectual, cultural, or social vacuum; there are always reasons for why new ideas (or the reexamination of old ones) necessarily take on a rhetorical flavor, why scientists, in the interest of accessible models or metaphors, must sacrifice precision in order to conceptually simplify their theories, and why other agents extraneous to the activities of science rely on the rhetorical dominance of scientific language to buttress their own arguments and agendas. Beyond this, however, we will see in the following pages how language as a whole is an intrinsically limited means of expressing *any* idea because it is by its nature a symbolic transformation of the thing it is intended to represent. We see this principle manifested in the ideas that dominate models of the mind, such as those in evolutionary psychology; we see it at work in the way policy decisions are made, especially those based on thin scientific knowledge or evidence; we see it in the misinformation that people have about some of the most well-established scientific principles that exist, like natural selection and evolution; we see it in the way that a scientific endeavor, like the space program, gets appropriated by the media and altered to make good copy or to side-step the scientific aspects of the events that are being described altogether; we see it functioning especially in the residual field that we are enveloped in when we try to make sense of issues or events that are remote from our experience and drowned out by the white noise of too many conversations taking place at once, all of which are fragmentary and susceptible to individual interpretive perspectives; and we see it in all the little dialogues we have everyday, whenever we debate an issue or establish a position.

1 A Culture of Science and Capitalism

This chapter will examine the intersection between scientific activity and consumer culture to argue that the face of scientific ideas has changed drastically in the last 50 years, especially concerning how science is viewed by the American public. Prior to World War II, science and technology had a significant, but not necessarily primary, role in American culture. It was only after the US emerged from World War II as a global superpower that advancements in science and technology became the top priority, and the resulting scramble to feed money into developments in science, engineering, computers, and weapons development was unprecedented. Federal acts such as the G.I. Bill provided opportunities for those who had fought in World War II (and for all subsequent veterans) in educational fields and training programs emphasizing science and technology, inextricably tying science to the military applications it was expected to enhance. Ideologically, of course, the goal was to preserve democracy against the spread of communism, but American democracy is (and perhaps always has been) analogous with capitalism and "free trade," meaning that the transformation of American science into a tool for national security would persist up until the present, post-Cold War age. The cycle is an interesting one, for it reflects a linkage between national protectionism, weapons development and other military applications, capitalistic economic concerns, and governmental support for science that meant the availability of increased funds for companies that pioneered scientific development. This, of course, meant jobs for those going into relevant scientific and technological fields, creating a "trickle-down" ideology that favored science because it: a) won the war for the US and its allies, lending the US its current "superpower" status; b) provided educational opportunities for veterans and other potential students of the sciences; c) allowed for the creation of highly skilled, highly paid jobs; d) consequently increased the standard of living for the American middle class; and e) came to parallel the political system that made scientific developments necessary to combat antithetical ideologies such as communism. It should come as no surprise, therefore, that science and consumerism have mutated into common goals in contemporary American society.

As the Cold War dragged on and finally fizzled out, the reasons behind supporting scientific development transformed as capitalism apparently won out. Throughout the last fifty years, part of the reason science became so profoundly important in American society was that, as military applications ebbed and flowed, the benefits of new technology, medicine, and employment were passed down to the consumer at a consistent pace, either when they became obsolete in their military role or never achieved the promise they had shown, necessitating another, civilian market. Now, developments in science and technology become less a matter of governmental imperative and more a matter of economic necessity, and the American public has become accustomed to being offered new high-tech products and services relying on cutting-edge science. This chapter will trace the history I just described to show how American attitudes about science guaranteed its primacy as an ordering system through the rhetoric of economics, psychology, and even religion with the goal of contextualizing the current status of science so we might determine how its language is applied and misapplied to garner, on the one hand, support from the American consumer, and on the other, misplaced suspicions about the nature and scope of science as an ordering system.

Military Concerns

Science and the military have always been strange but inseparable bedfellows. Nearly every technological advantage that civilians enjoy in our current state of capitalistic luxury can be traced to a military origin. Microwave ovens, radios, television, the Internet, the cellular telephone, many automobiles (like Jeeps and Hummers), and a host of other technological gadgets that we use everyday had military beginnings, only later to be marketed for civilian consumption, often with unprecedented success. The military/technological marriage is hardly unique to the 20th and 21st centuries; science has always been subservient to the military powers that fund and promote its research and exploit its progress. All warfare, one could argue, is necessarily a technological undertaking, whether the advancements sought are for new and more devastating weapons, for stronger and more effective armor, for creating surveillance and counter-surveillance measures, or for a more efficient means of transporting materiel, ordnance, or personnel. Perhaps the main distinction in recent decades can be seen in the transference of technological developments from a military to a civilian application. Following World War II especially, the conflict that was perhaps most reliant on scientific development of all the previous wars, we see an unparalleled reconfiguration of wartime technology to the public marketplace.

This arrangement, one where science plays a role of servitude to either the military directly or through the companies that contract with the armed forces, means that in order for science to have a "productive" outlet, it must ally itself with organizations whose interest in science is secondary to the usable knowledge and artifacts it can produce. Through this often imbalanced pact, scientists must philosophically sacrifice (or at least thoroughly modify) their primary motivational impulse to seek out new knowledge for its own aestheticism and for the ongoing improvement of the human condition. Most scientists, as with most young professionals, undoubtedly begin their professional pursuits as many of us in other fields do with a sense of idealism and hope, only later to discover that these naïve principles must be re-examined in order to maintain vocational stability. Pursuing most scientific interests is very expensive, and someone holding the purse-strings must have a strong economic motivation to fund research and development in the hope of having the investment translate into profit later on down the fiduciary road. Despite this obvious duty to economic "reality," we still often operate under the assumption that scientists are above all of the marketplace forces that bind the rest of us to economic servitude; the ethos of the principled scientist is powerful in our culture, to the extent that we sometimes forget that they are human, that they have families, mortgages, student loans, and all the of financial obligations that beset the rest of us. As the thinking most commonly goes for all of us in these situations, idealistic professional principles are wonderful to hold on to, but they don't pay the bills or provide for job security and a future in pursuit of the fields we have dedicated our lives to improving. What this often means, of course, is that many scientists must do work that they are highly qualified to do, but which may not be their first love. More importantly, the high degree of specialization of scientists and technicians in today's marketplace often means that they are not entirely certain of the ultimate purpose behind the research and development that they are contracted to complete. A scientist working for Lockheed/Martin certainly realizes that most of her research findings will go toward military devices—such a company relies almost exclusively on governmental military (specifically aeronautics) contracts for its survival. And since she is an intelligent and highly trained specialist, she may hazard a guess at what her research will be used for in the end, but the full picture is probably (and sometimes deliberately) elusive.

What this also means is that the minutiae of a given task provides for the dilution of accountability for any one scientist or even whole department. In other words, while a research and development department may be well aware that their work is going toward the completion of a new guidance system for a rocket designed to deliver a nuclear payload, their efforts are only one small contributor to the overall project. Also, such systems need not be used exclusively for the purpose for which they were originally designed. A guidance system for rockets can be applied to spacecraft for exploration and scientific data-collection of other planets. It can be used for rockets to send satellites that have civilian applications into orbit. If the military contracts for a system with tight specifications, the work done to complete those requirements can be used elsewhere later. So not only is the individual scientist often removed from the overall purpose of his work, he can also rest easy in the hope that his work will be used for peaceful ends in the long run. The government becomes a means through which work that would otherwise lack funding can get done. Given these conditions—a hungry post-doc scientist who needs a job, a government that is willing to fund certain types of research, a company that is interested in profit but needs highly trained and skilled scientific specialists to deliver its wares, the fact that specialists and departments are working on only one piece in a larger technological jigsaw puzzle, and the expectation that the work that is completed will ultimately have ethical and peaceful applications—the idea of an "indifferent," "cold," or "dehumanized" scientist, an attitude that some people assign to what they see as the impersonal self-interest of the scientific process, is overly simplistic and ultimately unfair.

At the very same time, people in general are more than happy to exploit the new developments in science so long as they benefit themselves in some tangible way. What results, then, is an interesting ambivalence about the nature of scientific inquiry and the advantages of scientific production: On the one hand, people venerate the austere body of knowledge that science has amassed and are perfectly willing to use its power to make their lives more comfortable; on the other hand, some people harbor a suspicion that science is behind many of the environmental, martial, and even political problems that we face in the United States. They wonder whether the keen judgment and pristine insight that scientists are reputed to have is really deserved. The climate in America is one of willingly embracing the practical and material benefits of science while concurrently looking for a source of intellectual authority to fault for some of the more unsavory conditions of the modern world. Typical of the American sense of entitlement, we easily construct false dichotomies when we assume there are things we can value, covet, and even possess without logistical and ethical cost. That is, we find it difficult to comprehend why science can produce so much that is positive for humanity while at the same time produce so much that can damage it.

Take oil, for example. Who benefits from the extraction, processing, and sale of oil? Simply put, those who sell it and use it. But this includes a wide range of people including those from a variety of socio-economic backgrounds, educational levels, and even nationalities. Certainly, Exxon and Shell benefit from our reliance on oil. But so too do those of us who rely on it. Idealistically, we loathe the idea of giving up the freedom that comes from owning and operating our own automobiles; economically, we dread the difficulty of making a decent living that surrendering oil would engender. This country does not encourage the use of mass transit much, mainly because, as a friend of mine who works for New Jersey Transit put it, "Mass transportation is always a losing proposition." No one gets rich from operating trains and buses, and the only reason they exist at all is to "keep the economy moving" in other areas. This is why mass transit systems are usually governmentally supplemented. Certainly Middle Eastern and Texan landholders (among others) benefit from our reliance on oil. Certainly the scientists whose job it is to uncover new ways to find, extract, and refine oil benefit if only because they are well-paid to do so. But who suffers from this arrangement? Thousands of miles of Alaskan coastline has seen environmental devastation because of crude oil spills. The Alaskan wildness as a whole is under threat from both pipelines and new policies under the Bush administration to open up land for oil exploration previously protected by environmental and conservationist laws. Those who fight in far-off Middle Eastern countries may think they are benefiting from fighting for the establishment of strategic occupation of oil-rich countries, at least until their tour of duty is prolonged, their veterans benefits are cut, their legs are lost, or their lives are cut tragically short. Greenhouse gases, smog and air pollution, and the general poisoning of our atmosphere are the inevitable outcomes of massive-scale burning of fossil fuels, regardless of how much we try to filter, convert, or otherwise retard the damaging effects of using petroleum products. This affects all of us, regardless of class status or wealth. And the final analysis is that natural oil is a non-renewable resource, at least in any practical sense of the word. New innovations for extracting oil from the earth will only stave off the inevitable for so long.

Scientists are therefore challenged with not only finding ways to supply the world with oil for its machines now, but with making the transition from oil-based engines, furnaces, and other machinery to "alternative fuels." Scientists are charged, ironically, with finding solutions to problems that science helped create in the first place. This may help explain the ambivalence some of us feel about the cost-benefit analysis of certain scientific enterprises, and may further explain why many people trust science as the only area capable of finding solutions to our problems while others are just as quick to condemn science for putting us in such a precarious position. Ultimately, it makes little sense to advocate a regressive philosophy; is anyone (except perhaps the most radical of environmentalists) really willing to return to an age when the great bulk of people's time was spent in the pursuit of mere survival? Science and technology have certainly improved our quality of life, and we wouldn't want to sacrifice the gains it has allowed us by pining for an idyllic, Utopian society that has shunned the advantages of technology (the Amish are an interesting an noble exception to this; their society has managed to function in peace and prosperity without the crutch of technology, but they have made many sacrifices in maintaining their highly devout and isolationist culture. Few of us, I doubt, have the same fortitude or will to accomplish what they have in the absence of technological aid). However, self-satisfied complacency is as great a danger as radical reform, so the question is: at what levels do we critique science and for what purpose?

Let us examine the military/scientific connection a bit closer. The November 23rd, 2004 edition of *The Washington Post* cited these numbers for research and development of a single weapon, a nuclear warhead "bunker-buster" bomb designed to penetrate deep underground enemy bunkers: "President Bush's fiscal 2005 budget contained $27 million to continue research on modifying two existing warheads for the earth-penetrator, or 'bunker-buster,' role, and it projected nearly $500 million over the next five years should a weapon be approved." While these figures may seem modest given the multi-billion dollar military budgets that have been approved and are still on the table at the time of writing this, $500 million for research of a single weapon type is significant, especially considering how and to whom this money will be distributed. The standard operating procedure for procuring a governmental contract for ordnance like "smart" weapons or aircraft is for the military branch interested in a particular type of weapon to put out a general call for proposals citing specifications for the parameters of the weapon they seek. Companies then send their proposals and, if they make the cut, construct prototypes that are subsequently reviewed by the military branch interested in the weapon. Usually, money is advanced for the research portion if the future prototype looks promising. This process translates into jobs for companies and the scientists, engineers, and designers who work for them. They are commissioned, in effect, to bring their scientific and technological know-how to bear on a military "problem," and the competition can be fierce. If, for example, the money in research alone is $500 million for a single type of weapon (we can assume that perhaps two or three companies will actually split this money in developing their prototype), then that becomes a significant incentive to win the contract, which might mean billions of dollars for the company. One major military contract can make or break a company's fiscal future for years to come. It is little wonder that weapons have become so amazingly advanced so quickly; if necessity is the mother of invention, then the push to make larger, faster, smarter, more powerful, and more destructive weapons fulfills a multitude of necessities—the military gets what it wants through hardcore competition between companies; the companies get at least enough money to continue research if they lose the contract and much more if they win, and the scientists, technicians, and engineers keep their jobs with opportunities for promotion and distinction along the way. A situation that pits some of the most talented scientists against one another for the creation of a better mousetrap guarantees that the result will be impressive (and dangerous) indeed.

And, again, the public reaps certain benefits from this form of military wrangling as well. When a prototype—or even a mechanism originally designed for it—proves unfruitful, American businesses, being the opportunistic and pragmatic institutions that they are, turn to the public sector to see if they can cut their loses. The Department of the Army, in fact, even created a "transfer agreement" that deals with the transformation of military technology to the public arena, suggesting that there exists a contingency plan for failed military R & D:

> The revised Department of the Army (DA) Regulation 70–57, Military-Civilian Technology Transfer prescribes DA policies and responsibilities for technology transfer with the domestic civilian sector. Specifically, it provides policies and operational guidelines for entering into cooperative research and development agreements, for the licensing of intellectual property, for the provision of technical assistance to state and local governments, and for other cooperative efforts in research and development necessary to provide new technologies of interest to both the civilian and military sectors. (Dobbins 15)

The correlative conjunction "both/and" in reference to both civilian and military "sectors" is telling because it indicates an important overlap between the technology designed to serve the needs of each of these social infrastructures. Our current standard of living, one could argue, would not exist were it not for the advancements procured during two major world wars, a number of smaller conflicts, and a Cold War that guaranteed the need to push military technology on all fronts. The exponential rise in high-tech machinery available to the public is directly connected to the wars that have facilitated the circumstances for that technological development. While this has always been true on some level, the fully industrialized and mechanized warfare of our two world wars created a technological momentum that has had far reaching influence on all aspects of modern and postmodern society.[1]

The connection between the military and the scientific does not, of course, end with hardware; biological and chemical weapons are also carefully researched, both as a means of offense and as a preventative strategy. The military has always been keenly interested in psychology, psychiatry, and neurology, and has even dabbled in the paranormal and the potentials of Extra Sensory Perception. These areas of scientific interest sometimes bleed over into the public sector, often in the form of cultural myths (consider, for example, the decades-long hype surrounding Area 51 and the elaborate conspiracy theories that have emerged out of it, spawning countless books, documentaries, and at least one fictionalized television series, *The X Files*). What is perhaps most interesting about these under-represented areas of science is that the military's interest in them belies the popular assumption of the conservative military complex. The military is, in other words, extremely open-minded when it comes to finding new ways to get an edge on its enemies, and the fact that its decision-makers are willing to consider scientific areas of study that are well outside the scientific mainstream indicates that they have faith in all forms of scientific inquiry. Of even more interest is the notion that the domain of science touches even the nether spheres of the metaphysical and that some of the most powerful people in our government are willing to entertain ideas that originate *from* those spheres. Here we see an example of the coupling of science and a kind of pseudo-spiritualism, a condition that is a rather frightening manifestation of military motivations. Behind this outlook, paranormal activity is only an as yet misunderstood component of our physical universe—it is science that will reveal and harness its power; biological and chemical weapons can be more devastating than their traditional counterparts, making them an extremely cheap and attractive (if wholly unethical—but what weapon is strictly "ethical"?) alternative. From a military perspective, the harnessing of power is perhaps its most important enterprise, especially when it comes to the design and manufacture of new weaponry, and new sources of energy that can be translated into mobility, fuel, and destructive power are welcomed. This almost obsessive preoccupation with new and better military devices—devices that, by definition, manifest themselves (either directly or indirectly) as instruments of war—promotes scientific and technological advancement in all quarters of society. Without the massive governmental funding necessary to explore new mechanisms of war, we would not be in a position to reap the advantages of the technology that it produces.

We should be mindful, also, of the sheer randomness of the military development we have seen in the last century. Had different circumstances dictated a different historical and intellectual path, the degree of technological convenience (and reliance) that we currently enjoy would simply not exist. Given the technological contingency that has unfolded for Western Civilization in particular, this means that the public must necessarily embrace the benefits that science has provided it while also holding in ideological esteem both the instrument of this success (science and technology themselves) and the principles under which it operates (rationalism, pragmatism, rigorous data collection, a tendency to favor hard evidence over more intuitive forms of knowledge, etc.). The trajectory of Western history is such that it has guaranteed a well-entrenched esteem for the means of its success, and this translates largely into the dominance of science and the technology that is its natural outcome. The specific historical events that contributed to this orientation deserve some attention, so I will discuss briefly several facets of American military, economic, and cultural history that have contributed to the primacy of science in American society in the next several sections.

Hot and Cold Running War

Favoring the intellectual and epistemological primacy of science has been swelling gradually ever since the Renaissance, but no single event marked the ascension of science as much as the First and Second World Wars. I refer to these conflicts as a "single event" because, as many historians rightfully argue, they are really two manifestations of the same struggle. It might, in fact, be more accurate to refer to World War II as "The Great War Part II," since without the conditions set into place by the outcome of the First World War, the Second World War would probably not have happened. Whether historians agree or disagree with this assessment is of little importance for this discussion, however; the fact remains that both of these wars saw unparalleled improvement (if one can call more efficient killing and destruction an "improvement") in weapons and the means to deliver them than in all the previous centuries of warfare combined. It is also ironic, however, that the advances in weaponry and defense technology were closely paralleled by the necessary advancement in medical technology; the incentive to improve medical science was, in many tangible ways, magnified by the need to treat the humanity that was being more efficiently mangled by the new weapons, not to mention the advances in medicines themselves that became a indispensable means for treating new and exotic diseases that cropped up in far-away campaigns.

One oft-cited reason for the unprecedented loss of life that occurred during the First World War is that, while great strides had been made in weapons like the machine gun, the airplane, the tank, chemical weapons, grenades, flame-throwers, and artillery, the tactics for warfare had changed very little from the Napoleonic days. Troops were expected to line up and rush a stronghold and overpower it with sheer numbers, but this method was useless against guns that could literally spray bullets at attackers at a rate of up to 500 rounds per minute. The machine gun, perhaps more than any other killing device, was responsible for more deaths than any other single weapon during this war, followed closely by improvements in artillery. Commanding officers and generals simply did not know how to adjust to the new machines of war, and they sent their soldiers on attack after attack and counterattack after counterattack throughout this conflict, certain suicide for the virtually defenseless soldier scrambling towards the opposing trenches across No-Man's Land in order to gain a few yards of precious real-estate. At the First Battle of the Somme, and later, Verdun, for example, the numbers of men lost are beyond staggering—they are almost impossible to imagine. Eric Margolis gives us the numbers in these stark terms:

> When [Imperial British Commander Sir Douglas] Haig finally gave up, in November, 1916, the Allies had gained a pitiful 125 square miles of bloody mud from the Germans at a cost of 600,000 men: 400,000 British Imperial, and 200,000 French casualties. The Germans suffered 450,000 casualties. At Verdun, 1.2 million men were lost on both sides.

To put this in perspective, a sold-out Major League baseball park will generally hold somewhere in the neighborhood of 50,000 people. Scan Yankee Stadium when it is full, and multiply the total of what you see by 57—that is the total number of men lost during these two battles. There was literally no place to put the dead. The carnage was absolute, and the primary machines responsible for the butchery were the machine gun and the high-explosive artillery shell. From our perspective, this seems not only callous and irresponsible, but downright immoral. But it illustrates a common problem that arises when technology outpaces our ability to properly apply it or our ability to ethically assess its consequences. As Margolis puts it, "In both titanic battles, military technology had far exceeded [Haig's] 19th Century military intellect. In both battles, the flower of German, British and French society were cut down, robbing these nations of their future." Moreover, two conditions conspired to make these battles the most devastating ones in world history: new and better ways to kill through technology, and old and outdated ways of military thinking. The human element must receive the condemnation it deserves in this example; without a stubborn, antiquated, perhaps even neurotic commander at the helm, such bloodshed would never have taken place. The folly of the first waves of mowed-down humanity should have been sufficient evidence that such tactics were ineffective. A good general would have reassessed the situation, if for no other reason than the loss of such huge numbers of troops was impossible to recover from. Simple empathy for the dying soldiers should, of course, have been the first guiding principle, but in the absence of this, practicality should have taken over. Yet, in the quest for glory and a "quick end," General Haig obstinately sent battalion after battalion to their deaths. The lesson here is that while new technology might come with operating instructions, it does not come with ethical instructions. In the hands of the ignorant, ambitious, self-serving, or merely foolish, the results of a cavalier attitude when dealing with the technologically disastrous can have dire consequences indeed.

But this also shows just how lopsided the rate of technological and scientific advancement was when compared to society's ability to understand and wisely harness its implications. Just 21 years after the Great War ended, a new war gathered momentum in Europe again. This time, the mechanization was better understood, as were the methods for employing it. While the same basic weapons were used during this war as in the First World War, the tactic of *blitzkrieg* had shown that they could be used as a more efficient means toward gaining military objectives. This war saw its casualties—and death on a scale that dwarfed even The Great War—but single battles where tens, sometimes hundreds of thousands of men were butchered trying to overtake their enemies' trenches were part of a different era. Now, technology was brought to bear in more precise ways; it reflected not only the weapons, but the counter-weapons. Radar, sonar, radar jamming equipment, remotely operated flying bombs, experimental aircraft, rockets, new kinds of guidance systems, bomb sights, etc. were all born during this war and developed into the arsenal that is standard equipment for today's military. Perhaps those who saw the greatest loss were the civilian populations who suffered most from both direct attack and from "collateral damage" from strategic bombing, a new concept that was possible only with the design and development of long-range bombers. (While bomber aircraft did exist during World War I, their deployment was limited and their success was negligible. They simply did not have the payload, power plant, or range necessary to make strategic bombing a viable tactical option.) Rockets were used for the first time during WWII, mostly against civilian populations. Hitler's V1 and V2 rockets (the "V" stood for, ostensibly, "vengeance") were launched against London, but were so imprecise that they could only function as weapons of terror. They did, however, lay the groundwork for what would become the Intercontinental Ballistic Missile (ICBM) capable of delivering a nuclear payload with precision against enemy targets thousands of miles away.

These two wartime developments, along with perhaps the aircraft carrier and the submarine (the latter was also used extensively—and effectively—during World War I) became the technological/military backbone of Cold War armies. As the US emerged from World War II as one of the new global superpowers, it realized the need to maintain and advance the science that had been birthed during the war. The main threat, of course, was the Soviet Union, which had also benefited from the technological progress that had been made during the war and had an even greater incentive to honor its hard-won victories (some estimates put total Soviet losses—military and civilian—during the war at 36 million, though this is impossible to verify. The US lost roughly 500,000 between both the European and Pacific campaigns). The technological race that followed manifested itself most ominously in the scramble to produce more destructive nuclear devices and new and better ways to deliver them on target. Compare, for example, the paltry 13 kiloton destructive power of the "Little Boy" bomb dropped on Hiroshima (equal to 13,000 tons of TNT) to the 50 megaton warheads that became commonplace in the hydrogen thermonuclear devices of the mid-1950s (these produce the equivalent blast of 50 *mil-*

lion tons of TNT). It is clear that the arms race was going to necessitate both policy and national defense measures, which was the central cause of anxiety for the general population during the early Cold War period—how do we protect ourselves from such a massive threat?

Nuclear holocaust and national defense were not the only causes of apprehension for the Cold War citizen, as we often assume. Other problems that fell under the sponsorship of science included overpopulation, food shortages, and the depletion of natural resources. While these issues were by no means new to the 1950s, they had certainly been aggravated by the so-called "baby boom" that had been the obvious result of the prosperity that followed the Second World War. Just as the U.S. population continued to produce offspring, so did they simultaneously worry if there was enough space, food, and resources to accommodate them all. Again, problems that had largely been created by scientific advancement—industrial pollution, medical advancements that allowed people to live longer (thus creating all sorts of shortages), and urban centers that required food, shelter, and waste management for high concentrations of people—required scientific solutions. These situations created an oddly unsure attitude toward science: While most of the population had great faith in science and scientists to address and resolve the problems it had created (even though this fact was rarely acknowledged), the general public did not appear to hold science accountable for the existence of the problems to begin with, an attitude that is quite understandable in the face of the progress and benefits that science had created (Tietge 98). If the public did see science as the originator of the problems that beset modern society, it was usually assumed that this was a natural hazard whenever new and progressive ideas were implemented. However, as we shall see later, technological advancement carries with it inherent hazards and by-products, what Edward Tenner refers to as "revenge effects"—the unintended consequences of a technological order that require further investment in the technological machine in order to correct. In general, the standard of living that was the happy by-product of scientific and technological expansion was enough to help people overlook the darker side of scientific progress. People mostly had faith in the science and the scientists who practiced it; both had been integral factors in the prosperity that America enjoyed following the Second World War.

This development is important in order to illustrate how the stage has been set for the *ideological* domination of science since at least as early as the 1950s. It has been frequently noted that ours is a technological society (Ellul, Tenner, Foucault, Dennett, and Buller, among others, have all made claims to this effect), but we often overlook just how a scientifically oriented society uses science and technology as the basis for a system of values that frames our experiences in pseudo-rationalistic terms. I mean by this that science functions as more than a mechanism for studying our natural environment and applying what we learn to it; science used in a social, political, and even theological context becomes a self-validating, self-perpetuating, self-rationalizing enterprise that goes beyond the observing, counting, and sorting that is at the heart of the scientific method. It has morphed into a decentralized institution that governs not only our efforts to understand and influence natural phenomena, but also directs our need to control all facets of human experience, from social interactions to education to personal activity. Those areas of knowledge that fall outside of science are often dismissed as frivolous and impractical (indeed, "knowledge" has become synonymous with "science," and the word "science" itself is etymologically translated from the Latin derivative *scientia*, a participle of *scire*, meaning "to know"). Art, philosophy, and literature are seen as expendable mental exercises that can be (and are) sacrificed in favor of more positivistic, practical pursuits. Other humanistic areas paint on a scientific face: History and politics, especially, are increasingly viewed as "social sciences." We turn to science whenever a difficult question needs to be answered, and our faith in science as an ordering system and as a problem-solving heuristic is applied to everything that touches our lives. The central argument here, then, is not one that hopes to discount or even challenge science in any overt way, but one that hopes to show the degree to which science and technology has penetrated our lives—not as a body of knowledge that all educated people possess, but as a *language system* that dominates our world view and is frequently appropriated by non-scientists for a-scientific ends, often with detrimental social, economic, and political outcomes.

Economic Considerations

Robert K. Merton, in his forward to Jacques Ellul's *The Technological Society,* summarizes Ellul's message about the nature of post-modern economics as an intellectual discipline that

> [...] itself becomes technicized. Technical economic analysis is substituted for the older political economy included in which was a major concern with the moral structure of economic activity. Thus doctrine is converted into procedure. In this sphere as in others, the technicians form a closed fraternity with their own esoteric vocabulary. Moreover, they are concerned with what is, as distinct from what ought to be. (vii)

This idea—that technological society has become an end in itself, not in service to a betterment of the human condition, whether that betterment be moral or political—is central to Ellul's point about *la technique.* Such a worldview about technology inverts the characteristic principle of agency that says machines are created to serve mankind—that they are invented to fulfill a previously defined purpose or to solve a problem confronting modern society. Instead, Ellul argues, technology becomes a locus of agency that becomes it own objective. As Mike Hübler notes in "The Drama of a Technological Society," "The technological drama manifests itself in the technological society as a rhetorically dominant narrative that treats human artifacts as if they were primary agents and human artisans as if they were passive agencies through which technologies acted." The "artifact" is the human construction of technology, whereas the "artisans" are the human creators and users and, in short, *we* end up serving *it,* and one of the principle reasons for this is economic. In describing *la technique* in *The Technological Society,* however, Ellul is careful to point out that the machine is only the purest form of this condition; it is more important to understand that *la technique* "transforms everything it touches into a machine" (4). This observation carries with it heady implications, because it suggests that *la technique* is more than just a condition that coexists with humanity—it is a condition that defines humanity in the "modern" world, constructing an inseparable marriage of identity between creator and created. Ellul puts it this way:

> When technique [*la technique*] enters into every area of life, including the human, it ceases to be external to man and becomes his very substance. It is no longer face to face with man but is integral with him, and it progressively absorbs him. In this respect, technique is radically different from the machine. This transfor-

> mation, so obvious in a modern society, is the result of the fact that technique has become autonomous. (6)

According to Ellul, the machine, which once eclipsed human autonomy, now found an integrating process in *la technique* that allowed man and machine to coexist without some degree of mutual disintegration. What he appears to mean here is that the mechanization that overwhelmed the social, economic, and political order of the 19th century required a synthesis of ends and means that found its manifestation through *la technique* (a term Ellul uses as a substitute for science proper because of the erroneous associations it can elicit), leading to "a more rational and less indiscriminate use of machines" (6). That machines had been used, we must presume, "passionately" and "indiscriminately" during the Industrial Revolution suggests that a continuation of that practice could only mean the eventual usurping of human will for the products and operations of the machine. Ellul's point is more subtle than this, and we will return to it in a moment. However, let's consider the implications of this premise. Today, even more so than at the time Ellul wrote *The Technological Society* in 1964, we have become attached to our machines almost literally at the hip. Our current economy relies on it, not only as a mode of information transference, but as the commodity that drives the economy itself. The great majority of products and services are both part of the technological infrastructure and the very products and services being sold.

With the ongoing influence of computers and the internet, the line between service and product commodities has become increasingly blurred. One of the most interesting developments along these lines is the proliferation of antivirus, anti-spam, and anti-spyware software. The software is itself an autonomous service that helps one maintain his or her computer system. It is, in effect, a virtual janitor for cyberspace. The product is ethereal in the sense that one can actually obtain the software online without ever physically possessing it (or, for that matter, physically paying for it). Such software becomes an electronic remedy that, like insurance, placates us with a sense of security. It is a "virtual" product, for while it may in fact do something that we want (prevent unwanted computer security breaches), it is a product that has neither the corporeality of traditional products nor the obvious benefit. If it is working properly, we rarely know whether or not it is performing the function it has been developed to perform. We may see occasional updates or lists of "ads blocked" or "viruses quarantined," but any conscious knowledge of the reality of these operations is eclipsed by our sheer faith in the system's integrity; when it works, in other words, we don't know it because our computer is doing what it is supposed to be doing. It is only when the protections fail that we have any cognizance of the software's operational existence. We notice it, in short, only when it *doesn't* work.

Perhaps even more interesting is the fact that never before has a product been offered that is specifically designed to prevent the customer from getting something (with the possible exception of medical inoculations or other preventative treatments). Think about this. We pay Symantec or McAfee a substantial fee for software that does not enhance computer operations (in many cases, it actually interferes with it) but is designed to keep it "clean" of outside infestation. We must also pay for periodic updates, tantamount to an ongoing cyberwar where antivirus software designers attempt to stay one step ahead of the computer geeks who exact their revenge on the world through programming sabotage that can have truly devastating effects. We, in essence, pay to get nothing, or, at least, to keep something away. Is this similar to home and auto security, insurance, or preventative medicine? In one respect, it is; antivirus software provides us with a product that guards against certain contingencies and gives us sense of well-being, whether deserved or not. However, the major difference lies in what is being proffered. Here, we are not getting hardware that deters thieves, a claim settlement that compensates for loss, or a fluoride treatment that helps us keep our teeth. We are buying lines of code designed to intercept other lines of code that could interfere with the lines of code we hope to keep intact. We are paying for information, but not information that we can apply in any traditional, practical way; it is information used solely as a means of snuffing out other information that might confuse our computers. And since this information is always in flux, we must constantly update the content of that information for it to remain effective. This notion, when broken down into its observable components, is bizarre on its face, yet it is something that every prudent computer-owner must have in order to protect his or her valuable information. Ellul alludes to this in his study, but today's computers are so far beyond the scope of technology he might have a working knowledge of in 1964 that it underscores his radical warning: that technique will actually assimilate humanity in a way that is dehumanizing and, short of a complete collapse of the technological infrastructure, irreversible. While this prediction may seem both reactionary and anachronistic, Ellul makes a strong case for the complete permeation of technology in postmodern society, and this is easily born out anecdotally. Beyond the PC, computers have gradually (though by technological standards, the shift has been incredibly swift) been introduced into everything from self-guided missiles to coffee makers. But these are only machines, and we would expect machines to get "smarter" as computer technology advances. The question is, rather, how does this affect our interaction with machines, our furthered reliance on them, and, ultimately, our perception that such machines are not only indispensable, but part of our very identity.

Technological Identity

My students are fond of forwarding the hyperbolic sentiment that they cannot "live" without their cell phones, an attitude that betrays a certain willful ignorance of history. I was born in the sixties, and cell phones were as remote a concept for me growing up as automobiles would be to someone reaching maturity in the Victorian Age. I use one now, but have never felt reliant on it for achieving my daily goals. I use it sparingly, usually as an information-gathering and communication-confirming device, not as a link to my social world or as a product of my identity. Yet my 18- to 22-year-old students see their cell phones as a personal channel to their friends and family, a device necessary to remain continuously "in touch" with those important to them. My students walk into class talking on their cells and walk out making new calls. The cell phone, like so many of our contemporary instruments of technology, has become more than a mere convenience; they are part of what makes us socially whole, to the extent that life without it seems actually incomplete, even frightening. Such a reaction takes us well beyond the idea that cells should be used for convenience (and "safety"); it suggests that it is an integral part of our personality. We even attempt to "personalize" our phones (like our computers) with special sounds, wallpaper, and icon imagery. They are, in fact, mini computers that have the capacity to organize our lives since they often include calendars, alarms, and reminders that keep our busy schedules on track.

The development of cellular communication, of course, carries with it a huge economic opportunity. The overhead for establishing cellular service is complicated, dealing as it does with things like satellites, relay towers, air time, FCC guidelines and infrastructures, etc., but it can certainly be classified as one of the more lucrative technological markets available today. Cellular service providers like Verizon or Sprint are firmly seated in an number of communications markets like land lines and internet service, and the sheer level of demand is enough to push development in these areas at a pace that is certainly a challenge. Software "improvements" and hardware upgrades are a constant factor in this business, and services and products are often introduced before they are ready. The public, in this respect, is a large segment of research and development for such services, functioning as a test bed for the efficacy of the new technology whether it is ready or not. But all the entrepreneurial details aside, the most interesting dimension of this market is the idea of "selling" communication in the first place. Obviously, this is not new; the ancestors of the modern cellular telephone have been around for well over a century. The telegraph, the "land line" telephone, and radio communication are all historical communications markets that have helped set the precedent for today's communication sales. So what makes cell phones so different from these earlier technologies? Obviously, the fundamental distinction is one of absolute access and mobility. In the days of the telegraph, one had to seek out a station like Western Union to send a message, and this was usually occasioned by a special need as far as the general public was concerned. Individuals did not routinely send telegrams unless there was an important reason to do so. The expense and inconvenience made such technology prohibitive for daily use. Even as the telephone developed to the point where there was one in almost every household, the user was always geographically tied to the permanent location of the device. The only alternative to home or business use was the pay telephone, and while in their heyday these were certainly ubiquitous, again, people tended to use them when there was a pressing reason to. As an example of one of the few monopolies in American history, divestiture of Ma Bell (American Telephone and Telegraph Corporation—AT&T) was completed by 1984 and opened up the industry for the sudden explosion of competitors that (it was argued) allowed market forces to drive both research and development and lower prices (AT&T). While new technology certainly became available subsequent to this change, it is debatable whether it had any positive effect on pricing for the general consumer. What did happen for sure was an increase in the number of people who eventually had mobile telephone service available to them, such that nearly anyone who wants it can have a cell phone. The mobility and access, while perhaps short of revolutionary, definitely had an impact on the numbers of people using cellular devices. This shift made cell use seem essential, not only for business purposes, but for personal use.

The trend just described did not, however, make life "easier" in the way that the industry mythology would like the American public to believe. Rather, it made contact with the user more readily accessible, on both a personal and professional basis, and this reality means that cell phone users are instantaneously within reach. From a personal standpoint, this may be desirable, since it allows, for example, parents to stay in close contact with their children (spouses may also be tracked down, to the chagrin of many husbands and wives). But professionally, people's jobs are not easier as a result of the cell phone. Like the computer, the fax machine, the Blackberry, and the Palm Pilot, the cell phone creates the conditions for greater productivity. Predictably, greater productivity is expected. Strangely, this fact is frequently overlooked, even vehemently denied, by the most ardent users of these devices. They are convinced that their ubiquitous communication devices not only make their work production more convenient, but also that the machines are indispensable components of basic business operations. This is another example of the failure of the current generation to consider the historical conditions under which people lived prior to the introduction of such tools. Business may have operated differently in the past, certainly at a different pace, but it did operate, and with a saner momentum than people experience today. The only reason that business cannot function without the cell phone, the computer, and the Blackberry today is because these devices have been made foundational to the stability of the business infrastructure. Business, in an effort to increase productivity and efficiency, has anchored its daily process to the very machines that make this possible, but it certainly does not follow that they make life easier as a result. Edward Tenner points out that one of the unintended consequences of this heavy reliance on computer-based technology is constant maintenance. Without the performance of daily janitorial tasks—deleting, organizing, and arranging email, data input, scheduling maintenance, software upgrades, etc.—computer usage would quickly become unmanageable. Computers and their smaller technological cousins would become clogged with so much irrelevant information that their function as data-sorting and analyzing devices would soon become paralyzed. It is the human element, still, that must ultimately decide what happens to the information these machines receive and store, and neglecting that role can have rather messy consequences. The result is an expectation of immediate action. If, in fact, one can receive information, requests, questions, or make basic decisions via technology, then that is what is expected. Far from making life easier, computer-based technology tethers us to the machines of our trade, and we are expected to take care of business whenever and wherever business may happen.

Gone too are the days when we could harness our own machines and fix or maintain them ourselves. While we do have to perform the basic data moving operations that are part of using computers, most of us are ill-equipped to deal with the technology itself. Anyone who has tried to retrieve drivers for a video card installed in an obsolete computer knows the frustration of having just a little technical knowledge. In most cases today, warranties are voided if anyone but a "qualified service technician" attempts repairs on most household electronics. This industry-serving policy has been taken to some extremes, especially by automobile manufacturers. Even attempting to change the oil yourself can in many cases void the manufacturer's warranty, and this can have dire financial implications if anything major goes wrong. In some cases, a "qualified service technician" is enough; if you own a Ford, then the car must be serviced only by a "Ford Specialist." This, of course, is a carefully orchestrated way of gouging customers who become legal hostages of warranty policies, and it is a clever way of generating further profits, at least for the short term. However, such policies are industry standards now, meaning that there is little in way of competitive market options for the consumer. We simply accept the idea that we cannot fix our own machines, so as a result, we either trundle them into the nearest dealer (usually at their convenience) or, an even more encouraged response, throw the broken device away and buy a new one. The result of this cavalier wastefulness and submission to the machines we use (or as Ellul implies, to the machines that use us) is a pervasive dependence on the culture of the machine and those who have intimate access to its fundamental operations.

Selling Scientific Ideas: The AI Example

Not everyone sees machines and our reliance upon them as starkly as Jaques Ellul, however. Many scientists, engineers, and designers (as one might imagine) feel that machines will surpass us not only in their physical and mental capacities, but also in their *spiritual* ones as well. Explaining this strange idea to the general public is not an easy task, and Ray Kurzweil, 1988 MIT Inventor of the Year and author of *The Age of Intelligent Machines* and *The Age of Spiritual Machines,* uses a disruptive technique of "dialogue" to forge understanding with the reader on his most complicated ideas. As Kurzweil describes the wonders of today's machines, he reinforces the culture of technological identity and dependence on which so much of our economy rests. The following passage, taken from *The Age of Spiritual Machines,* is intended to create the illusion that we are having a conversation with our mentor, Kurzweil himself. While it is taken out of context, the important thing to note is the phrasing of the questions and the patient narrative responses:

> NOW ON THIS TIME THING, WE START OUT AS A SINGLE CELL, RIGHT?
>
> That's right.
>
> AND THEN WE DEVELOP INTO SOMETHING RESEMBLING A FISH, THEN AN AMPHIBIAN, ULTIMATELY, A MAMMAL, AND SO ON—YOU KNOW ONTONGENY RECAPITUALTES.
>
> Phylogeny, yes.
>
> SO THAT'S JUST LIKE EVOLUTION, RIGHT? WE GO THROUGH EVOLUTION IN OUR MOTHER'S WOMB.
>
> Yes, that's the theory. The word *phylogeny* is derived from phylum . . .
>
> BUT YOU SAID THAT IN EVOLUTION, TIME SPEEDS UP. YET IN AN ORGANISM'S LIFE, TIME SLOWS DOWN.
>
> Ah yes, a good catch, I can explain.
>
> I'M ALL EARS. (47)

A "conversation" such as this one has a couple of rhetorical advantages, the first being that the author can manipulate the direction of the lesson in a way that makes us feel like actual participants. We sense that we are supposed to "get" the ideas that are being volunteered, and that for the more obscure elements of the theoretical argument, we, as discerning pupils, will have certain questions, seeing as we do apparent contradictions in the thread of the theory. These conceptual conflicts, however, are obviously manufactured by the writer, and may not reflect the questions we have at all. In fact, the questions showcased in the dialogue sections of Kurtzweil's text are more likely the sorts of questions a scientist is likely to have, yet we feel as though these should be the questions we would be asking if Kurzweil were really there to hold our hands through the assemblage of the theoretical puzzle. The effect is, for this reader at least, a combination of confusion and impatience; the disruption to the primary text, while designed to be helpful, produces a conceptual disturbance that muddies the theory rather than clarifies it. The dialogue sections, though formatted to be separate from the main text—they are set against a gray background and bordered on their own pages—can be skipped, though readers may feel they are missing something important if they opt to move over them. This experimental format presentation is, apparently, designed to mirror a Windows operating system or a Web page, separating different segments of information in different fields and frames. It assumes that the reader interested in technological topics is also one who is more comfortable with computers than a traditional book layout. This arrangement may also be an attempt to push book formats in the direction of computer layouts, seeing it as a more navigable and efficient form of information delivery.

Another rhetorical advantage to this method is that it presumably counters possible objections before they can arise. Because Kurzweil has presented what he anticipates to be possible questions, he constructs a fictitious sense of discourse about the arguments he is making. As we will see, Stephen Hawking uses a much more subtle form of this technique by presenting timely scientific debates in their historical context, showing contrasting ideas as part of the natural development of what eventually gels into accepted scientific wisdom. Hawking notes, for example, the competing ideas surrounding black holes from scientists such as John Wheeler, John Mitchell, and even Isaac Newton. This practice is far more reflective of the actual process than Kurzweil's, and having ongoing debates really is an important aspect of advancing scientific knowledge. Kurzweil, however, has disarmed opposition in a preemptive discursive strike; by anticipating certain objections, he can create the impression that the debate has already taken place when, in fact, it hasn't. In a typical positivistic move, Kurzweil has forwarded a theory in a way that leaves the reader with the sense that the matter has been settled, that all the important objections have been dispensed with. Any rational person, we can infer, would come to the same conclusions, especially under Kurweil's benevolent guidance. The truth of the matter, for the average reader, is that we are not in a position to accept his implicit claims at face value. We haven't the training or the critical equipment to make determinations about the status of the state of the art in artificial intelligence. We must trust that our authorial guide will give us an honest sense of what stage the technology and the theoretical underpinnings that drive it are in.

However, if we turn to other scientists in the field, such as *Flesh and Machines* author Rodney Brooks, we see that there is serious disagreement about the nature of artificial intelligence and even the time scale on which the basis of such development is possible. More importantly, we see strong evidence of the human factor propelling scientific fields like cybernetics. Kurzweil, beyond his interesting predications about the time scale necessary for "true" AI, is really interested in the notion of extended life through the use of downloaded wet ware—that is, implanting human consciousness into a computer. Brooks says of Kurzweil's theory on cybernetic immortality that he has "succumbed to the temptation of immortality in exchange for [his] intellectual soul" (note the religious metaphor and Faustian literary reference) (205). According to Brooks, Kurweil has made absurd prognostications regarding science's temporal proximity to solving the problems of AI, much less the capacity of transferring consciousness into an electronic receptacle (206). Moreover, of those scientists Brooks says have made similar predictions, this anecdote sheds some light on their real motives:

> In 1993, I attended a technology and art conference, "Ars Electronica," in Linz, Austria, where my former postdoctoral student Pattie Maes gave a talk titled, "Why Immortality is Dead." She took as many people as she could find who had publicly predicted downloading of consciousness into silicon, and plotted the dates of their predictions, along with when they themselves would turn seventy years old. Not too surprisingly, the years matched up for each of them. Three score and ten years from their individual births, tech-

> nology would be ripe for them to download their consciousness into a computer. Just in the nick of time! They were each, in their own minds, going to be remarkably lucky, to be in just the right place at the right time. (206)

Kurzweil himself, also not surprisingly, has predicted that "a singularity in which computation makes us all powerful" will happen around the year 2020 (206). Guess how old Ray Kurzweil will be in the year 2020.

Whereas it is easy to dismiss Kurzweil as just a self-absorbed kook with delusions of grandeur, this is perhaps an impulsive position to adopt. A more thoughtful look at Kurzweil's predictions reveals that a mythos exists to create the conditions not only for the lofty claims that Kurzweil makes, but also the public receptiveness to those claims. Kurzweil's book is not aimed at the scientific community; on the contrary, it is clearly written to enlist the support of the layperson, a move that speaks to Kurzweil's interest in garnering public allegiance for the purpose of creating a new set of popular legends about the likelihood of transmuted consciousness in the near future. This is, at least, one possible motivation. Public support translates into funding for research because it suggests a potential market, and if Kurzweil hopes to cheat the grim reaper in time for his seventieth birthday, time is of the essence. But there is also the possibility that Kurzweil is simply looking for an audience, a set of readers who will accept what he says and buy into his ideas, however unlikely. The title of the book is, interestingly, religious in nature, dealing as it does with philosophical and even theological notions of what machines are really capable of in terms of their "spiritual" potential. Kurzweil goes so far as to say that machines, if they reach a certain level of consciousness, will "naturally" develop a form of spiritual identity. He argues that humans themselves possess a "God Spot" in the brain that drives spiritual impulses, "a tiny locus of nerve cells in the frontal lobe that appears to be activated during religious experiences" (152). This "module" was discovered by University of California neuroscientists who noted this "neural machinery" in epileptics during seizures: "Apparently the intense neural storms during a seizure stimulate the God module. Tracking surface electrical activity in the brain with highly sensitive skin monitors, the scientists found a similar response when very religious nonepileptic persons were shown words and symbols evoking their spiritual beliefs"

(152). Kurzweil establishes this apparent fact so that he can then argue that machines, if they become conscious in the future, would have a corresponding "God Spot" and become spiritual entities themselves. According to Kurzweil,

> Just being—experiencing, being conscious—is spiritual, and reflects the essence of spirituality. Machines, derived from human thinking and surpassing humans in their capacity for experience, will claim to be conscious, and thus to have spiritual experiences. They will be convinced that these experiences are meaningful. And given the historical inclination of the human race to anthropomorphize the phenomena we encounter, and the persuasiveness of the machines, we're likely to believe them when they tell us this. (153)

This claim is at once bizarre and seemingly, at first glance, unnecessary. Why would it be important for our machines to have a "spiritual" dimension? How does this forward Kurzweil's argument? From a standpoint of popular support, it is rhetorically efficacious to suggest that machines possessing artificial intelligence are (potentially, at least) another form of life, that they have "sentience" which allows them the same emotional responses as humans. However, the spiritual argument may not have the effect that Kurzweil is attempting to establish; while a certain segment of his audience may see this as a compelling reason to continue researching the prospect of AI that can become spiritually aware, there is another segment of the population who would view this effort as blasphemous in the extreme. The religious right is unlikely to accept the idea that anything besides humanity can have a spiritual consciousness because to suggest otherwise is to place humanity in a different "God spot," a place that gives us the seat of power and the right to be worshiped. Would a machine have a spiritual affinity for a human god, or would they be more likely to worship, as we do, their creator? These are not new questions, believe it or not. They are at least as old as Mary Shelley's *Prometheus Unbound* (a.k.a. *Frankenstein*)[2] and have been a fixation of science fiction writers like Isaac Asimov and Arthur C. Clark. It has been a question of some preoccupation for Western thinkers, and is perhaps one of the central ideas used to define human existence and free will. Are we, ultimately, to remain subservient to our creator(s), or does our "natural" human inclination toward new knowledge and continuous inquiry about the structure of the universe give us a sovereignty that is our birthright as humans? These are tough philosophical questions, and, far from being mere intellectual exercises, they have significant practical implications. A cavalier push to do something because we can without regard for whether we should has been a key ethical concern ever since science became the dominate intellectual force in Western societies. This is, in sum, the scientific hubris that many fear could be our eventual undoing as a species, and it is one of the reasons that science experiences a strange ambivalence from the general public. Kurzweil's tactic might have the effect of preaching to the converted, but he is not likely to gather new members into his flock on the predication that machines will someday be spiritually equal (or even superior) to humans. This is an example of a scientist who seems to have failed in his rhetorical task, at least with the audience he is apparently addressing.

Kurzweil seems willing to risk the alienation of one sector of his readership in favor of another, more sympathetic audience. The rhetorical carrot is, as I've already mentioned, immortality. This particular issue is far too important to dismiss, so some discussion of its significance will, I hope, help prepare us for understanding how it is that scientists are able to cast their "spell" over the public. If we consider that the primary motivation for most religions is not the impulse to do good, but the promise of everlasting life (whether that be "life" in heaven, through reincarnation, or through enlightenment), then we get a sense of how it is that religion has held such sway over civilizations for so long. Science gives us an opportunity to have control over our own mortality, but currently, only in a very limited, temporal sense. Advances in medicine, diet, and surgery certainly have allowed us to prolong life (sometimes under dubious circumstances), but these are far from the sort of groundbreaking discovery that will give us everlasting life—unless Kurzweil is right. This may, in fact, inform his need to emphasize the spiritual aspect of machines, for if we are to cybernetically infuse ourselves with hardware in order to achieve immortality, it is only an appealing prospect if we can be certain that the machines we inhabit possess the same qualities necessary for a "soul" that humans do. Otherwise, we are little more than the machines themselves. Kurzweil is performing an interesting, if deeply troubling to some, infusion of science, religion, and magic in order to pull off this rhetorical performance. He insists that technology is "evolution by other means" (14), and argues that the human capacity to keep records is the same process used by DNA in other organisms:

> [Technology] involves a record of tool making and a progression in the sophistication of tools. It requires invention and is itself a continuation of evolution by other means. The "genetic code" of the evolutionary process of technology is the record maintained by the tool-making species. Just as the genetic code of the early life-forms was simply the chemical composition of the organisms themselves, the written record of early tools consisted of the tools themselves. Later on, the "genes" of technological evolution evolved into records using written language and are now often stored in computer databases. Ultimately, the technology itself will create new technology. (14)

Is the similarity between a circuit schematic and a strand of DNA really that close? Both contain "information" in the sense that they contain the necessary "data" to reproduce either a new circuit or an organism, respectively, but does the comparison go beyond this? This fallacy is an example of the "false analogy," the attempt to connect two different phenomena using a comparison that breaks down after analysis of the superficial similarities is revealed. In this case, the information connection may in fact exist, but the enormous difference is one of agency. DNA needs no extraneous interaction from an outside agent in order to make it reproduce. It does so because it is in its very make up to perform that task. Tool records, conversely, require a human agent to reproduce the tools they describe, and they can lay fallow forever if no one is there to exploit the data. Even more significantly, the record is not *part* of the machine itself; it does not embed itself into the machine and reproduce along with it. The exception to this is, as Kurzweil implies, the computer, which can contain the information to create another computer, but still cannot do so autonomously. In order for Kurweil's analogy to work, he must admit a "prime mover" for DNA which, by all definitions of and evidence for evolution, does not, and has never, existed. Can a computer, if left to its own devices over the eons, "evolve" independently to reproduce itself? The answer is clearly no, because it does not have the capacity to instinctually survive, and therefore has no basic motivation to adapt, assuming it had the intrinsic equipment to do so in first place.

This is a truly unusual argument for it infers a god in the original creation of life, but uses the theory of evolution to do so. According to Kurzweil, we, like the prime mover of the "first" DNA strand, are the godlike interveners who will eventually be obsolete, as machines become conscious and can do for themselves what humans once had to do for them. While it is true that there currently exist machines that make other machines, self-replication in any autonomous sense seems far off. The other question, of course, is whether or not such an ability is desirable in machines. A fully autonomous, conscious, and self-replicating machine could certainly qualify as "life" as it is presently defined, but is this, in itself, something that is advantageous? Kurzweil seems to imply that it is inevitable because it is, according to Kurzweil, the logical extension of an evolutionary process, but he makes little effort to address the question of whether we *want* machines that, of their own will and accord, can accomplish a feat heretofore reserved for living organisms. This is an especially important question if such machines are, in fact, "intelligent." If human history is any measure, we know that higher intelligence breeds higher ambition and an increased willingness to assert one's will. This has not always resulted in positive outcomes. Unlike other evolutionary processes, we have control over the future of machines, whether Kurzweil believes so or not. The unlikely assumption that Kurzweil makes is that our own autonomy has no bearing on this contingency.

For example, he quotes Arthur C. Clarke, a man interested in inventing "laws" (like Isaac Asmiov and his laws of robotics) by which the philosophical dimension of science and technology ostensibly operate. Clarke's three laws of technology are:

1. When a scientist states that something is possible, he is almost certainly right. When he states that something is impossible, he is very probably wrong.
2. The only way of discovering the limits of the possible is to venture a little way past them into the impossible.
3. Any sufficiently advanced technology is indistinguishable from magic.[3] (qtd. in Kurzweil 14)

As interesting, and even plausible, as these claims are, they are hardly "laws" in any conventional sense of the word. For one thing, qualifying phrases like "almost certainly" and "probably" are at odds with the concept of a law to begin with. For another, scientific "laws" are not established unilaterally—they must proceed through a long series of processes and tests until they become mainstream truisms for the entire scientific community. In terms of the first problem, Clarke seems to be conceding within the "laws" themselves that some things are impossible, establishing within the definition of technology the very exceptions that the law would normally be meant to challenge. Isaac Newton, for instance, did not say, "For every action there is a reaction that is *probably* equal and opposite." In order for a law to have any credence, it cannot hedge it bets through the use of innate qualifiers that overturn the very law that is being established. Yet Kurzweil forwards these "laws" as if they are universal and accepted by the scientific community at large, a claim that would certainly have no validity. To posit these as laws gives him license to make many claims that might otherwise be dismissed, and to pursue ideas that are beyond the scope of what we might consider "possible" is, in itself, not a problem. But the move is rhetorical, not scientific. His goal is to impress upon the reader a series of postulates that mislead his audience into believing his claims are a mere extension of these laws. The definition of "law" in scientific parlance, if the *Webster's New World Dictionary* is to be trusted, is "*a*) a sequence of events in nature or in human activities that has been observed to occur with unvarying uniformity under the same conditions (often **law of nature** *b*) the formulation in words of such a sequence / the *law* of gravitation/ the *law* of diminishing returns" (812). It is not my intention to split semantic hairs here, but when a scientist (and a reputable one in his field at that—Kurzweil is not merely some amateur garage scientist) is attempting to put forth a "law" in order to rationalize a theory that is at best untenable and at worst downright "impossible," one must question what motives would drive the casual use of such a serious claim as to have proffered a "law" a la Arthur C. Clarke (Clarke, by the way, is known more for his science *fiction* than his acumen as a scientist). This suggests that Kurzweil either "believes"—or wants us to believe—that the inevitability of machine sentience is as unstoppable as the gravitational pull of our own sun. This is simply untrue, and is both bad science and sloppy rhetoric. However, it may very well be effective nonetheless.

Kurzweil may be used as an extended example of a failure of scientific rhetoric—the same rhetoric that, as I will show, other scientists like Stephen Hawking have mastered. Whereas Kurzweil's motives for educating the layperson in the field of artificial intelligence are almost immediately suspect, Hawking is a trustworthy teacher because his intent and his method seem far less self-serving than Kurzweil's do. Hawking's project and the success of *A Brief History of Time,* which will be discussed in more detail in Chapter Four, are based on the public indication "that there is widespread interest in the big questions like: Where did we come from? And why is the universe the way it is?" (vii). These questions are central to our cosmic identity, and they represent a very old preoccupation with understanding the nature of humanity and with trying to predict our ultimate purpose. The origins of the human species are closely related to our notion of ourselves because we want to know not only where we came from, but also where we're going. Kurzweil reflects a hubris about our godlike stature while at the same time relinquishing this role to our own creations; Hawking manages to put or tenuous existence in more humbling terms, and it is this quality that gives his position a more spiritual significance. While we crave to be the center of the universe, we also like to believe that there is something bigger than ourselves. In this way, we have an incentive for moving forward, improving our condition, and envisioning our own future.

An examination of Kurzweil's argument helps us see that scientists, far from being above the human desires and ambitions that beset us all, are in fact susceptible to allowing their yearning (in Kurzweil's case, for immortality) to color their objectivity. What is important to note is not that this is unexpected; rather, beliefs, imagination, and motivations for self and others are part of the scientific discovery process. Without such a dream, no important innovations in science would be forthcoming. It is, in other words, the very *subjective* nature of personal ambition and the longing to make a dream a reality that both influences scientific activity and provides a motivation for discovering ground-breaking new theories that can, ultimately, become reality. The problem is one of self-image: Scientists and advocates of the scientific methodology have created an unnecessary fiction about the objective, detached, neutral orientation of the scientist. The reason for this is obvious enough. By downplaying the emotional source of scientific activities, one can easily claim that reason, rationality, and a dedication to the "facts" are all that drive scientific projects and discoveries. In a social climate that responds more readily to quantification—to the illusion that numbers are fixed and data is stable—the ethos projected by the "objective" scientist is more convincing than admitting that the real reason a scientist might be so interested in the future sentience of AI is that he wants to live forever. In this sense, computers, robots, and virtual reality are the postmodern alchemy, but what we should remember is that scientists are influenced by the same fears, misgivings, and impulses as the rest of us. Kurzweil practices his alchemy—and there is no reason to doubt that some of his predictions may in fact come true—but the most important lesson to draw from the Kurzweil example is the absolute humanness behind the scientific motivation.

The Economics of Specialization

Returning to Ellul, as a scholar interested in the trajectory of modern science, he indicates that true "scientists" are rare these days, and he further asserts they have been supplanted by technical specialists. The distinction is an important one because it means that the general scientist—though he or she may indeed have a specialty—is a nearly extinct species; the sheer depth and breadth of the technology that infiltrates our lives is so omnipresent and complex that it requires specialists of many different stripes to maintain the technological network.[4] The other main implication of this distinction is one of maintenance over research. The research scientist has a very specific field of study, and the goal is to obtain new knowledge. The technician, conversely, is the starter and stopper, the fixer and oiler of the machine. It is through the specialist that the engineering details are maintained. The general public may not have regular contact with scientists unless we happen to be one or we associate with academics, but nearly everyone has regular contact with his mechanic, his physician, his computer support person, all of whom qualify as technicians under Ellul's definition. Farther-reaching in its implications are those people ". . . who are unable to do research at all without the help of machines, large teams of men, and enormous amounts of money" (*Technological Society* 9), a body of technicians who could easily be mistaken for scientists in our research universities and hospitals. Ellul sees this distinction as important because what we often view as "science" and who we view as "scientists" is really the maintenance performed by highly specialized mechanics—people who may in fact do research but who mostly apply their tightly-focused craft in new ways. These are what Thomas Kuhn calls the "normal science" scientists, those who perform routinely applications of *la technique* and lack either the will or the imagination to pursue ground-breaking science. Extremely rare, in fact, are the scientists of the so-called "Golden Age" of science—the Galileos, Newtons, Darwins, and Einsteins of the last few centuries, men who not only practiced a method of inquiry, but also stood the world on its ear with the results. The practice of science today, according to Ellul, most often manifests itself as *merely* technique without imagination, skills without enlargement, rote activities without real purpose. This impression is interesting because it means that we most often put our faith in a scientific community that, in fact, has a very limited scope in terms of it area of expertise. Any glance at an itemized list of hospital charges, for example, illustrates all too clearly just how specialized people are. There are different technicians for anesthesia, radiology, post-op, etc., etc., not to mention the doctors and various nurses, and each of these technicians has their specific role and their specific price. How all this is coordinated is something of a mystery to the uninitiated, but it does reflect the need for a complex bureaucracy to bring the various components together. The system, of course, has its flaws, and the fate of many patients is placed not so much in the hands of the doctors themselves, but in the efficiency of the administrators who may or may not bring the network together in an efficacious way.

Obviously, the bureaucratic dimension of contemporary science is a key factor in how economics play a role in science's power and influence. As with most workforce environments, the administrative tier shoulders the decision-making burdens, and this means that the administrative branch of hospitals, universities, and corporations is larger than the actual practitioner branch. This also means that the financial decisions are left up to people who are often neither scientists nor technicians, and the clear consequence of this arrangement is that scientific projects are weighed against fiscal feasibility and institutional priority. Scientists, in other words, are often not in a position to make suggestions for the direction of research; they are bound by the needs of the institution and the availability of funds. The result of a system that places its experts on the lower rung of the organizational ladder is a precarious (and often distorted) fixture of the command structure. Anyone who has worked at a university or other institution of higher education has probably experienced the frustration of an administration that does not use its specialized talent in the most productive way. Advice is often ignored in favor of a seemingly more prudent (but usually short-sighted) financial dictum, and this can result in a deep-seated mistrust between administration and the people in its charge. More importantly, the face of administrative public relations is often not in line with the reality of the work actually being accomplished. Scientific advancement, under these conditions, is almost always exaggerated, and certainly the information shared with the public is selective and fragmentary. One of the interesting residual effects of a democratic society is that people have a tendency to voice concern over projects that have soaked up a lot of money without any discernible progress. What is often overlooked is that, especially in science, success is never guaranteed. Sometimes, the hopeful prognosis that was articulated at the beginning of an enterprise is not possible in the face of new discoveries, new data, or unexpected results. However, corporations, foundations, health associations, and universities tend to offer a different façade in the hopes of retaining old benefactors and recruiting new ones. The public's impression of what takes place in the laboratory or the field is, for this reason, a deflected one.

Technology, of course, has the greatest direct impact on (and the greatest received influence from) economic forces. Technology is the reason for our current standard of living and has been a major contributor to the current direction of economic trends. Everything that is produced relies, to some greater or lesser extent, on the state-of-the-art of the technology that is available for that product, service, or industry. However, economics, as a field, is considered a social science. As such, it is bound by the formulae that are created by economists, making for a complicated interrelationship between the forces of economic "reality" and its reciprocal reliance on modes of technology. Economics, then, is both a scientific study of market environments and is firmly anchored in the state of the technology that drives its growth. My point, of course, is that science today could not exist without the terminology for nor the conditions provided by economics as we understand it today. These two modes of knowledge (science and economics) are more than bedfellows; they exist in a symbiotic relationship where the health of one greatly impacts the well-being of the other. According to Ellul, "economic science" has seen two definitions in history that have framed this relationship for us:

> The first was given in 1850, the second in 1950. In the first, economic science was defined as the "science of wealth." Its object was primarily acquiring wealth and disposing of it. It was therefore an individual and private matter. [The more current definition of economic science is through the] objective of political economy [and] is conceived in such a way that it is virtually impossible to encompass it in a formula. As Marchal shows, we have the problem of satisfying the needs of humanity, co-ordinating [sic] the available means of production, modifying existent institutions, and even transforming human needs. These problems must all be studied not on the plane of the individual, but on the plane of the social group, and an effort made to disengage the laws of these social groups. (158)

The economic intersection with science can be seen, then, in the 1850s notion of a "science of the production of wealth" and the more contemporary problem of a "science of the administration of scarce goods" (158). This is another key distinction set down by Ellul, for it drastically changes the *rhetorical* framework that terministically informs our understanding of what both science and the economy are designed to do. The first glance, a worldview of scientific economy in the production of wealth, is a uniquely American brand of economics—one that puts the pursuit of the individual above the needs of the social order in which he lives. We see this all the time. Individual interests, from a fiscal standpoint, are not only tolerated, they are encouraged in American society. This is not to say, however, that economics are a driving force only insofar as the individual is concerned. On the contrary, according to Ellul,

> [m]ore and more, the economic fact covers all human activity. Everything has become function and object of the economy, and this has been effected by the intermediacy of technique. To the extent that technique has demanded complete devotion of man or brought to light a growing number of measurable facts, or rendered economic life richer and more complex, or enveloped the human being in a network of material possibilities that are being gradually realized, it has

> transformed the object of the economy. The economy now becomes obliged to take into account all human problems. The development of technique is responsible for the staggering phenomenon of the absorption by economics of all social activities. (158)

Ellul suggests here that technique, as Mike Hübler puts it, forwards a ". . . sociological drive to transform every human purpose and practice into a systematic and preferably quantitative method that can be measured in terms of its efficiency." This is especially true of our economic pursuits, which on a formal basis are described through terms such as "cost-benefit ratio" and "market capitalization rate"—terms that expose the scientistic nature of the economic formulae that guide financial thinking. Economics becomes, through technique, a mode of operation and a manner of thinking. Thorstein Veblen would describe this specifically as "trained incapacity," a function of corporatized educational thinking that allows conditioned businessmen to miss key opportunities through the blinding lens of a single economic orientation, an idea Kenneth Burke would later adopt and apply to all orientations, business related or otherwise. Economic language cast in scientific terms functions as more than a convenient vocabulary for describing market forces; it creates a worldview that defines our reality. All things, then, can be described in terms of an economy of technique. Hübler sums up nicely the implications of Ellul's take on the economy of technique (and the technique of economy) this way:

> The enduring value of Ellul's perspective on technique is that he is less interested in the particular artifacts of an industrialized society or information economy (since these artifacts usually change faster than an editor can publish a critical essay on them) and is more concerned with the general worldview arising from exponential growth and change in technical praxis.

The overriding question for our purposes here is how the conditions I have described above contribute to the rhetorical environment in which scientific discourse is disclosed, received, processed, and acted upon by a public that accepts scientific and technical language as both valid and, often, indisputable. More specifically for the purposes of this chapter, how are the dominant discursive forces of science and economics forged into an ideological alloy that frames our worldview, and hence, our reality? There are several prospects to consider:

1. Science has evolved into an unshakable epistemological foundation, such that almost all "legitimate" knowledge is validated or dismissed on the basis of a scientific standard.
2. Economics (in the case of the United States and much of the rest of the world, Capitalism) drives the practical concerns of daily life, from the purchasing of coffee in the morning to retirement funds to world trade, so it too is used as a sounding board for all ideas.
3. Science and economics, given their epistemological place in post-modern American society, are the ideological lenses through which all judgments must pass.
4. Given these conditions, the merger of science and economics was inevitable. As a result, Americans respond most readily to issues and ideas when they are expressed in these terms.

Science and economics, especially in the form of capitalism, which relies heavily on the advancement of new technologies, are inseparable as mutually defining bodies of knowing; the evolving nature of one impacts the development of the other in a manner consistent with the understanding of the general public. Americans, as a group, both respond to and articulate a scientific/economic vocabulary, a fact which is easily illustrated through any advertisement that depends on scientific language to sell its product. Beyond that, however, people are swayed by econo-scientific discourse in all areas that touch their lives—education, politics, health and therapy, personal finances, even domestic concerns.

It is important to understand this rhetorical dynamic in order to discern the level at which a scientific vocabulary influences and is influenced by economic factors. The details of such an analysis could be, and has been, the topic of book-length studies in itself. For our purposes here, however, it is important to establish the notion that the two most influential domains of modern life have a discursive mutual reliance and corresponding rhetorical implication at their base. By understanding how it is that people respond to language based in scientific and economic terms, we can more easily unpack some of the ways that science has achieved such rhetorical currency in contemporary society.

It is clear, for instance, that Americans are preoccupied with finances, an understandable condition considering the nature and ramifications of capitalism for the well-being of American citizens. We admire those who have attained a level of wealth that allows such people to enjoy a shift in status and privilege, essentially providing for them the means to live under a different set of guidelines than those of us who have not achieved this happy position. We not only have a separate set of rules for this segment of our population (despite what legal idealists may claim), but also believe that they have earned it through their ability to play the game with skill and shrewdness. We do occasionally show contempt for those who have played with an unfair advantage (consider the Enron scandal, for example), but this traditionally liberal view is quickly losing favor in the face of a much more conservative and binary America. People with wealth are "winners"; people living in poverty are "losers." The control of wealth is directly linked at the very least to the control of technology and, I would argue, to the favoring of the scientistic *Weltanschauung* that governs the rules of the economic game. The rationalistic, hyper-efficient preoccupation with the best ways to make, move, and keep money is at its core a scientific practice. More than that, as Ellul has certainly shown, postmodern economies cannot thrive unless they both use and sell science and technology efficaciously in a mass market. While this may not be an earth-shaking conclusion, the critical thing to remember is that, given our propensity for econo-scientific ways of thinking and seeing, the language that drives this mindset will spill over into every facet of our lives whether the use of such a terministic template is appropriate or not. As we shall see, the over-reliance on and misapplication of such terminology can have dire, indeed irrevocable, consequences.

Cashing In on Science

We are led again to the strange bedfellows mentioned earlier that are science and industry. To review what was discussed in this chapter: The intersection between scientific ideas and its fiduciary support is not new, but there are a number of contemporary developments that make such an arrangement a necessary component for pushing scientific progress. Where the Cold War military-industrial complex that Dwight D. Eisenhower warned about helped establish the need for scientific specialists and innovative technology, the post-Cold War world possesses less of a globally political motive and more of a globally economic one. In fact, these two areas of American ideological identity have become inseparable. In many ways, the factors that propel a strong American economy are reliant upon a sturdy technological base; in order to compete, companies must either develop technologically or perish in the wake of those who can. Not only does technology help traditional markets to compete, but also new technology itself is a valuable commodity. Computer and communications markets are at the leading edge of technological industries, and new products and services are being endlessly offered, often even before they are ready for the mass market. Furthermore, well-established products are constantly improving, new formulas become available, and consumer choices for items as simple as floor cleaner or toothpaste seem infinite because of advances in chemistry, for example. Given these conditions—not to mention competition for grants, research funding, military, and medical development—it is clear that "science for science's sake" is no longer a tenable prospect, provided it ever existed at all. A scientist, like everyone else, must make a living, and to do so, he or she must play the mercenary game of selling to the highest bidder. What, we might ask, does this do to the integrity of science and the discoveries it makes, and how does the general population know where the science ends and the public relations begins?

Consider the ways in which we "sell" illness and remedies. The culture of therapy is an offshoot of the idea that human behavior can be scientifically cataloged, but even assuming that science is the best way to address our personal conflicts, we should probably concede that modern psycho-therapy has less to do with science and more to do with skilled listening and responding—what is loosely known as "talk therapy." The psychiatric approach of the past (one that many still adhere to today) was a perspective that embraced a physiological cause for many of our most common psychological ailments. While it is entirely likely that serious mental illnesses like schizophrenia or manic depression have an underlying biological cause (I am not a neurologist nor do I pretend to understand neuroscience), most other common maladies probably are more a product of environment than a strict "chemical imbalance" in the brain itself. In a recent Zoloft commercial (Zoloft, for those who don't know, is an upwardly mobile antidepressant), the argument reads like something out of a Dick and Jane primer: "While the cause of depression is unknown, it may have to do with a chemical imbalance in the brain; Zoloft works to correct this imbalance so you can be your old self again." This fallacious line of reasoning is then accompanied by a cartoon graphic (ostensibly meant to represent neurons fighting each other until the super-hero/drug Zoloft steps in to break it up) that pushes the point home: Drugs are the answer, even if we don't know the cause.

Based on this example, we can assume a number of things about how the reciprocal arrangement between the corporate world and the scientific world has come to pass by examining the Zoloft commercial as an interface or access point separating the pharmaceutical developments that made Zoloft possible and the consumer population that will apparently benefit from the product. First, it is common knowledge that the pharmaceutical industry is one of the highest-grossing markets in America today. Even a casual examination of stock performances for many of the nation's leading pharmaceutical companies shows just how much money is involved in the drug industry. The economic value of these companies has increased with the recent easing of advertising regulations for prescription drugs and the corresponding jump in sales that results. Commercials like the Zoloft ad mentioned above encourage self-diagnosis because the more people "asking their doctor" about the benefits of a given drug for a malady that they have determined, based on the vague list of "symptoms" usually accompanying such an advertisement, that they do indeed have, the more likely it is that the drug will be prescribed—whether or not it's needed. Disclaimers that urge people to seek the advice of a physician assume that all doctors are meticulous in the screening and cross-referencing of the drugs they prescribe their patients and that they are equally diligent in remaining up-to-date on the latest available drug treatments. The reality is that medicine, like nearly every other professional field today, is populated by specialists who rely almost exclusively on patients' honesty to discern which drugs they are already taking. It is not uncommon for people, especially the elderly who may be seeing as many as half a dozen different doctors at any given time, to be taking several prescriptions from each of these specialists. While it is not my intent to condemn the private medical practitioners for carelessness in their prescription practices, the point here is that given the current medical and social climate—a climate that encourages the consumption of drugs for everything from heartburn to "male erectile dysfunction"—it is not hard to see just how much *business* is at stake in the pharmaceutical industry.[5] This industry relies on a wide range of scientific specialists and technicians to research, develop, and produce its drugs, and the relationship between economic need and scientific research is a tricky one. On the one hand, amazing strides can be made in the discovery of life-saving drugs for diseases ranging from AIDS to cancer; on the other hand, pharmaceuticals, like any other product today, *are* products—products that must be developed, marketed, and sold. It requires clever marketing, shrewd advertising, and incentives for the best scientific talent to succeed.

In a best-case scenario, these conditions create the possibility for enormous advancements in medical innovations; in the worst case, they set the stage for corruption. As corporate-friendly administrations make the likelihood of drug industry deregulation more probable, the market is wide open for competition in an area where such a doctrine has more dubious consequences. While a *laissez faire* approach is always popular with the corporate world—and even the middle class it relies on most heavily for its wealth—it promises serious fallout in areas that really have few businesses regulating themselves. While it is true that a litigious society might provide for adequate fail-safes in a free-market drug industry, the opportunity to cut corners and create cost-efficient shortcuts is too tempting. For the purposes of this discussion, we should consider the effect on the corporate-science association and how this relationship carries over to the private sector. The ubiquitous drug ads that we have all become accustomed to (in some cases, jaded by) have become just another part of our consumer culture. The difference is, however, that most products do not require anything more than to follow the principle of *caveat emptor,* and the consequences of a poor choice are usually not detrimental to one's health. With the increase in drug "choices," though, we necessarily require a third-party advocate, one who understands the nature of the product and the appropriate administration of it. This again obliges the consumer to have faith in his or her physician, the pharmaceutical companies that produce the drug, and the government agencies that regulate them. More than that, a faith in the image of the company is tantamount to having faith in the scientists and technicians who both develop and regulate the production of the drug. Moreover, we are forced to trust those in positions of scientific and/or medical know-how with our health, and we usually do not make a distinction between the company and the people who work for it because we can't; the company, for all practical, product-development purposes, *is* the scientists and technicians who work for it.

Let's examine this rhetorical intersection a bit, since it is through the veneer of the corporate trust image that we can and do exercise our own brand of rhetorical shorthand. It has always been interesting to me, for example, that some people won't buy a generic brand of aspirin (or any other drug), even if the difference in cost is significant. I have seen people who would rather pay seven dollars for a bottle of 150 200 milligram tablets of Bayer Aspirin than to buy the "off-brand" equivalent for three dollars a bottle. Even when I have pointed out that aspirin, by FDA law, *must* be the same (the only discernible—and legal—difference involves additives or buffering agents *added to* the drug), they respond that they "trust" Bayer (this has been, in fact, a slogan for Bayer aspirin—"The brand you trust"). We consume the corporate reassurances as readily as we consume the product, to the extent that aping their slogans becomes part of our everyday vernacular.[6] This is only possible through shrewd marketing, but the basis of that marketing is a carefully cultivated public-relations image that allows us to substitute the brand for the company, and by extension, the people behind the company. In many cases, of course, so-called "off-brands" are really just name brands that have foregone the expense of advertising and flashy packaging, the rhetorical interface that helps reassure consumers that they are "certain" about what they are buying. Brand recognition has always been an important component of American capitalism; it is the discursive template, the product's attire, which provides a convenient access point to all the associations that we have with that product. Brand recognition breeds customer "loyalty" and helps guarantee the longevity of a company. Most significantly, it functions as a symbolic motivator, carrying with it the values and anticipations of the consumer to the point of purchase, the ultimate goal of the brand's icon.

But the expectations we assign to brands because they carry with them the familiar images, icons, and symbols that have been part of our household as long as we can remember are usually fictitious; they are the superficial reminders of why we trust or prefer a particular product even if those feelings are strictly psychological. It is undoubtedly true that some products are in fact better that others, but that is not the purpose behind encouraging fidelity to a particular brand. In fact, brand loyalty is designed to keep consumers from even considering other brands, thus making any comparison of this sort is impossible. Science, or what looks like science, is often invoked to underscore a claim or to emphasize the features of a given product. Pseudo-statistics become a primary point of "proof" that a product can out-perform its competitors. But the "statistics" are often dubious, as when a brand of chewing gum claims that "four out of five dentists surveyed recommend sugarless gum for their patients who chew gum," which of course is not the same as saying that four out of five dentists recommend chewing gum—sugarless or otherwise. The apparently quantifiable nature of the four-fifths fraction speaks to the audience's tendency to respond favorably to simple numbers. Numbers are stable, we believe, and are therefore less prone to misrepresentation. But, when carefully qualified with disclaimers like "for their patients who chew gum," all the statement really means is that, according to 80% of the dentists surveyed (we have no idea how many this was; it may have been as few as five), if you must chew gum, it is better for it to be sugarless. When considered in terms of what the claim is really saying, the endorsement is far less ringing concerning chewing gum and its qualities as a dentifrice.

Product invocation of scientific "facts" or "data" is, ultimately, at odds with the intended purpose of advertising. Commercials are designed to persuade the audience to do one thing: buy the product or use the service being advertised.[7] Science, quite conversely, is normally seen as a means for weeding out unlikely outcomes, data, or hypotheses from likely ones; hence, the ubiquitous use of "science" in an effort to convince the audience of the merits of a given product are not compatible goals, in most cases. Nevertheless, scientific "authority" is a frequent tactic for selling everything from automobiles to razor blades. One notable example is for an "air purifier" that uses ionization to capture and collect contaminates from the air we breathe indoors. Called the "Ionic Breeze," it is a product that not only uses science as a marketing tool, but is ostensibly a scientifically researched product—that is, it uses (rather simple) technology to perform its claimed function. In this case, the technology amounts to series of electrically-charged plates within a housing structure that "ionizes" the air around it to capture contaminate particles on the plates. When the plates become sufficiently dirty, one can remove them from the housing and simply rinse them off, meaning the unit requires no replacement filters. It is of course necessary to first establish that such a device is a necessary component in your home, which may be the product's biggest challenge. However, given the current climate of environmental paranoia, it isn't difficult for the product's manufacturers to convince people that they are breathing horribly contaminated air that must be filtered in order to maintain optimal health in the home. According to the television version for this product, the only authority cited to establish the level of pollutants in one's home is, of all things, *The Wall Street Journal:* "According to *The Wall Street Journal,* airborne pollutants in your home can be up to five times greater than those outside" ("Ionic Breeze" Ad). The claim is accompanied by a brief glimpse of a graph that apparently shows the correlation between indoor air pollution and that of the outdoor variety. This is the extent of the "proof"—not only is the source suspect (is *The Wall Street Journal* an acceptable authority on the dangers of air pollution?[8]), but the claim is carefully, if quickly, qualified using the flag terms "can be" and "up to," suggesting that not only is indoor air pollution not necessarily greater than that outside, but that at its worst, it is five times greater. There is, of course, no discussion of which pollutants are worse or how they might be greater, nor is there any comparison between which homes are more likely to contain greater contaminates inside as opposed to their outdoor environments. It is entirely likely, for example, that a home located in the mountains away from the usual carcinogens of an urban setting that contains much tobacco smoke, uses harsh chemicals regularly, and reflects an otherwise generally unsanitary state would in fact be more polluted than the outside. But this is likely to be the exception, not the rule, a ratio that is carefully inverted for the purpose of establishing the need for the Ionic Breeze.

From the Internet version of a similar product, The Fresh Air Machine, we get a better sense of how science is invoked to sell this type of product:

> Here's how it works. Negative ions are invisible particles which carry a negative charge. Dust and pollutants have a positive charge. Since opposites attract, the ions quickly bond with pollutants. This makes them too heavy to float, causing pollutants to be knocked out of the air. The Fresh Air Machine produces high quality cleaning ions using a process that "simulates" nature: Natural Fiber (NF4) Technology. This technology is longer lasting, and is capable of producing

millions of ions at the press of a button. While you can't see the invisible ions, you'll easily see the results... less pollutants in the air you breathe. And there's more. FIVE technologies scrub your air clean The Fresh Air Machine is packed with air cleaning technology. Its advanced ionic cleaning technology will overcome many types of pollutants in your air. To go even further, it also has 4 additional, powerful cleaning technologies at work to help make your air even more clean. These include particle screening, smoke and odor absorption, H.E.P.A. deluxe, and harmful gas reduction technologies. Here's how this all works together to help super clean your air:

First, the Fresh Air Machine circulates millions of healthy negative ions using its powerful air circulator. Alongside this powerful technology, the following components further purify and refresh your air:

1. The unit's particle screen quickly captures moderate-size particles.

2. Smoke and odor are rapidly absorbed by the unit's absorption component.

3. The High Efficiency Particulate Air (H.E.P.A.) component traps difficult-to-get particles. (HEPA is the **only** type of technology recommended by the U.S. government for emergency preparedness.)

4. Harmful gases are reduced using an exclusive multi-catalyst gas reduction component.

5. Finally, air is **enhanced** with a trace amount of beneficial ozone, which assists with **control of airborne bacteria**.

The result is cleaner, **fresher air.**

> With **all of these technologies** working together, you'll have **everything you could want** to purify your home's air. Why settle for a limited-function air purifier when you can have **all this** for the same or **less cost?** There is simply no better overall value today than the **Fresh Air Machine.**
>
> **Discover what it's like** to breathe **fresh air**. Click here to enjoy the Fresh Air Machine risk-free in your own home. See for yourself how much **better** your air can be. We guarantee you'll be amazed at the results ("How the Fresh Air Machine Makes Your Air So Clean")

The site from which this advertisement was taken is named after its parent company, Science Air, a label that lets the reader know from the outset that this is a scientifically state-of-the-art air filtration device. The ad is typical in many ways of the testimonial/graphic method of introducing scientific information for the purpose of selling a new gadget. In this case, enthusiastic claims liberally punctuated with bold key words help create an effect that draws the viewer toward the visually attractive key words while concomitantly distorting selective information designed to produce the desired consumer effect (namely, enough awe to get out the credit card to purchase the thing). The structure of the information is carefully offered to be as accessible and simple as possible while retaining the key scientific features that give the claims authoritative credibility via the ethos of science. The reader is not overwhelmed with technical information, but at the same time, technical words and phrases are introduced at strategic points to create an impression of a product that has been scientifically researched (which, on some level, it probably has), but also to underscore that this device is the state-of-the-art in air-purification technology.

The brief physics lesson on the behavior of ionic particles is cursory at best, and the description of the scientific function of The Clean Air Machine is tellingly incomplete. Simply stating that "opposites attract" and that ions have a positive charge while pollutants have a negative charge does not go very far in describing how, exactly, The Clean Air Machine produces the positive ions necessary for the device to function. All that is said is that it "simulates nature" and that the process is called "Natural Fiber Technology." The technologized acronym attached to the filter, NF4 (for "Natural Fibers 4"), gives the reader an impression of technological innovation, as if the unit is just out of the experimental stage (the number following such designations usually indicate the number in the series of prototypes, i.e. this is a "fourth generation" design; though such info may not be readily known to the average reader, the technological designation is adequately impressive). It is a cutting-edge piece of equipment. However, the audience is left to make the leap that this technology is both appropriate and adequate (and logically consistent) for fulfilling the function the device claims to perform. The appeal of a product that works "naturally" relies, of course, on the buzz-word "nature," even if this is contradictory to the notion of "technology." Technology, by definition, is a rearrangement of nature for predetermined, artificial purposes. Yet, people respond to the idea because, as we have been conditioned to believe, that which is natural is always desirable over that which is not. "Natural," as a central concept, suggests a deliberate choice of words for the most likely audience; that is, those interested in home air purification probably respond well to anything that might remove synthetic contaminates to create a more breathable environment, like the air that is found in nature. A return to a "natural" environment, ironically, requires the use of new-generation technology—The Clean Air Machine is sold as a crucial piece of life-support equipment.

Also of interest are the grammatical modifiers that the ions produced by this machine are not only "cleaning" ions, but "high-quality" ones at that. None of these claims is independently verifiable by the average consumer, so even if they are true, there is no way for a purchaser to know for sure whether the machine is really doing what the ad writers say it does. Do ions vary in quality, and if so, how? Does it have to do with the strength of the charge? The marriage of traditional marketing terminology to that of authoritative scientific language makes for a very shrewd presentation of the product's assets. The advertisers can reap the benefits of a recognizable glitter-phrase like "quality" while also bringing scientific integrity to bear on the assertions that are most likely to persuade the viewer that his or her home would be incomplete without The Clean Air Machine.

As an added appeal to authority, the ad writers collate the scientific nomenclature with a governmentally-sanctioned safety approval rating. The HEPA filter concept, which is by now well-known to most American consumers as High Efficiency Particulate Air (used frequently in conjunction with vacuum cleaner systems), is both technically recognizable and scientifically appealing—another reinforcement that this is the cutting-edge of filtration technology. Add to that the government's endorsement of the product as a *safety* device, and you have a winning marketing combination. The governmental approval standard for "emergency preparedness" is deliberately vague and, I would argue, dangerously irresponsible, implying (without directly making a claim the advertisers can't possibly rationalize or which might be legally actionable) that it not only cleans your air, but also prepares you for "an emergency." What sort of emergency would require a high-grade air filter? Two possibilities leap to mind, both of which capitalize on terroristic implications: a catastrophic structural failure which would require survivors to live in a debris-contaminated environment; or a biological/chemical attack or even a natural disaster that might release contaminates from damaged buildings, etc. Under none of these emergency situations would a mere air-filter suffice to make the environment contaminate-free, and the ad writers cleverly avoid such a claim by allowing the viewer to connect the diversionary dots. The use of the phrase "Emergency Preparedness" is an especially reckless attempt to suggest that any responsible survivalist must have The Clean Air Machine in his arsenal, along with his duct tape, two-part epoxy, and heavy-gauge plastic wrap.

While some of the claims about The Clean Air Machine may in fact have some validity, it quickly becomes difficult for the viewer to distinguish legitimate information from that which is carefully delivered to both obscure the incompleteness of the testimony and to hook the viewer into buying the product. Advertisements in general are designed to encourage impulse rather than sustained critical analysis, since a potential buyer exercising any level of critical scrutiny is less likely to concentrate on the gut level response to get out the credit card and dial the toll-free number. What complicates this for the ad using scientific tactics is that the ad writers *do* want the audience to think about the legitimacy of the information, at least on a superficial plane. However, over-analysis would surely be a disadvantage since it is at the critical juncture between accepting the claims and thinking about them too much that the audience might see how fragmentary and loose (and sometimes irrelevant) the claims actually are.

Government scientists: "Our studies show results"

> Leading **scientific research** supports the technology behind the **Fresh Air Machine**
>
> Two decades of research**:**
>
> This powerful technology originated more than 2 decades ago. It was initially developed for use by the U.S. Navy, aboard its advanced submarines, where clean air is essential.
>
> Since that time, millions of dollars has been spent refining and developing the technology. In addition, we've added 4 of the best supporting purification technologies available today. The result is a **superior**, highly **advanced air purifier** that you can easily afford to place throughout your own home.
>
> The following scientific studies provide additional information about the many benefits of this promising technology. ("How the Fresh Air Machine Makes Your Air So Clean")

Just as interesting, perhaps, is the "questions and answers" section of this site, which poses "common" or "frequently asked" questions that the advertiser wants us to consider (as opposed to the questions we might actually have). It is not, obviously, an open forum that allows the potential buyer an opportunity to ask about the product and have a dialogue with a salesperson/technician, but rather, a series of pre-formulated questions to which The Fresh Air Machine people have provided carefully constructed answers. A favorite is, "Should I be skeptical of filters that supposedly last for years?"(http://www.freshairmachine.com/air/faq.htm). The question is typical of polls and surveys that deliberately frame a "discussion" by limiting the extent of the terms supplied or the questions that can be asked. "FAQ's" are, in fact, a kind of linguistic aggression designed to channel thought into preordained spaces. Anyone who has ever attempted to gather extra information from, say, a telephone company only to have the befuddled "customer service representative" prattle off scripted irrelevancies has some idea of the limitations of the FAQ approach to customer satisfaction. Rhetorically, the method aids in channeling thinking into carefully prepared dialogical domains, limiting the likelihood that someone might ask a question for which there is no satisfactory answer. The question about "being skeptical" is basically a stroking device, suggesting that the potential consumer will be intelligent enough to ask the important scientific questions. Consumers are often rightfully skeptical about product claims, and this rhetorical device helps to validate that skepticism, to feed into the consumer's conditioned sense of healthy doubt about products overall, and to provide a serviceable indictment of the competition in one shrewd "rhetorical" question. Other questions are similarly syntactically crafted, as in "how easy is the unit to operate?" or "how does the Fresh Air Machine help with tobacco smoke?" These questions presuppose that the unit *already* performs these functions, not *whether* it does, so that the answer relates to the degree of that performance, not whether it is in fact capable of these operations.

Moreover, products like the "Clean Air Machine" have the ostensible endorsement of science behind it ("Doctor recommended"; "Underwriter Laboratories Approved"), even if the information offered about its air-cleaning properties is superficial and not a little questionable. It is important to note that these devices are not inexpensive (they range anywhere from $300-$500 apiece and can only "clean" an area of about 10' x 10'), so the importance of establishing a seemingly incontrovertible environmental need for the machine is that much more crucial to the success of its sales. The ad uses a variety of tactics to forward its claims, and science under-girds them all. It is especially significant that the recent hysteria about air quality is at the heart of the rhetorical strategy. Without a climate (no pun intended) of concern about the increasingly receding domains of fresh, clean air, The Clean Air Machine would be a difficult sale. Here, the intersection between media exposure to repeated warnings about the waning quality of what we breathe and the science that provides the data for this conclusion is where the rhetorical efficacy of this marketing tactic dwells. People are deeply troubled, for example, by the existence of "second-hand smoke" because they have been told that it is nearly as dangerous as smoking itself. Politics intervenes to prohibit smoking in public places (even bars) in cities nationwide. The irony, of course, is that urban geographical environments are among the most polluted in the country—if not the world—in terms of air quality, making the prohibition of second-hand smoke a bit like cleaning a highway with a toothbrush. Yet, science is on the side of the righteous, at least so far as we know, and there is no reason why this can't also be capitalized upon by the free market.

One notable feature of this particular product is the conspicuous absence of actual scientists to endorse it, but this is not the case for many other products. "Alternative Medicine" has been a popular "field" for science-marketing, and while herbal therapies have been particularly under fire by the medical community and the FDA (at least insofar as the latter refuses to allocate resources to research the validity of claims made by the manufacturers of herbal supplements), they have also attracted many "supporters" from both medical researchers and practitioners—especially when monetary compensation is used as an incentive. Stephen Barrett, MD, notes one well-documented case:

> The most notorious "endorsements" I have seen involved United Sciences of America, a multilevel marketing company that sold various vitamin products with claims that they would protect against many diseases. In 1986, the company proudly announced that its products were endorsed by a prominent 15-member scientific advisory board that included two Nobel Prize winners. However, what actually happened was something else. The board members had been offered a yearly retainer and promised that a percentage of product sales would fund research grants for which they could apply. They were not told that their names would be used for marketing purposes. Most resigned when they found out how they were being used. That plus government regulatory action quickly drove the company out of business in 1987, but its total sales probably exceeded $50 million. ("Endorsements Don't Guarantee Reliability")

Beyond merely attaching their names to products that have dubious medicinal or convalescent properties, many scientists and physicians openly promote such products, often in misleading ways. It is not unusual for advertisers and promoters to use visual backdrops for their marketing materials that are designed to evoke a picture of competence and professionalism, such as the mainframe computers that are visible in the United Sciences of America promotional brochure (note, too, the convenient acronym that parallels United States of America—an effort to conflate the scientific with the "made in America" directive). This is part of the scientific and technological ethos that is mined to create an image of the importance of the work, the capability of the personnel, and the efficiency of the technological equipment. The reader must beware in such cases, however, because, as in the United Sciences of America case, one can never be sure if the claims are authentic or staged to deliberately give the impression that the company has access to resources it does not in fact possess. It makes one wonder whether, given the inaccuracy of the setting Barrett describes, the expertise of the personnel is equally exaggerated. But I should note that no claims are made beyond the assertion that USA has a database of dietary supplements—it never says, for example, that the rows of computers visible in a brochure for USA behind USA representative Dr. Wise (a fortunate surname for the company, no doubt) are the ones that actually store the information. This is a common rhetorical tactic for advertisers and marketers: Set a stage and let the audience draw reasonable (if unsubstantiated) conclusions. The company avoids "false advertising" liability while still achieving the desired result, a carefully forged image of the company's professionalism, expertise and access to high-tech equipment.

There are many such examples; so many, in fact, that it is unnecessary to burden the reader with them here. Any casual search on Google will reveal just how many doctors, scientists, and technicians prostitute themselves for pecuniary gain. I do not, however, wish to give the reader the impression that this is the rule, not the exception. The truth of the matter is, it would be very difficult to establish what the "rule" is, but like any other profession, greed can and does become a factor towards sacrificing integrity, and the sciences are certainly no exception. My point is less to provide a scathing exposé of corruption (or just plain charlatanism) in the sciences and medicine as it is to give the reader a cautionary stricture against the faith we might reasonably have in the science and scientists behind the endorsement of products. Advertising and product marketing are, at their very core, rhetorical practices; this does not mean that they are necessarily "bad" or "empty" or "false," but it does mean, to coin a favorite media quip, that they will always have a "spin" (the media's use of this term is misleading insofar as it suggests that this is a new phenomenon in mass media broadcasting—as if, until recently, all reports were "objective").

In subsequent chapters, I will provide some further examples of and possible remedies to these tactics in the hope of giving the reader a vocabulary and a method for dealing with distorted, deflected, or downright debased scientific information.

The purpose of this chapter has been to examine the access points at which science is disseminated to the general population, both in "legitimate" forms and in those manners that originate beyond the sphere of direct scientific activity. I have covered several possibilities because, as I hope the reader is beginning to see, the areas where science and technology touch our lives are as varied as they are, often, undetectable. One common thread is the presence of the economic factor in scientific activity. Like any other field or professional area, science and scientific activity are tied inextricably to the purse-strings that allow it to exist. This nearly always means that, in order to survive individually and collectively, scientists must also be rhetoricians, sometimes selling their wares to the highest bidder, other times convincing a patron that their wares are worth buying in the first place. And by "wares," I am speaking figuratively; I am not referring strictly to the physical products of science (though that is ultimately what benefactors want to see), but to the enterprise—the *practice*—of science as well. While science has a privileged position in today's marketplace, it, like anything else, must maintain its status through constant vigilance and advancement. The adage, "but what have you done for me lately?" applies to science as much as it does to business.

This being said, it is important to understand that the public's relationship with science is a capricious one. While the overall population responds enthusiastically to all that seems scientific, it also harbors a certain level of suspicion about the skepticism science advocates. As odd as this seems, the notion of suspicion about skepticism captures a prevalent attitude regarding science, because while the public enjoys the fruits and wonders that science offers, it does not like to accept that there are limitations. If galaxies can hold hundreds of billions of stars, why can't I talk psychically with my dachshund? If medical science can produce drugs that enhance my sex life, why can't ginko biloba make me smarter? Hence, the scientific pundits have an odd role in today's hedonistic society—the most responsible ones try to educate the public on the actual status of scientific projects using the appropriate zeal tempered with a cautious optimism, or, if warranted, the proper degree of skepticism. Irresponsible ones will gladly sensationalize anything that promises what the public wants for reasons that have little or nothing to do with science *qua* science. The key, as we shall see in subsequent chapters, is developing a method for discerning the difference. Before we do, however, it will be necessary look at some of the scientists who have helped popularize science and to see just how the media (in its varied forms) have used these personalities to promote science. The next chapter, then, will help give us both a theoretical and a personal framework for discovering how science has risen to the status that it enjoys today.

2 The Creation of Media-Ready Science

The celebrity scientist has a substantial tradition, going back at least as far as Benjamin Franklin (who was really only an amateur scientist but represented the practical side of American innovation) and, later, high-profile figures like Thomas Edison. Neither Franklin nor Edison were "scientists" in the sense that we have come to think of the university-commissioned, professional research scientist who specializes in a certain field; rather, in the humanistic tradition of Leonardo de Vinci, both men were simply well-trained recreational inventors with highly inquisitive intellects who applied scientific concepts to create new methods and devices, many of which show their influence in designs that survive to this day. At this they were highly successful, for many of our modern conveniences can be traced back to rudimentary prototypes devised by these men and others like them. Edison alone is credited with some 1093 inventions ranging from a practical light bulb to the phonograph to the film projector, along with other notable contributions to existing designs like the stock ticker, telephone, and telegraph, and it is difficult to imagine what modern life would be like without the offspring of these ubiquitous devices. Edison earned the moniker "The Wizard of Menlo Park," and for good reason: No one before in history had enjoyed such a string of successes with his new gadgets, and Edison came to be seen as more than just a highly adept and innovative inventor—he would become one of the first media celebrity scientists. A biography of Edison compiled by Rutgers University describes his celebrity status this way:

> In the last two decades of his life he became the nation's inventor-philosopher. Reporters sought his opinion on any and all subjects, from the role of inventions in the Great War and the technologies of the future to questions of diet and the existence of God. The Sec-

> retary of the Navy appointed him head of the Naval Consulting Board in 1915 to review inventions submitted for the nation's defense, and Edison conducted his own defense research after America's entry into the war. He received a special congressional medal and socialized with presidents. In 1929, the fiftieth anniversary of the electric lamp, Henry Ford staged a ceremony attended by President and Mrs. Hoover and broadcast across the country. On Edison's death in 1931 the president asked the nation to dim its lights in his honor.
>
> Edison lived the second half of his life in the glaring light of modern celebrity, under a spotlight he welcomed and sometimes directed. But he earned that light. The bankers who financed his first great undertaking, the electric light, were buying his accomplishments as a telegraph inventor, as the man who made Bell's telephone a practical instrument, and as the creator of the marvelous phonograph. Even more, they were backing the work of the man most responsible for what Alfred North Whitehead called the greatest invention of the nineteenth century—the invention of the method of invention. ("Detailed Biography")

Inventing at a time when the wonders of science had become something of a preoccupation with both the press and the public, much of Edison's celebrity status can be attributed to simple good timing. While he dwarfed other important inventors of the time like Elisha Gray, Emile Berliner, and Edward Weston as a major figure in the public spotlight, it was perhaps a combination of brilliant new technological ideas and sound business sense (and considerable success) that made him a favorite among the public. It was also clear that he cultivated his own image, actually inviting reporters to his Menlo Park and West Orange laboratories to see what his latest brainstorms were producing. His labs became a regular stop on the reporting circuit, and newspapermen would drop by on slow news days because they knew Edison could give them a quick story. He was a manifestation of all that America valued ideologically, the American Dream personified. Reflecting Yankee know-how with a capitalistic flair, and, perhaps just as importantly, hailing from modest beginnings, Edison was an icon of the best that America could produce and was viewed as a paragon of what hard work, smarts, and a good head for commerce could accomplish.

However, science and celebrity have never been closely linked. Except for Edison, most people who enjoyed fame through science did so under the banner of adventurer rather than strict scientist. Charles Lindbergh and Chuck Yeager, for example, while possessing a good degree of training in the sciences, likely would not be considered scientists themselves. They made their mark by fearlessly piloting some of the most sophisticated (and experimental) machines of science. Perhaps Wilbur and Orville Wright can be considered celebrities, but this too is a bit of a stretch since much of their widespread fame came about only very late in their lives and after their deaths. Other celebrities were explorers *and* scientists, like Jacques Cousteau, who enjoyed considerable fame for his brilliantly documented excursions aboard the ship *Calypso.* Famous women scientists have often been showcased only in naturalistic environments and are also considered adventurers as much as scientists, like primatologist Jane Goodall and anthropologist Margaret Mead. Science and celebrity in American history—when it has existed at all—has traditionally emphasized the link between the Romantic notion of the renegade researcher who has risked life and limb to further knowledge and the willingness to pursue unconventional ideas in unforgiving environments.

This legacy of the star-scientist has perhaps faded today and been replaced by battalions of "experts." One possible reason for this shift may be that, as a generation so comfortable with the technology that dominates our daily life, we simply aren't as awed by the wonder of new technological innovations as we once were. The novelty of scientific advancement has worn off to the extent that we expect technology and science to continue its break-neck pace. We have become jaded by new scientific capabilities for the simple reason that they are commonplace. We have grown up with the understanding that technology is almost instantly obsolete, and we even cynically speculate that such obsolescence is orchestrated to flood a market only so greedy companies can introduce the technological replacement for the gadget we have just purchased in order to keep the great money-making machine well primed. The last true celebrity scientist was Albert Einstein, and the closest thing we have to a "star" now is Bill Gates, a man hated by as many as he is revered. Gates complicates the media scientist role because he is not technically a scientist either, but a computer programmer (and, as is often pointed out, one without a formal degree). Gates is also something of a recluse, rarely granting interviews or allowing pictures in the mainstream media. Many don't trust him for this aloofness and for the suspicion that he has completely taken over our technological lives. He is a figure of immense power, and we are as likely to level at him our oaths of frustration as we are to hold him up as a paragon of technological genius. He is the face we put on a marching army of nerds toward whom we react with a mixture of jealousy and revulsion. It is as if we are experiencing a petty high school rivalry of cliques written on a massive scale, one where the geeks are exacting their revenge for all the injustices heaped upon them by the "popular" crowd, and Gates is leading them into the battle over the control of information upon which we all rely so heavily with a skinny, bespeckled smirk. He is not, even by the most generous analysis, a particularly attractive media personality; we sense in him a creepiness, a feeling of wizardry not exactly of this world. His reclusive habits only add to his persona of mystery, an image that is no doubt carefully cultivated for maximum rhetorical effect. Microsoft itself is an entity of considerable dominion and almost unlimited resources. It has effectively crushed or subjugated all competition not only through its unambiguously monopolistic practices but also through its dominance as a "software giant." While there are some important exceptions to this practice, the reality of the industry is that Microsoft has and does still control the most lucrative components of the information war.

The above portrayal of Gates is intended to illustrate not that the man is some evil magician casting his spells maliciously against a helpless public that is at the whim of his technological and capitalistic wizardry, but to suggest that this is how he is often painted for and viewed by the general public. He is an example of how capitalism and a free-market system have, in fact, utterly failed. Under Gates's careful control, the cornerstone of market-force theory is removed, for his company has seized control of the information superhighway and undercut all possible alternative avenues. While it is true that others are getting rich on the Internet, particularly those who are constructing important search engines like Yahoo! and Google, none of this would be possible without the omnipresent (and seemingly omnipotent) Microsoft and its implicit endorsement. Microsoft is the Ma Bell of our times, but it goes a step further than the other communication giant in that, besides providing a service (and the hardware that goes with it, as was the case with the early AT&T), it also introduced a new product, one that was "smart" in the sense that it could be programmed to accept only compatible products from potential competitors. From almost the very beginning of the computer revolution, software programs, especially operating systems, unlike telephones or the infrastructure to make them practically useful, could monitor, manipulate, configure, and reject any other programs that were brought into contact with them. What this meant for Microsoft was that it alone possessed a corner on the software market; those programs that were not Microsoft compatible could not be used on PC's or, if they were, risked system crashes, software conflicts, and a host of other troublesome, irritating, and ultimately damaging operational problems. The final result was that consumers either used Microsoft endorsed or manufactured products or they became well-trained enough in systems analysis to overcome the problems through their own programming knowledge. Of course, this second ability is possessed by only a very small fraction of the computer-using public, and most had to surrender willingly to the demands of the Microsoft system or opt to avoid computers altogether—a decision that is becoming increasingly difficult in today's working environment. One either has a set of computer skills, or one risks being pushed further down the corporate hierarchy, doomed to a position of economic servitude.

Philip Eubanks adds another interesting rhetorical layer to the conversation about Gates: his place in the technological and scientific American narrative. In a poetics/narrativity study conducted by Eubanks, he observed in both a *Playboy* and a *Time* interview of Gates that what the computer mogul really wants is to be part of the "story" of the American myth of the self-made man. Responding to charges of unethical business practices, Gates provides in both interviews his own version of the narratives that circulate about him. In this way, narrative becomes more than just stories spun for entertainment purposes or to reinforce cultural values; they become arguments and rebuttals. For example, Eubanks notes that "Gates is irked most by a version of his personal success story that begins with his having a million-dollar trust fund while he attended Harvard" (39). Gates's denial of the trust fund but confirmation of the success of his family indicate an effort to recast an appropriate myth (39), but why Gates would consider this important is the interesting question. In his own cleverly rhetorical way, Gates recognizes that the public support of a pampered trust fund baby who becomes the richest man in the world is far less appealing than the more popular American myth of the man from humble beginnings (like Edision) who makes a fortune through hard work, innovative ideas, and perseverance. Such romantic heroes of yesteryear are hard to find indeed. The Carnegies and the Edisons of the past—men who had faith in the American Dream and sought to make their own—no longer exist. Gates makes his narrative argument at the intersection between myth and reality, where he wants to acquire enough of the features of the boot-straps image to be appealing but to concurrently show that the corporate hero of today is a different figure. However, Gates may be what he fears the public really sees: an advantaged youth who had a jump start on the competition and seized it—unapologetically and even, at times, ruthlessly. Such a picture is not nearly as romantic as the alternative, but Gates is not a romantic figure. He wants to reinvent the myth just as he wants to monopolize the software market.

Viewed this way, it is perhaps no exaggeration to say that Bill Gates and his company have a rhetorical and practical stranglehold on the working public, and the workforce is, of course, only one facet of the growing computer industry. Recreational computing accounts for almost as much computer activity as professional applications do, and between the gaming market, the Internet, digital photography, personal websites, and a wide range of other computer-based diversions, the bubble shows no signs of bursting anytime soon. So while on the one hand Gates pioneered the information revolution, on the other, he became its most dubious proponent, for it seems that his entire project has been, from the outset, less than altruistic. His contribution to technology and its place in the public arena was arguably every bit as influential as Edison's, but he has certainly not been received by the public with the same enthusiasm and reverence as The Wizard of Menlo Park. Gates's name has, rather, become synonymous with Silicon Valley (though Gates himself is from Seattle), an exceedingly adolescent place that does not always evoke the most sympathetic associations, but instead conjures images of yuppiedom, inflated egos (and inflated real estate, bank accounts, and identities), excess, technological monstrosities, and rampant capitalism. While such a portrayal may not be fair, Gates garners little in way of the type of reverence that Edison enjoyed, largely because his financial maneuverings have always been a bit below board. Microsoft has been—and is going—through several antitrust cases, and there are many conspiracy theories surrounding his sovereignty in the computer world. Whether or not Gates is really staging a full-scale monopoly or trying to take over the world and cast it in his own nerdish image is beside the point; the public in general has an impression of the man that separates him from the great inventors, innovators, and thinkers of yesteryear (like Edison, Ford, or Einstein) who enjoyed a stardom that was widespread, reverential, and, perhaps, even deserved. The Bill and Melinda Gates Foundation notwithstanding (some argue that this foundation was established only as a way of repairing Gates's public image as a malevolent, vengeful opportunist), Gates simply does not command the same respect or status as our earlier technological pioneers, even though he is rich (by most estimations, the richest man in the world) and famous (or infamous, depending on one's point of view). This may be in part because people are simply not as easily impressed by technological innovators as they once were, in part because they are jaded by pubic figures who seem only to disappoint, in part because Gates is not as charismatic (or even likable) as some of his predecessors, and in part because the media itself downplays its attempt to glorify such personalities in favor of movie stars and athletes who have far more selling power and public appeal than inventors and captains of industry in this role.

A more focused examination of how such people and their projects are portrayed in the media might help answer the question of how our society is informed about the science and technology that is dispersed to it and how the talking heads of science are projected. It is also important to note that the issues attached to the media-created scientific persona has an impact on both how we view the issue and the person who represents it. An area of considerable anxiety in recent years has been, of course, the environment, a rather sweeping word insofar as it reflects a concept that is applied to a broad scope of potential problems and concerns. Strangely, many of the most "popular" environmental crises are far less significant to the health of our planet than other, far more pressing, environmental problems that receive little or no attention at all.[1] The Ozone Layer, Global Warming, and the erosion of rain forests, while important problems to consider and, we hope, find solutions for, are probably not as dire as other, sparsely-covered topics like overpopulation, air and land pollution, and simple resource depletion.

The erosion of the environment, one is compelled to point out, is not in fact so neatly categorized; the problems with the atmosphere, for example, are a direct result of larger environmental issues. Perhaps the precursor of all environmental concerns should be overpopulation, for if it weren't for the sheer number of humans on this planet, problems like pollution, global warming, ozone weakening, increased levels of carbon dioxide, and a host of other environmental ills would be far less disheartening than they currently are. Overpopulation, as a topic of concern, was in fact discussed far more seriously in the past than it is currently, and the primary reason that this issue has fallen from favor is because solutions to such problems always trespass upon areas of human rights that spring the lid on a Pandora's Box that we are neither technically nor ethically equipped to deal with. About the only measures that can be practically taken to stay the swelling of the earth's population have to do with control of reproduction, a highly charged subject that would certainly have social repercussions far beyond the strictly empirical observation that there are just too many people in the world. A statement from the Hampshire College Population and Development program illustrates how politically unpopular this topic actually is, especially among academics harboring a tendency to read any such discussion of these problems as an attempt to oppress the weak or to marginalize minorities, as Anne Hendrixson does in her article "U.S. High School Social Studies Textbooks: Perpetuating the Idea of Overpopulation":

> Too often references to population are full of stereotypes of the "poor South" versus the "affluent North." [The high school textbook] *World Cultures: A Global Mosaic* is laden with judgments about differing levels of development in the North and South. The authors use words like "failure" and "condemn" to describe Southern levels of development . . .. Accompanying these perspectives on the South and North are racial stereotypes that can engender misunderstanding and dislike. For example, the author of *World Geography and Cultures* promotes the stereotype of Africans as sexually promiscuous. She charges that the out-of-control population growth rate in Africa is causing the spread of the desert, and that African people have to take responsibility for their actions (i.e. use birth control) for

> the process of desertification to slow.. . . The texts hold women as largely responsible for population growth because of their role as child bearers. The majority of texts present government-enforced population control, rather than user-controlled family-planning methods, as reasonable measures to curb population growth, without critical discussion of the effect of population control methods on women's health and rights.

For Hendrixson, the introduction of this topic is an occasion to further her own pet political agendas and interests, typical of the gender, race, and class inclination found so often in today's scholarship. "Overpopulation" for her is a phrase (and an idea) used to promote oppression of minorities—the poor (this group is not technically a "minority"), the black, and the female, specifically. She concludes that the textbooks she cites present a "simplistic, incomplete" overview of the issue, and that their discussion of overpopulation "perpetuates misunderstandings." The texts she cites may indeed be misguided in their overview of overpopulation issues, but the significant point here is that overpopulation, as an environmental topic, has become appropriated by other political platforms, decidedly leftist and contentious ones at that, effectively shutting down dialogue on the subject for fear that any discussion of remedies for a very real problem will be viewed as an expression of intolerance, or worse, seen as the first step in the repression of the individual rights of minority citizens. The issue is, as a result, simply too politically hot for the mainstream media to handle, and there are safer rhetorical ways to appear environmentally responsible and "raise public awareness."

Given this kind of topical out-of-bounds, therefore, the media has chosen its own fixation on politically innocuous environmental concerns that only require us to use non-aerosol products or to drop a dollar in a 7–11 jar to "save Fluffy, The Local Homeless Cat" for the SPCA. Such measures are inoffensive and promote the illusion of widespread social consciousness—a win-win situation. No reasonable person would argue (which is not the same as saying that the argument wouldn't be made) that avoiding the use of CFC's will somehow rob a minority group of their collective identity and economic opportunity or that we should allow unrestricted cat abuse because it is so central to the livelihood of people in underdeveloped countries. We want to be environmentally responsible, but not if it means we will also be vilified for our choice of causes or, perhaps more hurtful, unduly inconvenienced. The failure of early efforts to recycle should be a good indication of this lesson about American environmental consciousness—only the most dedicated among us were willing to sort, stack, and bundle our recyclables in those days when this cumbersome task was required before our plastic, metal, and paper would be hauled away to the processing station. It was only after many communities were willing to put in place fully-equipped recycling centers that would sort our refuge for us that we were willing to make an effort to have a "garbage can" and a "recycling bin." Now, recycling programs are hugely successful, but only because the process has been made more convenient and accessible.

Media representations of environmental issues have been largely disarmed in this way, and those who control media outlets have, as a result, opted for a superficial treatment of environmental concerns that affect us all. Science fits into this equation, but only on the shallowest level, as a source of authority but without too much superfluous information—a kind of ongoing *Idiot's Guide to Environmental Concerns* program. The dilution of scientific information and the oversimplification of the complexities of environmental problems stem from the basic assumptions held both by the scientists who produce the data and the media that delivers it. According to Peter Weingart in "Science and the Media," the "traditional" view of scientific popularization holds that the audience of such information is passive and unspecified. He notes that the general public is "perceived as purely receptive" and that it is "excluded from the production and validation of knowledge" because "by implication [it] is considered to be incompetent to judge the transferred knowledge" (869–870). The result of this "asymmetrical image" is a further implication that such audiences "have no selection criteria of their own when processing scientific knowledge" (870). This approach, according to Weingart, was in part the result of the training scientific journalists received, while later (post 1970s, by Weingart's account), media coverage of scientific issues came under critical scrutiny of science when the adequacy of information was in question (870). Despite this, it seems that the media controls the locus of power in this discursive dynamic; science possesses the apparatus to scrutinize media coverage in positive or negative ways, but the media has by far greater influence. Perhaps because of some of the assumptions Weingart cites—i.e., that the general public is too "scientifically illiterate" (to coin a common educational term) to adhere to sound scientific arguments regarding the inadequacy of the media's representation of scientific activities—the media holds sway over public opinion on scientific matters. The media can, for example, be selective in the types of science covered and in the scientific personalities it uses. Weingart has critiqued the tendency to seek sociological validation of the obvious through his examination of R. Goodell, who makes the final declaration that the way scientific figures are chosen for media representation have to do with criteria like "extraordinary personality, high level of communication and self-representation, [and] the attractiveness of the topic of his/her work with reference to problems and fears of society" (871).

This list of criteria is, of course, no different for science than for any other area of interest covered by the media. It is rare that ordinary personalities, poor communication, and revulsion at the work someone does are going to make good copy (this "traditional" set of requirements may be shifting, however, if reality television shows are any indicator, but "the media" in Weingart's essay are the media of news and features, not of the amateur-hour circus that passes for entertainment these days). For Weingart, as for myself, the question becomes a central one for rhetoric and science: how much and in what way does science become reliant on public consent, and how might this fundamentally change both the way science is conducted and how science is represented to a public possessing certain informational needs that may be in contrast to how the science is being done and what it ends up saying? The notion of scientific "reputation," for example, has a far different meaning through media exposure than it does within traditional scientific communities. Reputation, it has been said, is everything to the scientist, and reputation is often hardly a matter of strict scientific judgment. One's reputation is also one's professional meal-ticket; tenure, grants, job offers, and simple renown are among the motives for achieving a solid reputation among one's peers. But Weingart notes that "the media have their own internal reconstruction of scientific reputation. Media prominence of scientists is a media-specific construct" (874). For those scientists who have had other avenues to establishing a reputation cut off for whatever reason (not the least likely is plain mediocrity), media recognition becomes an attractive alternative. Weingart further suggests that

> [f]rom the perspective of science it is theoretically possible that the attention of politics is directed to media-prominent scientists and their topics and that this attention is translated into decisions of resource allocations even though the scientists' prominence is not in line with their reputation; from the perspective of the individual scientists, it is likewise imaginable that the path to success which has previously been blocked in internal evaluations may be found via prominence in the media; from the perspective of science policy, this creates the potential problem of focus being directed at topics whose state of research does not justify the support suggested by media attention. In other words, the media coupling may theoretically lead to a competition between media and scientific criteria of relevance and validation. (874–875)

Public reception of the issues, moreover, is what drives much media coverage, especially those issues that seem to have a direct effect on the quality of life that Americans have come to enjoy. One such recurring discussion is climate change, specifically, the "Greenhouse Effect," a metaphorical representation of a phenomenon (as we are all perfectly aware now) that contributes to global warming. Metaphorical explanations, especially dominant images or comparisons that define an issue, serve to both clarify and distort the science they are meant to illuminate. In the case of the Greenhouse Effect, we see the conflation of a well-understood phenomenon—the retention of heat and other climatic determiners—within the glass structure of a greenhouse, a process that is used for climate control for the artificial reconstruction of plant and animal habitats, to the atmosphere which acts as an increasingly unyielding trap for sunlight (and therefore heat) the denser the atmosphere becomes.[2] While this is an important contributor to maintaining a suitable living climate on earth, it can be environmentally problematic if the atmosphere becomes so thick with man-made pollutants that the effect becomes unnaturally amplified (hence people often confuse the Greenhouse Effect with global warming, even though the former is a necessary component of a stable environment and the latter is an artificial acceleration of a natural process). The result is a rise in the global temperature that can have unfavorable long-term effects on the biosphere. Without a basic understanding of how a greenhouse works, this phenomenon would be abstract to most casual recipients; it is therefore often accompanied by visual aids that demonstrate the metaphor graphically, though "realistically," through an image of the earth and its atmosphere and the corresponding arrows, lines, and labels that make the a demonstration of the greenhouse effect more complete. Sometimes this analogy is simplified even further, foregoing any image of the earth at all in favor of the central metaphor itself. This is especially true when the process is explained to children, as in Figure 1.

Figure 1. "Greenhouse Effect." *Climate Change Kids Site.* The United States Environmental Protection Agency. 2006.

The treatment of issues such as these often lead to what Weingart refers to as "catastrophe discourses," as those that may occur when science is called upon to provide an overview of a particularly pressing environmental scenario and to offer both an explanation of the causes of a specific problem and to forward possible solutions for it. When this happens, the discursive interface between the scientific community and the public must shift rhetorically to capture the severity of the problems and impress the need for immediate action, and to do this effectively, the issue must often be stripped of its technical aspects and condensed for an appropriate public understanding of its gravity. It is not a matter, according to Weingart, of whether scientists themselves provide the public with deliberately erroneous information to unduly dramatize the importance of addressing a problem; rather, he says, "scientists adapt to the media when addressing policymakers and the public. The results are simplified, dramatized pronouncements and prognoses calling for immediate action which are taken up and amplified by the media" (876).

Weingart's insight here illustrates the process well. As with all commercial media, the end result is driven by a corporate mentality. In many respects, media outlets seem less concerned with whether the information they are offering is technically "correct" than they are with how well their representation of the problem lends them authority on the issue at hand or underscores the severity of a problem that is deemed too important to trivialize. This is more than a simple "bottom line" mindset; in fact, it reflects a rhetorical need to appear to be on the cutting edge of society's most immediate problems. While this approach may ultimately translate into higher ratings and more money for a network or a newspaper, its real value is in maintaining a media source as a central, trustworthy, and progressive resource for important and apparently accurate information. Scientists are sometimes naively caught in the middle of a media tempest that simply exploits their status as "experts" and "spins" an issue in a particularly sensational way. For example, cable stations like Discovery, The Weather Channel, The Learning Channel, and their subsidiaries like the Discovery Science channel must compete with other programming just like any other cable network. To accomplish this, such stations often take wholly natural phenomena like tornadoes, hurricanes, or other atmospheric forces and give the viewing public the impression that these are brand new occurrences and lend them the "human element" necessary to construct a dramatic narrative. This provides the audience with an odd sense that he or she is witnessing catastrophic natural events of epic proportions and further lends the station credence as a reporting instrument of important, opportune information. Educational channels have resorted to "edutainment" in order to survive in the highly competitive cable station market, reducing such natural events to an undignified series of ostensibly "extreme," "intense," or "freak" occurrences. The overall effect is one of novelty, a feeling that such natural phenomena are brand new blights against humanity, regardless of their obvious stature as a mainstay of the natural world.[3] Despite our best efforts, floods, tornadoes, hurricanes, and earthquakes do happen, and there is not a thing we can do to prevent them. Nature is indifferent to its own contribution to our suffering, and in this respect it should be clear that we are no better nor worse than any other organism that has struggled for survival in the very long history of life on this planet. Dramatizing a natural event for improved ratings and anthropomorphizing it so that we can make meaning out of the loss and destruction it may cause does not change this harsh fact.

Unlike the punctuated effects of natural events like hurricanes, global climate change has the advantage of being a phenomenon that can be tracked slowly, and data can be collected systematically providing much time for careful interpretation. This is also its disadvantage, from a scientific point of view; since the data are cumulative, they are also ambiguous and can be interpreted in more than one way. Is climate change, which does naturally occur, being artificially accelerated by human activity, and if so, what are the implications for the long-term "health" of the planet, and what should be done to correct the negative influences humans have had on it? What are the specific human activities that are contributing to these effects, and at what rate are they amplifying their natural progress? These are not easy questions, because collecting data on trends in climate change leaves little to compare them against in geological terms. We can extrapolate certain patterns, but only for as long as the information has been accurately collected. In many cases, this means less than one hundred years, a veritable nanosecond in meteorological time scales. Scientists can only observe past climate change indirectly, through evidence gathered from its effect, like the analysis of glaciers. The reality of climate change undergoes a morphing process when it enters the very human realm of media and politics, creating what Sheldon Unger refers to as the "knowledge-ignorance paradox," which he describes as "the processes by which the growth of specialized knowledge results in a simultaneous increase in ignorance" (297). This interesting notion reflects what Ungar sees as a public trend to "militate against the acquisition of scientific knowledge" whereby he argues that discourse concerning the ozone hole could command a different level of attention than could garden variety climate change because, as an issue rather than a simple scientific fact, it is at once a topic of concern and could rely on "easy-to-understand bridging metaphors" that made the concept far less elusive than the more mundane descriptions and impact of climate change (308). Ungar concludes that the general population is motivated only through a "need to know" mentality, and this is often how scientific information is offered. Beyond the immediacy of a real or perceived need, most of the public, Ungar says, is neither equipped to deal with nor particularly interested in obtaining a higher degree of scientific literacy (308).

This position is, admittedly, difficult to argue against. It seems clear that science is not alone in its inability to foster increased awareness or its inability to supplement knowledge about its own pertinent activities—many other areas of specialized knowledge that might have useful applications in a democracy are also seen as irrelevant or ignored altogether—politics, history, and civics, to name a few. The atmosphere of instantaneous information and the erroneous expectation of absolute certainty about scientific claims exacerbates this problem. Stephen C. Zehr has argued that when scientific uncertainty about an issue (in this particular case, global climate change) is acknowledged by the media, the general public views such uncertainty as a scientific failure rather than as a necessary component of further research and inquiry. He further suggests that this very admission of uncertainty about global climate change "was used to help construct an exclusionary boundary between 'the public' and climate change scientists." Such a boundary was rhetorical in that it "delegitimated lay knowledge by suggesting that the public did not hold appropriate reverence for scientific uncertainty and the need for more research" (85). Such a relationship can create a great deal of tension and ill-will between scientists and the public, the former seeing the public as merely uninformed masses who don't "get" the process of science, and the latter seeing scientists as a group of self-important grandstanders who aren't even sure about their own findings. One general conclusion we can draw from this regarding the public/scientific community relationship is that the public has a clearly misguided view of science and its capabilities. People seem to view science as infallible, an attribute that it simply doesn't possess and an access to certainty that it doesn't always have. This may stem in part from the cultivated image of science and the scientist as almost magical; the wonders of science are so impressive—and so ubiquitous in our society—that we have been lulled into a false sense of comfort about its process and its capabilities. We have come to expect perfection, and I would argue that, contrary to Zehr's conclusion, this does not indicate "a lack of reverence" toward science but an all too profound reliance on science to supply all the answers all the time with absolute certainty and exactitude. This state of mind can only come into being if it is encouraged by agents not concerned with reflecting the genuine nature of the scientific process.

Determining whether we are, in fact, ushering in new "ice ages" or are experiencing "global warming" seems as much a matter of media capriciousness as meteorological and geological reality. According to Weingart, media hysteria regarding these issues was not exclusive to the United States, and even certain scientific journals showcased studies that predicted imminent doom:

> The debates on climate change ran more or less parallel in the USA and Germany. [. . .] In the 1986 an article in the journal *Physikalische Blatter* predicted a climate catastrophe that would render the earth uninhabitable. The prognosis came with a relatively precise date: 'Irrevocably in the next 50 years.' The authors made explicit what interest had motivated them. Atomic energy should be developed rapidly in order to lower CO2-emissions. In the same year the German Physical Society published a 'warning of impending worldwide anthropogenic climate change' with estimates of the rise of the sea level (5 to 10 m were believed possible) which, in turn, triggered the media to dramatize. *Der Spiegel* provided the icon of the debate on its title page with the Cologne cathedral standing in water. The first proclamation demanded an immediate worldwide regulation that focused specifically on Germany. The scientific uncertainties of the prognoses were played down in view of the claimed urgency and feared every irreversibility of the change. (876–877)

A rise in ocean levels of five to 10 meters worldwide would indeed be catastrophic, but twenty years after this dire prediction, sea levels have not come anywhere close to rising at the rate needed to realize this prediction. According to Anny Cazenave of *Science* magazine, "Remote-sensing data suggest that ice sheets currently contribute little to sea-level rise. However, dynamical instabilities in response to climate warming may cause faster ice-mass loss" (1250). Despite recent arguments about the dangers of melting ice sheets by politicians like

Al Gore in his *An Inconvenient Truth,* other equally compelling arguments are being forwarded by scientists in venues that demonstrate how media fixation on the more sensationalistic predictions gets airtime over more conservative estimates, but also provokes the question of how much responsibility the media should shoulder in providing evenly presented information, especially for issues on the scale of Global Warming. As seen above, the German sources have a solution for suppressing the increase of CO2 emissions which lead to a heightened greenhouse effect which causes accelerated global warming which melts icecaps, raising the sea level too quickly: nuclear power, which is a cleaner energy source in terms of the atmospheric emissions it creates (though is also generates other unsavory environmental outcomes which should not be overlooked). One must wonder, then, how politically motivated such studies are, who's paying to have them conducted, and how the results are selected and to whom they are distributed. There is more than strict science going on here, in other words, and it is not unreasonable to suggest that, in the interest of professional stability, some studies are simply unreliable. Bring in the media, with its own motives and desired outcomes, and the result is less than conclusive; it is downright suspect.

To illustrate this rhetorical intersection further, consider the June 17, 2005 issue of *The Chronicle of Higher Education,* in which research science was revealed to be far less sterling in its professional ethics than most of us assume. According to a survey of more than 3000 research scientists, "a third of participants [in the survey] acknowledged that they had engaged in actions such as overlooking others' use of flawed data, failing to present data contradicting one's own work, and circumventing minor aspects of human-subject requirements" (Monastersky, "Scientific Misbehavior," A11). More significantly, of those who admitted to these types of ethical indiscretions, "15.5 percent reported that they had changed 'the design, methodology, or results of a study in response to pressure from a funding source'" (A11). Scientists are pressured to produce results for certain types of projects, and those results can sometimes have as much to do with the agenda of the funding source as with any interest in actual scientific advancement and new knowledge. There are many cases where researchers working for corporations (or university professors funded by them) are expected to pursue a line of inquiry that will lead to a predetermined result. If this result cannot be achieved, then either the researchers' funding is cut or the data that are available are skewed for public dissemination. This is not a conspiracy; it's business, and funding sources today want (and demand) results for their investment. And a large investment it is. Research and development of many products or potential technology is extremely expensive, requiring as it does massive capital commitments to equipment and personnel. Scientists working for "big tobacco," for instance, may have extreme pressure placed on them to produce data that contradicts more common sense conclusions in an effort to aid tobacco companies in both costly litigation and in the image they hope to improve for the media, all because the companies see the role of scientists (just as they see the role of lawyer or accountant) as one that aids them in furthering their business ventures.

Smoking is, incidentally, a topic of scientific/rhetorical interest for a number of complicated reasons, because the fabrication of scientific evidence cuts both ways. While the motives behind tobacco companies encouraging data favorable to them is obvious, less clear is why there seems to be such a blatant maliciousness against smoking itself. The subject has become at once a social and an environmental issue that has made it an interesting scapegoat for a sweeping array of health issues, ranging from lung cancer to breast cancer to increased incidents of ear infections in small children. There is a huge and powerful movement to ban smoking in all public places and even outside of certain public buildings. Most states have already adopted such legislation, all because, we are told, "second-hand smoke" is as detrimental (some even argue more so) to non-smokers as direct inhalation is to smokers. These claims are verified by a number of oft-cited scientific studies, but note that the media is quite selective in how it chooses the studies to showcase and how it delivers the information in a socially acceptable way. There are a number of credible studies that, in fact, do not support the politically popular notion that second-hand smoke is the cause of so many health problems. One of those was conducted by James Enstrom and Geoffrey Kabat, and the results appeared in the *British Medical Journal* in May of 2003. Its objective was "to measure the relation between environmental tobacco smoke, as estimated by smoking in spouses, and long-term mortality from tobacco-related disease" (1). 118,094 adults were enrolled in the American Cancer Society cancer prevention study between the years 1959 and 1998, and the focus of the study was to determine whether 35,561 of the participants who never smoked were affected adversely by spouses who were known smokers. The results are interesting, because they are in direct contradiction to the inherited wisdom the public has been given by media outlets representing other scientific findings:

> No significant associations were found for current or former exposure to environmental tobacco smoke before or after adjusting for seven confounders and before or after excluding participants with re-existing disease. No significant associations were found during the shorter follow-up periods of 1960–5, 1966–72, 1973–85, and 1973–98. (1)

The conclusion was just as strident in showing that there was no significant association between environmental tobacco smoke and death by tobacco:

> The results do not support a causal relation between environmental tobacco smoke and tobacco-related mortality, although they do not rule out a small effect. The association between exposure to environmental tobacco smoke and coronary heart disease and lung cancer may be considerably weaker than generally believed. (1)

Surprising as the findings of this study may be, they are not isolated. In the Sept./Oct. 2001 issue of the *Journal of Exposure Analysis and Environmental Epidemiology,* researchers conducted a similar study on the effects of second-hand smoke in the workplace. Below is the abstract for the resulting paper, titled "Environmental tobacco smoke in an unrestricted smoking workplace: area and personal exposure monitoring," included in order to establish its methodological credibility, though the final claim is the most important for our purposes:

> The objective of this investigation was to determine the extent of a real and day-to-day variability of stationary environmental tobacco smoke (ETS) concentrations in a single large facility where smoking was both prevalent and unrestricted, and to determine the degree of daily variation in the personal exposure levels of ETS constituents in the same facility. The subject facility was a relatively new four-story office building with an approximate volume of 1.3 million cubic feet.

The exchange of outside air in the building was determined to be between 0.6 and 0.7 air changes per hour. Eighty-seven area samples (excluding background) were collected at 29 locations over the course of 6 days of sampling. Locations included offices and cubicles occupied by smokers and nonsmokers, common areas, and the computer and mail rooms. Twenty-four nonsmoking subjects wore personal sampling systems to collect breathing zone air samples on each of 3 days in succession. This generated a total of seventy-two 8-h time-weighted average (TWA) personal exposure samples. In all samples, respirable suspended particulate matter, ultraviolet light-absorbing and fluorescing particulate matter, solanesol, nicotine, and 3-ethenyl pyridine were determined. With the exception of a few locations, tobacco-specific airborne constituents were determined in all samples. Not surprisingly, areas with the highest ETS constituent concentrations were offices and cubicles of smokers. Median and 95th percentile concentrations for all area samples, excluding background, were determined to be 1.5 and 8.7 microg/m3 for nicotine, and 8.2 and 59 microg/m3 for ETS-specific particles (as solanesol-related particulate matter, Sol-PM), respectively. Personal exposure concentrations of ETS components were similar to those levels found in the area samples (median nicotine and Sol-PM concentrations were 1.24 and 7.1 microg/m3, respectively), but the range of concentrations was somewhat smaller. For example, the 95th percentile 8-h TWA nicotine and ETS-specific particle (as Sol-PM) concentrations were 3.58 and 21.9 microg/m3, respectively. Intrasubject variation of daily concentrations ranged from 20% to 60%, depending on the component. Self-reported proximity to smokers was supported by higher ETS concentrations determined from the personal monitors, but only to a modest extent. Although smoking was completely unrestricted inside the main office areas of the facility, ETS levels, either real or from personal exposure measurements,

> were lower than those estimated by Occupational Safety and Health Administration to be present in such facilities. (375)

It is not my intent here to crusade for "smoker's rights" (a fictitious "right" indeed) or to forward a conspiracy theory against overly-zealous political activists who have taken aim against smoking in any form, but to demonstrate the misinformation (or, at least, highly selective information) that is doled out to the public, not in the interest of "consciousness raising" or scientific accuracy, but in the interest of imposing partisan views on the public using science as the foundation for its authority. Smoking is harmful, and most of us don't need scientific studies to validate this reality—a hacking cough, decreased immunity, and shortness of breath are all the personal evidence most smokers need to attest to that truth. It is also not news that smoking is addictive; anyone who has tried to quit knows just how true that also is. "Second-hand" smoke, however, falls in the much fuzzier area where one person's activity may have an impact on the comfort of another—but "comfort" has been generously extended to "health" in this matter, an interesting transformation indeed. But why is the fervor against second-hand smoke so disproportionate to the apparent damage it causes? Why would groups lobby so stridently against it, as if it is truly a life or death issue? Surely, there are other carcinogens in the modern urban and suburban setting that are more damaging—gasoline and diesel exhaust leap to mind—but no one seems zealously driven to outlaw automobiles or buses. No one is campaigning for the prohibition of driving "in public places." There is a movement to increase research in renewable fuels, but this has more to do with a feared shortage of fossil fuels and personal economy or the effect on the biosphere as a whole as it does with the personal health risks associated with vehicular air pollution. The benefits of lottery-like court settlements for the families of dead smokers seems a likely culprit for the incentive propelling the anti-smoking crusade, at least on an individual level, but one could just as easily argue that the real reason has to do with worker productivity and the cost of insurance for smokers to employers, not with any genuine concern for our personal well-being.

And the point is that science and the public, ostensibly the two most important agents in this drama, become incidental playing pieces in a much more intricate economical/political process. At the media center of the popularization of science is the "science reporter" who has the rhetorical advantage of controlling the discourse of the scientific information that is released to the public. Jay A. Winsten offers a list of the complaints and problems surrounding this arrangement, including

> . . . the limited, incremental advances in research [that] are portrayed inaccurately as major developments or breakthroughs; that the risks posed by putative health hazards are frequently exaggerated; that there is often a striking imbalance between the amount of attention accorded a piece of research and its actual scientific importance; that many medical news reports shamelessly exploit the emotions of desperately ill patients, their families, and the public at large; and that science news coverage often is plainly inaccurate, due to errors of commission, omission, or both. (6)

While Winsten admits that these criticisms are sometimes justified, he suggests that more often the onus for the blame in science reporting inaccuracies can be attributed to the scientific research community itself. He further notes that, in order for the press and other media outlets to receive the information that is central to the reporting of science news, the research community must first have an intimate understanding of how such reporting agencies work within the confines of its own public role and responsibility (7). One aspect of this understanding is what Winsten refers to as "constraints on quality," which include the following 19 restrictions in accuracy and clarity:

1. Time
2. Space
3. Competition for page one
4. Competition among news agencies
5. Editor as gate-keeper
6. Editing processes
7. Constant demand for something new
8. Degree of access to news sources
9. Degree of disclosure by news sources
10. Bias of news sources
11. Difficulty in explaining science
12. Bias and style of reporters

13. Experience of reporters
14. Structure of news stories
15. Headline writers
16. Need for a newspeg
17. Prior reporting
18. Nature of the publication
19. Tyranny of bandwagons (8)

While all of these factors contribute to the distortion of scientific findings and outright misinformation regarding the content and significance of scientific reporting, certain statements by science reporters in Winsten's twenty-seven interviews offered particularly revealing insights into the reasons behind this situation. One interviewee, identified as "the science editor of a leading newsweekly," admitted that a central problem facing science reporters is the fear of "losing the story": "That's the danger," he said, "if you are too low key. Science, relatively speaking, is allocated much lower priority compared to other aspects of news. To get into print, you do have to beat the drum and sometimes that leads to a little exaggeration" (9). Science reporting, as a "low priority," suffers from the ethical dilemma of either becoming sensationalized, in which case the news may lead to alarmist reaction on the part of the audience, or from the prospect of being ignored altogether. To weigh the impact of a particular story, the reporter must decide whether it is better to push "the boundaries of truth" (an interesting relativistic phrase by any account) or to let the information fade into irrecoverable obscurity. Like so much of the information available to us, we rely heavily on the judgment and integrity of the people reporting it. Science is not sexy, in most cases, and therefore requires a different rhetorical touch than, say, celebrity news. Ethically, this means that a reporter must rely on his or her sense of professional propriety and decide whether a science story is important enough to warrant embellishment if only to give it print space or air time. This is, of course, a best-case scenario; it should be clear that some reporters, in the interest of professional or personal ambition, promotional acknowledgement, or simple vanity might resort to anything from mild embossing of the details to complete fabrication of the results in the interest of furthering professional and/or political aspirations. Another interviewee, also a science journalist, put it this way: "There is always the tension of what is the strongest thing I can say about this story and still have it be accurate. It is not how wishy-washy, how cautious, how moderate I can make it and get it buried way back in the paper. The fact is, you are going for the strong" (9). Science news, like any other story, must reflect audience interest, contextual relevance, and readability. However, the scientific process is often at odds with this very set of requirements, demanding as it does careful, often long-term analysis, tedious data collection and organization, and dry subject matter.

It should not be surprising, therefore, that medical advancement, new studies in medicine, and health news are frequently featured in print and broadcast news sources. Sometimes regular columns are dedicated to these topics. Such cutting-edge information is often the closest thing news sources have to science-related stories with which readers/viewers can identify. Personal health is a popular topic for a variety of reasons, but audience interest in appearance probably tops the list. In a culture that emphasizes health and beauty (often for the most superficial of reasons—sex appeal), medical news is likely to emerge as the only scientifically-oriented type of news feature that has widespread audience interest. It is also one of the few scientific topics that seems to have a direct effect on the people who are being informed about it. While environmental issues and stories on the latest technology certainly contain information that has an impact on us all and to which the audience can "relate," they can still be seen as distant to the everyday encounters one has with the self. In other words, medical news taps into our sense of vanity and self-preservation; it speaks to our most deeply-rooted needs to look good and live longer, higher "quality" lives. The question becomes one of how seriously we should take every new reported "development" in medicine or health research. We become saturated with advice about our "lifestyles," and as soon as we adopt one mode of behavior at the behest of a series of reports concerning a particular danger or a specific regimen, along comes a new "development" to contradict the old one. One irony is that if we become too obsessive with following each new health trend, diet, or drug in order to live longer and be healthier, we have little time for the activity of actual living. Common sense seems to dictate that a moderate diet of low-fat, low-sugar, high-fiber, and protein coupled with regular exercise is probably the most effective way to maintain health (other factors like not smoking and moderate drinking also contribute), yet media outlets have managed to manufacture so many different variations on this particular theme that the results cannot really be called "heightened awareness" but must, rather, be described as utter confusion.

Perhaps the most important thing to realize about the media as a source of scientific information is that it is itself an industry, and a for-profit one at that. This is why understanding the motives of everyone involved becomes so central to understanding the rhetorical phenomenon as a whole. While providing advice on diets and exercise generally does not also promote misinformation that can be considered dangerous (there are exceptions, especially with "fad" diets that can actually lead to malnutrition and even drug dependence), other topics have ripple effects throughout society that can lead to panic and hysteria. The AIDS "epidemic" is one such topic. Before the early 1980s, acquired immune deficiency syndrome was virtually unknown, but this soon changed with a series of news stories and news magazine broadcasts that catapulted over the boundaries of acceptably "strong" journalism. The stories reporting the onset of an AIDS epidemic were, strangely enough, validated by official statements from normally trustworthy authorities like the Public Health Service. Winsten explains that what "galvanized public attention was preliminary information which raised the possibility that the disease might escape the confines of high-risk groups [i.e. homosexuals and intravenous drug users] and threaten the public at large" (11). He further asserts that news of the disease was "needlessly and harmfully exacerbated" because of statements by the Public Health Service and media sources like *The New York Times* and ABC's news magazine, *20/20*:

> On May 24, 1983, influenced by political pressures, senior federal health officials declared AIDS to be the "number one priority of the Public Health Service." This proclamation, meant to reassure the public, sent the opposite message, because it permitted—indeed encouraged—people to conclude that the risk of contracting AIDS was far greater than health experts had been willing to admit. In addition to undermining efforts to quell the public's fears, this proclamation officially certified a major news story for the media—and resulted in the first page one story on the epidemic to appear in *The New York Times*. Then, on May 26, ABC's "20/20," in its second program on AIDS, told 19 million television viewers, "There is now steadily

> growing fear that the nation's entire blood supply may be threatened by AIDS . . . The safest thing to do is to store up your own blood." The story's factual content fell just within the boundaries of "truth." The tone of the story could not have been better designed to provoke massive public hysteria. (11)

The fallout from this report was potentially disastrous; in the weeks following the broadcast, the Red Cross reported a severe decline in blood donations across the nation, leaving blood banks virtually decimated in the wake of this sensationalized exaggeration of the real threat of AIDS. In another *Times* article run shortly after the *20/20* broadcast, the journalist writing the story carefully organized the information to begin with a string of frightening scenarios in the first several paragraphs followed by the actual numbers of the AIDS cases in the latter half of the sixth paragraph, revealing that, nationwide, the number of cases "mounts beyond 1500" and the number of deaths "nears 600," hardly an epidemic of global proportions. One's chances of getting hit by lightening are almost as likely, but the carefully chosen words "mounts beyond" and "nears" is a clear attempt to tweak the numbers' implications for more than they actually warrant, giving the reader a sense of gravity for numbers that are, at best, modest, and in truth, statistically insignificant.

Winsten cites a number of such stories, including those that promise advances in medical treatment for serious diseases like ovarian cancer, generating an end result of scientifically unfounded hope in the readership. Such reporting goes beyond mere embellishment and transgresses the borders of journalistic responsibility, for to manipulate a vulnerable readership like those who make up the audience of such features, namely sufferers of cancer and their families, establishes a false sense of optimism regarding the treatment of diseases that are much further away from a cure than they are represented. But science itself can shoulder some of the accountability for such exaggerations. A number of professionally ambitious scientists have recognized that the press can work in their favor, and when accomplishments (or personal perceptions) fail to sway one's peers, one is provided with a career advancement opportunity with the public. Many popular scientists, as we shall see, while usually respected in their own fields, have capitalized on the advantages of being in the public spotlight. High public visibility can translate into better odds for promotion, greater chances for securing large grants, and increased opportunities for pursuing pet projects. The attitudes among scientists regarding the exploitation of the media fall generally into two camps: those who willingly (or even enthusiastically) pursue media attention for the afore-mentioned reasons, and those who view both the public and the media with disdain, preferring to remain aloof and avoid complicating matters by not acknowledging the added rhetorical attention needed to succeed in the media game (14). Many scientists in the first category are quite adept at this game, some even to an unscrupulous degree. One of Winsten's interviewees admonishes those "scientists who manipulate the press, who exaggerate their accomplishments, who neglect the contributions of their peers. And they often get away with it. Often the press is blamed for exaggerating something, and the scientists who gave them the story are just as much to blame" (14). According to Winsten, media relations center upon five main groups within the scientific community, and each have their own idiosyncratic rhetorical and professional requirements: the academic medical centers; the individual researchers; the biotechnology firms; the private health care practitioners; and the universities (14).

The details of the media relations with these discrete groups are interesting, but the important issue here is the recommendations Winsten provides as a result of his study. Beyond the problems he has already indicated, he sees eight areas of improvement that can be addressed when reporting on scientific studies, and they are worth mentioning here in abbreviated form as a way for the reader to understand the constraints of science reporting and to keep an eye out for instances where they are violated:

1. Acknowledgement of the limitations of research and/or studies.
2. Less reliance on single sources.
3. Greater emphasis on research trends (as opposed to isolated single studies).
4. More attention to socially relevant areas of study.
5. Explanation of clinical obstacles (i.e. the practical application of research).
6. Acknowledgement in health-related stories of differing interpretations of data.

7. Appropriate balance in health-risk stories (i.e. is treatment worse than cause).
8. Illumination of available choices to allow for informed decisions. (21)

I would add that the informed reader has a responsibility to understand the necessary limits of science reporting—limits that are benign in themselves but that can be manipulated for rhetorical gain when one of the agents involved in the reporting process uses the venue for either unscientific ends, or for dubious journalistic motives, or both. Just because a "new study" has found something compelling does not mean that we are obligated to act upon those findings. In so many cases of media reporting, findings are preliminary, individual, and isolated, and may have no significant bearing on our lives, but we may be led to believe that they do because to present it that way simply makes better copy and more riveting edutainment. Science operates in a way that is quite often at odds with our habits of immediate gratification. Solid scientific research takes years, sometimes decades, to produce practical and applicable results; it is a slow-moving process because it is so important to avoid drawing hasty conclusions on the existing evidence. This means that scientific "fads" are a media invention more often than a scientific one. We should practice the same degree of skepticism that scientists pride themselves in when receiving new information about scientific activity. The informed audience will weigh reports with critical scrutiny, and if he or she is interested in the development of a given scientific project, issue, or topic, he or she will seek out further information on it. This is perhaps one of the more useful applications of the Internet technology that we now have available to us; it is packed with information that we can use to verify or dismiss scientific findings from other sources. However, we must caution ourselves when using alternative forms of information, since we are all aware that the very same media restrictions that affect scientific reporting in its traditional forms can in fact be amplified when using source material from the Internet.

The two most influential institutions in this drama of ideas and attitudes are education and the media. I have attempted to provide a basis for further inquiry for those disposed to examine the media juggernaut and its impact on scientific information in greater detail. There are many excellent articles and books in a variety of fields for those interested in a more complete account of specific areas of discourse theory, cultural studies, rhetoric, and communication as they relate to science and society. My intent has been to suggest a pervasive and swelling cultural condition that, if not addressed in some significant and tangible way, will mean a serious blow to the democratic responsibilities of American citizenship. Education will be examined with more scrutiny in the final chapter, but as a preliminary premise, I think it is important to keep in mind that students, administrators, teachers and professors, and popular media sources are all responsible for improving literacy in both science and media analysis, for if they don't, it means much more than just the college graduate who claims merely to have been a bad student in biology; the issues science raises regularly encroach upon the public sphere, forcing us and our leaders to make key decisions based on important scientific information, of which the popular media is a primary source. If this information comes from non-scientific sources and is pushed in a certain direction because of a political agenda or is "spun" in a certain way to sell newspapers, then we are destined to make poor, even catastrophic choices on topics that affect us all.

3 Two Popular Representatives of Science

As I discussed in the last chapter, the most flagrant perpetuators of the distortion or outright misrepresentation of scientific ideas are that ubiquitous and highly elusive entity known collectively as "The Media." But what is the working definition of this rhetorical agency? Is it Time-Warner, Fox, CNN, or the national broadcasting companies that we have come to associate with large (and constantly shifting), bottom-line corporations that sensationalize the trivial and trivialize the sensational for corporate gain? Is it the print media such as *The New York Times* (or any other major city newspaper), *Time, Newsweek,* or *US News and World Report* that sacrifice depth for shotgun coverage aimed at the basest elements of their reading audience? What about alternative sources of information, especially special interest publications or underground newspapers? While these may indeed provide alternative interpretations of issues to those of the mainstream media, their journalistic integrity is often in question as much as it is for their more popular counterparts. The journalistic orientation of presses with smaller circulation is often based in political partisanship or special interests that slant their coverage (*The Village Voice* and *Mother Jones* are prime examples). An environmental publication, for example, is not going to champion the wonders of science for creating toxic chemicals, nor is it likely to trust science to construct strategies for cleaning up an environmental hazard, at least not without the close scrutiny of watchdog groups. In the eyes of publications such as these, scientists are often part of the problem, not the solution. The sudden explosion of the Internet as a source of information is another factor in the level of misinformation that is available for consumer use, and one must question whether the common citizen is equipped to discern credible sources from crackpots for issues that affect her most directly. Examination of the form that our information takes is important because, looming behind the information glut that characterizes the "Information Age," the problem is not *whether* the information is available; it is whether we can find *trustworthy* information, interpret it properly, and sort out the valid from the merely opinionated, a process which raises the very complicated question of what constitutes legitimate knowledge.[1] The saturation effect favored by modern mass communication contributes to the increase in scientific authority because we are generally willing to allow scientists to do much of our thinking about the hard questions for us. Their ethos as creators and organizers of knowledge means that we can turn to them for the answers, and when all other sources are questionable, scientists still maintain that sense of integrity and honesty absent in more dubious personalities and organizations (like politicians and the government). "The Media" recognize this as a way to lend hot stories an air of credibility, using scientific expertise to define and interpret an issue. Who does the speaking for science (as both authorized and unauthorized representatives)? How is science represented by those who speak for it? This chapter addresses these critical questions.

If we intend to discuss The Media with any regularity and consistency, we should have at least a working definition. Many scholars have provided definitions and established parameters for The Media, but Robert W. McChesney is perhaps the most vocal about the relationship between a strong media information source and the health of a functioning democracy. In his short book, *Corporate Media and the Threat to Democracy,* McChesney notes that the idealistic hopes of thinkers like James Madison and Jurgen Habermas (who both saw the media as central to any true democracy) have been eclipsed by the corporate-controlled media that he says has its own conservative vantage point (both in content and partisanship). McChesney quotes Madison as saying that "[a] popular government without popular information, or the means of acquiring it, is but a prologue to a farce or a tragedy, or perhaps both" (6). Habermas was seeking to reestablish a "public sphere" that he had identified as key to any successful revolution. In both examples, the intersection between political awareness and action and reliable and honest information sources was predicated on an understanding of the need for an informed constituency. For McChesney, in fact, the American system of media information (i.e. "the news"), based almost exclusively on a handful of media giants like Viacom,

Time-Warner, and Disney, is highly resistant to the politicization of the news because its success relies so heavily on the maintenance of the status quo. Such media corporations have a far-reaching investment in the established system, and often argue that "market forces" are the best self-corrective since they determine "what the people want" in their news (8). This maintains the illusion of democracy while serving in reality to keep people, in the words of Walter Cronkite, awash in "a stew of trivia, soft features, and similar tripe" (qtd. in McChesney, 24). The Media, for McChesney, must therefore serve the public interest while keeping its own interests at arm's length, as much as that is practically and realistically possible. Its primary purpose is to provide reliable, politically neutral, thorough coverage of issues and events that are ultimately answerable to the voting public.

Jacques Ellul, who was discussed earlier for his contribution of the concept of *la technique,* provides further insight, this time in *Propaganda: The Formation of Men's Attitudes.* In this 1965 book, he describes media as an agency for information that must operate under two conditions in order to be successfully used for propagandistic purposes: it must be subject to centralized control, and it must be diversified (102). I do not wish to give the reader the impression that I feel there is an orchestrated and conscious political effort to forward scientistic manipulation on the population. The effect is more subtle than outright propaganda, it does not contain the overtly insidious political indoctrination, but it does present an allegiance to an epistemological ideology. Ellul's observation is useful because it points out that in order for The Media to function as an effective instrument of propaganda, it must be both centralized and diverse. While certainly the latter can be considered true—truer today than when Ellul was writing—the former is questionable. We do have huge mega-media outlets like Time-Warner and the Turner Network, but our choices today are the product of market forces that in some ways pit competing networks against one another, such that the careful coordination of rhetorical messages is more incidental than deliberately orchestrated for a single political or ideological purpose. There is, in other words, a sometimes maddening consistency in the types of programming imposed upon us (the relentless stream of "reality TV" programs is a good example) and the methods used to promote both the networks and their sponsors, but the effect demonstrates a lack of imagination and fiscal adventurousness on the part of programmers—an unwillingness to make television networks or news sources distinctive from one another because it might constitute an investment risk—not so much an effort to push a particular propagandistic agenda. That is, the amorphous effect of numbingly unimaginative programming is not the result of centralization, but of competition that paradoxically supplies homogenous programming because that's what competing networks provide (and apparently what people watch, read, and listen to). The notion of the Liberal Media, for example, is a reactionary fiction promulgated by conservatives who, these days, seem to have the ear of the media more than their liberal counterparts.[2]

Moreover, The Media is often spoken of as one ubiquitous entity, but it is really a composite of many smaller informational sources and entertainment forces. To suggest that the media networks that exist have the will, the motivation, or the ability to synchronize material that could gel into something like scientific propaganda is to give the media more credit than it deserves, though McChesney's points about the media monopoly has much merit for the mainstream media. We do, however, see a new and interesting (and not altogether understood) media agency in The Internet, which could be viewed as an equalizer in the vast sea of public opinion. Through it, anyone can not only access information on almost any subject that can be imagined, but users can also post their own insights about those subjects. More and more, people prefer to get their information from online sources, and electronic devices that make that possible are accessible virtually anywhere. Even for those who do not have personal computers, cell phones, or Blackberries, it is difficult to find a venue that does not have a television screen bleating out information or advertisements. Even standing in line at the grocery store now means being subjected to television and Internet "news." From an infrastructural standpoint, there are so many ways to access information (or have it foisted upon us) that centralization is elusive and, perhaps, undesirable. What it does mean, however, is that the means by which information is conveyed is inescapable; unless one were to make a conscious effort to avoid civilization and its technological machinery, there is no way to completely sidestep the media influence.

One workable definition of the media is that it is a decentralized mechanism for mass-marketing, information, entertainment, and editorial, a communication tool that comes in many forms, from print media to satellite radio to the World Wide Web and is closely linked to the state of the art of the technology available for delivery. It therefore has ripple effects that touch nearly everyone worldwide, and has a concomitantly significant impact on our understanding about what is going on in the world and what our world means to us. This last detail—that media outlets have much control over meaning—is perhaps the most important rhetorical component of media influence. It serves as a source for the stock responses people have to all sorts of issues, and it should be noted that issues are not confined merely to news and news magazine programs. (In this respect, it may often be mistaken as having propagandistic properties, even though it is questionable how conscious this effort is on the part of media coordinators.) Current events and the issues that help define them seep into all areas of the media, including entertainment. Many of our "prime time dramas" deal routinely with current issues, and this fact contributes to how they are framed and received by the viewers. It is common for people to adopt the prepackaged opinions that are laid out by these programs, and even sit-coms often make flaccid attempts at being "socially conscious" in order to "raise awareness." The media sources we use the most, then, function as more than conduits for communication; they construct our ethical and social identity, making education in the decoding of the media's rhetorical instruments an important part of a person's cultural and critical training. Science and scientists, likewise, are frequently showcased in the media, performing roles that range from expert commentator to cameo appearance (Steven Hawking and Carl Sagan have been favorites, as we shall see). Science itself is a constant presence, and it too takes on an eclectic array of forms. Science fiction programs are among the most popular, but shows like "Nova" and stations like The Discovery Channel (which even has a "Discovery Science" version), and The Learning Channel have a substantial viewership as well. To understand how these sources operate rhetorically, we will first look at some of the popularizers of science, then discuss how their presence contributes to the prepackaging of science for effortless public consumption to create an audience receptive to the promise and credibility of science.

Popularizers of Science: James Burke and Carl Sagan

Recently, the Discovery Science Channel showed reruns of classic science programs like "Mr. Wizard," "The Ascent of Man," and "The Secret Life of Machines." These programs, ranging in their original airing dates from the 1950s through the early 1990s, each had their own appeal and idiosyncratic attraction. *Mr. Wizard*, perhaps the archetype of science programs aimed at educating young people, featured children who helped demonstrate scientific principles, emphasizing the fun, mysterious, magical world of the physical and biological sciences. *The Ascent of Man*, with host Jacob Bronowski, gave popularized science an air of sophistication and academic relevance.[3] His presentation was much like an on-location lecture, thereby utilizing the benefits of both a classroom atmosphere and the wonders of video reproduction to create a truly unique learning experience. *The Secret Life of Machines*, hosted by two disarmingly dead-pan English engineers was campy and funny as the hosts demystified the most common devices used in the modern household: televisions, video cassette recorders, refrigerators, vacuum cleaners, telephones, automobiles, quartz watches, etc. One of this program's primary attractions was the audience's realization that, while all of the items discussed were incredibly commonplace in the modern home, most of us knew absolutely nothing about how they came into existence or even how they worked on the most basic level. All of these programs—and others still relatively popular today, like *Scientific American Frontiers* with Alan Alda—serve an important public relations function in the furtherance of science: In order for science to have power and relevance to the contemporary layperson, producers of science programming saw the need to educate the public about what science did, how it operated, and what, specifically, it was capable of explaining.

The popularization of science in the United States can be traced back at least as far as the periodical *Scientific American,* which began publication in the mid-nineteenth century. However, the real push to educate the general public and to recruit scientists and technicians did not begin in earnest until after the Second World War, when the growing need to compete with the Soviet Union necessitated programs designed to draw people toward science and to explain U.S. foreign policy using the language of science.[4] Since then, television programs, magazines, radio shorts, and books have become increasingly commonplace. Without a venue to promote scientific projects, many such projects would be seen as extravagant expenses that were both politically untenable and financially unwise. Scientific spokespersons like Steven Jay Gould, Carl Sagan, Brian Greene, David Suzuki, and James Burke became household names (at least for any moderately educated, middle class household), and their wisdom, confidence, and stage-presence (relative, at least, to other scientists) made them and science increasingly popular.

The same Discovery Science Channel that has been rerunning old science programs also featured a highly unusual, often fascinating show known originally as "Connections." On this program the host, a would-be English teacher named James Burke, traced links from a variety of scientific discoveries to one another, showing an amazing intellectual tapestry of interconnectedness between such unlikely events as the development of quinine and the invention of radar. The sheer tedium of scholarly dead-ends this must have produced seems, at least initially, to be prohibitive as a premise for an educational series and as a research project. It would require at minimum some hunch that two temporally remote events or functionally disparate discoveries had some link, however unlikely. Admittedly, the connections at times seemed forced, yet Burke seemed to pull off the plausibility of these intersections episode after episode; in fact, the premise was so successful that it generated two "spin-offs," entitled (unimaginatively enough) "Connections 2" and "Connections 3" with exactly the same conceptual framework. Burke was (and still is), therefore, an important figure in the popularization of science because, of course, he made it entertaining. Just as the accomplished teacher of history makes past events come alive with real people in a compelling narrative, so too did Burke manage to capture the historicity of science—the "story" of it—in a way that was both appealing and instructional. In short, Burke established that he was a good teacher because he was able to successfully engage his media "classroom." The very fact that "Connections" began as a PBS program (which, we are loathe to admit, does not have the highest ratings compared to its mainstream, commercial brethren) to the cable network Discovery is a testament to its popularity and mass-appeal.

When one actually sits down to view the program, it is also important to note that Burke's *modus operandi,* his rhetorical mode for the program, is the narrative. This is clearly a deliberate rhetorical choice, knowing as he does that the success of his show is determined by the number of viewers he can get to commit to watching it. On a playing field where one must compete with the sensationalized television spectacle and the prime-time dramas that enlist a masterful use of the narrative for entertainment purposes, casting his show in the form of a narrative drama is a logical and effective choice. The connections he makes really comprise a story, and the story must be engaging and entertaining in order to teach successfully. With so many entertainment alternatives, Burke and his producers realized the need to create a program that was at least as compelling as the other programs in the same time-slot. This is the corporate reality of the popular media; ratings are everything since without them sponsorship quickly dries up, and from a rhetorical standpoint we must admit that without a significant number of viewers the message, no matter how important, will not be heard. That Burke's show was a success—at least by educational programming standards—means that he hit upon a winning formula. But the logical academic question is, what must be compromised in terms of content in order to make the narrative entertaining? Can one be completely true to the historical spirit of events while also maintaining a degree of academic integrity? It is an old teacher's question. Knowing as we do that we must vie for students' attention with diversions that are far more stylized, bright, flashy, faced-paced, and, admittedly, entertaining, how do we capture our students' interest and attention while staying true to the subject matter we are required to teach? A skilled teacher will strike a balance between entertainment and the delivery of material, whereas less successful teachers will sacrifice delivery for content only. Many of us have endured the stuffy lecture hall where we listened to one of our professors drone on in monotone about matters that we had no connection to and less interest in. But this does not make for effective television, and if the goal of scientific popularization is to get people interested in science and the history of scientific discoveries, we must perform. Burke does, using everything from costumes to props to on-site field trips, and while we might envy him for the resources on which he is able to draw, we must also admit that he has achieved his goal: to make the scientific connections provocative and entertaining. Here are some typically glittering accolades given on Burke's behalf:

> Hailed by the *Washington Post* as "one of the most intriguing minds in the Western world," James Burke takes audiences on a creative journey through the history of science, technology and social change. He is the creator, producer and host of many award-winning TV series including *Connections* (PBS), *Connections2* and *Connections3* (Discovery), *The Day the Universe*

> *Changed* (PBS), *After the Warming* (PBS) and *Masters of Illusion,* for the National Gallery of Art. With a unique and entertaining perspective on how people and institutions change, Burke has been a popular speaker for such companies as IBM, NASA, Procter & Gamble, Microsoft, Cisco and countless universities and museums. He has also produced customized videos for companies seeking to explain new technologies and ideas. (Royce Carlton, Inc.)

Clearly, Burke has made his mark on both the educational programming world and, based on his association with a number of major corporations, the business world. He is a successful model of the marriage between knowledge and commerce. It is interesting to note that the biographer here, an unacknowledged public relations writer for Royce Carlton, Inc.—a company that manages, promotes, and provides agents for company speakers—would consider NASA a "company" in the same category with Microsoft or Proctor and Gamble, further evidence that education is most palatable if it is also profitable. Also interesting is how his credentials are underplayed in one respect and amplified in another: "Burke was educated at Oxford and also holds honorary doctorates for his work in communicating science and technology" (Royce Carlton, Inc.). But there is no mention of what his degree was in or whether he, in fact, ever *received* a degree from Oxford (only that he was "educated there"), nor is there explanation of what an "honorary doctorate" actually is (information that might not be readily known by the average consumer). Still, one cannot argue with his success, and if he can establish a rapport with the public more effectively than his academic counterparts, then more power to him. His appeal is his eclecticism, as well as his willingness to take the time to crouch down before the masses for a detailed explanation that is, in Erik Davis's words, "part avuncular professor, part oddball autodidact, part Monty Python."

The real reason for mentioning James Burke here, however, has to do with a monologue he tacked onto the end of the first episode of the first season, an episode called, "The Trigger Effect" (1978). After an especially dramatic finale to the recreation of the connections made in that particular show, Burke posits an idea that he admits some of his audience might "disagree with violently," namely, that the movers and shakers of the early modern, modern, and postmodern world were scientists and technicians, the Einsteins, Oppenheimers, Galileos, and Newtons of history, not the Beethovens, Shakespeares, Michelangelos, or Nietzsches. The reason, he says, is that the latter group had a fatal flaw in the gathering and organizing of their knowledge—they were "interpreters" of the world, not observers of a natural, organized pattern of order. A work of art, for example, said more about the person who created it than it did about anything tangible in the world, and therefore, scientists, whom Burke claims are not afflicted with this interpretive distortion, were able to describe the world "as it really is." The sublime poem written by Coleridge does not possess the same practical power that the new household invention does because it can never "move forward" a way of living. All of the revolutionary rhetoric in the world, he suggests, is useless without the machinery and the scientific efficiency to put it into motion. According to Burke, the artist can produce, at best, a "second-hand knowledge" that is further altered by a third party viewer, reader, or listener who in turn interprets the interpretation (*Connections*, "The Trigger Effect"). The assignment of meaning, it seems, is a detrimental condition that has kept us at an intellectual impasse, wondering, ever wondering, without hope of uncovering the unifying Truth that will make all things clear. That is, until modern science came along. With this tool, the distortion is lifted, the empirical evidence rules, and the universe becomes finite, understandable, and ordered. The cosmos is carefully patterned, and science merely reveals those patterns to mollify chaotic human thinking. With this teleological tool, human beings are able to move, to change, and to "progress." Without it, we become, in the words of another Burke, Kenneth Burke, beings who must huddle together, nervously loquacious at the edge of an abyss.

While the assertion is not necessarily one that the general viewer would challenge, much less violently disagree with, it might be more important to ask *why* we *might* challenge it. Most of us readily accept the notion that science creates the conditions under which change can happen. But what "change" are we referring to? While it is certainly true that a symphony cannot put a man on the moon, it is equally true that if you randomly fed all of the notes and principles of music into a computer, you wouldn't get the 1812 Overture. Technological change (that which James Burke seems to be alluding to) has made our quality of life *different,* but has it made it *better?* Burke concedes that we rarely acknowledge, much less understand, the technology we rely on, but the implication that "progress" is measured only by the functionality of the things we create is a distressingly limited view of progress. And such an overarching assumption may be a primary reason why humanistic wisdom has not kept pace with technological advancement. If the latter is what we principally value, then the former will be seen as the thought absent the action. In countries like the United States, action is valued over words, and this is a principle we often emphasize unthinkingly. Why do we see as inferior, say, a pursuit of music, art, or literature? Perhaps because it does not pass the test of legitimate knowledge judged through the lens of econo-scientific viability. Science and technology, on the other hand, enjoys a prominent space in the marketplace because it produces commodifiable products. But is our current reliance on computers, cell phones, fax machines, and satellite television an "improvement," or have these devices simply shifted our attention to, what are in most cases, banal, superficial concerns, making us not "better" in any significant sense of the term, but rather, lazier, more dependent, and more distractible?[5] It's a debatable prospect, but the more important question for our purposes here is not whether it is better or worse, but how we came to the ideological conclusion that scientific advances and knowledge have primacy in the first place.

Equally important, perhaps, is a fundamental flaw in Burke's argument, one that I find it odd he should have made given the very premise of his program. He suggests that science (and, by implication, the scientists who practice it) is divorced from the bewildering clutter of language and interpretation, as if science had discovered not only the best symbolic expression of the world (mathematics) but had also so thoroughly trained its users that interpretation becomes unnecessary. Scientists, he implies, don't interpret the world; they merely present the "facts" which speak for themselves. But we know that facts alone are untrustworthy because they don't by themselves contain meaning. We cannot expect facts to function as the timekeeper of knowledge, as if the known facts "speak for themselves" to tell us what time it is. It requires a knowledge of the workings of clocks, an impression of what function a clock has, tools with which to assemble the device, a desire to know the time, and a reference to check the time against to even begin to have the individual parts make sense. Even this analogy reveals a penchant for mechanistic metaphors that suggests only order and precision and ignores the artistry, the need to know the time, the vanity, in short, the *meaning* behind a clock. A scientist without interpretive acumen is a poor scientist indeed. In fact, one cannot call such a person a scientist at all, and the idea that Burke's very project relies on just such interpretive abilities—the capacity to make connections where none seemed to exist before and translate them into an accessible, entertaining, and educational form—makes his argument that much more bewildering.[6]

But few would question that science has a "precision" or an "order" that makes it more reliable, and therefore, more useful as a predictor of phenomena or an instrument for altering our surroundings to suit our needs than, say, that of the poet. To compare the two is, perhaps, unfair; after all, poetry doesn't attempt to accomplish what science does or vice versa. One may very well rely more heavily on emotive outpourings and human-centeredness than the other, but that hardly makes it inferior. Artistry and science do, in fact, share many qualities, but they also differ profoundly in their intended goal. Where poetry's primary function may be to express the sublime or the profane using a necessarily limited human vocabulary, science's is to discover consistent and natural principles so that they may be understood and manipulated (or at least materially applied) in some way, all the while using a different, but equally limited, vocabulary. Perhaps one of our most persistent mistakes is to fail to make a distinction between what we create and what we observe. Science is a method; humans are the interpreters. Art is a catch-all phrase for a variety of forms (painting, music, poetry, sculpture, literature), whereas science is a catch-all phrase for a variety of functions (experiment, engineering, design, observation, prediction). At the same time, it is hard to imagine science without the human element of imagination, that strange capacity we have for considering possibilities that we have not ourselves experienced. In short, art can live without science, but science cannot survive without art, or at least an artistic temperament. It needs it to turn raw data into something else, lest we are left with nothing but raw data, a fairly useless jumble of numbers, letters, measurements, and "facts," which by itself stands meaningless.

Burke's monologue is probably not entirely genuine, and this is part of the point. His effort to maintain science's degree of intellectual primacy is an important component of the success of his and others' popularization project. The scientific enterprise, like any other human endeavor, has that important rhetorical layer, the area of discourse within which the all-important attitudinal dimension resides. Put bluntly, science cannot exist in any meaningful way without its supporters and detractors; it needs both to function as a social process, and the effort to popularize science is nothing if not a social process. Whenever we have discursive banter and "wrangle in the marketplace," as Kenneth Burke puts it, we have rhetoric. Science is no more above the need for this banter than any other language system, but it does have at least one advantage over the others: It is intellectually favored, and it can therefore suggest that it has qualities that puts it out of the reach of other human endeavors that are more obviously language-centered. The equally obvious question, though, is what human endeavor *isn't* language-centered? Whenever we need to create meaning, we have language; whenever language is called into question, or there are spaces in which meaning is either ambiguous or not wholly agreed upon, we have rhetoric. James Burke's rhetorical move is to construct a fictitious dichotomy, one that suggests there are "artists" and there are "scientists," and the two have not only fundamentally different methodologies, but they also have fundamentally different objectives. It is important to maintain this illusion, lest we allow the gray-areas of humanistic knowing to contaminate the immaculate domain of the scientific process. To admit that all things scientific are not in fact cut and dry, that there are uncertainties, that the quest for truth is actually messy, laborious, and even counter-intuitive, is to damage the public ethos of the scientist. All good scientists know this, and while Burke may appear to speak with the zealousness of a convert, being humanistically trained himself, it is just as important to consider the market-value that his show represents. Herein lies the important consideration of media influences and how scientific PR is constructed for public consciousness.

We should remind ourselves that "the media" refer to any broadcast instrument designed to reach a large audience at once. It is, of course, through the advances in science itself that new media have gained influence, and it is for this reason, if for no other, that training in media decoding skills is an important aspect of any contemporary education. This task has been left largely to the humanities which specialize in interpreting cultural phenomena (though of course social science has its dubious influence as well). The development of radio, television, and computers (not to mention reproduction technologies like video recording, compact disks, digital video disks, and memory technologies used in computers) is the province of science and engineering schools. These technological strides have led not only to the ability to reach a wider audience in less time while sending more information, but have made us reliant on such technology for nearly everything we do in our daily personal and professional lives. Hence, I may risk belaboring the obvious when I say that it is hardly surprising that media vehicles like television and the Internet have a huge impact on how we understand and organize the world. The interesting thing to note, however, is the irony that this swift advancement in media technologies has not made us more discerning observers of our world, generally speaking. We have become intellectually lazy because we have plenty of media personalities or information centers that fulfill the interpretive role for us. In this may lie the most troublesome aspect of media technology of the twenty-first century, so it is crucial to repeat this mantra: *More* information does not mean a *better* informed public.

The issue of quantity over quality in regard to the information explosion will be dealt with in more detail. For now, it is necessary to establish both who is afforded the privilege of advocating scientific information, ideas, and prognostications, and how those people are invented for optimum media effect. James Burke is a good example of the archetypal media "scientist" because he is not, in fact, a professional academic, at least not in the traditional sense of the word. His M.A. in English from Jesus College, Oxford in the late 1960s led him to Italy where he was director of the English School at Bologna. Later, he went into broadcasting, where he developed a number of documentaries (some having to do with more humanistic themes like Renaissance art and the history of Western knowledge) until he hit on the winning combination that produced *Connections* (Palmer). In fact, in the sketchy biographies that are available on Burke, nowhere does it say that his educational background has anything to do with science. He is, for all practical purposes, a broadly educated layperson, and it is perhaps this quality that has given him the necessary ethos to be a successful popularizer of science with a general audience. A professional scientist—one with a specialization and a vocation that utilizes it—would have great difficulty relating to a general audience because it would be difficult for such a person to present language that was accessible to a wide range of viewers, compromising the authoritative ethos necessary for popular success. Burke, by contrast, represents in many ways the model popular scientist. His personality, while apparently affable and approachable, conceals a hubris regarding the epistemological superiority of science that has left an indelible mark on the mainstream public. He is knowledgeable, clever, and ultimately, convincing, and this triad of abilities converts his mild-mannered, even playful, professorial demeanor into the persona of the modern Western scientist. In this capacity, his claims about the primacy of science have currency with the public, and it is through such scientific pundits that the scientific world-view dominates. He is only one symbolic representation of how media images—at once tangible and ethereal—have such a major impact on the mindset of the community at large. It is through such spokespersons that science maintains its secular piety to a method and a goal that is, in American society at least, largely taken for granted.

In many ways a contrast of Burke, we have Carl Sagan. While Sagan was a professional scientist, he did not possess the same stage-presence as Burke. Despite this, he was remarkably popular, mainly because his topic (articulated broadly as an exploration of the "Cosmos") seized the imagination of the public unlike that of any previous advocate for scientific education. According to the back cover of his most famous popular book, *Cosmos,* the PBS series of the same name which aired during the early 1980s was viewed in sixty countries worldwide by "approximately 3% of the people on the planet Earth" (as opposed to the Martians who forgot to tune in). Likewise, according to the back cover, the hardcover edition of *Cosmos* remained on the *New York Times* best-seller list for 70 consecutive weeks, "and appears to be the best-selling science book in the English language of the twentieth century." As a planetary astronomer, Sagan was able to put into perspective the significance of the human animal in the larger (much larger) framework of the universe in a way that was at once persuasive and unsettling. His famous illustration of the "calendar" of evolution is perhaps the most memorable example of how a complex scientific concept can be made accessible through the use of a commonplace analogy. Sagan explained that if we think of the development of the planet earth (to say nothing of the universe as a whole) in terms of a single year, modern human beings (i.e. *Homo sapiens*) did not come on the scene until 11:59:59 p.m. on December 31st. While the analogy is expressly dramatic, and therefore, somewhat imprecise, it did give the public a scientific sense of how recently we appeared in the great evolutionary date book. Such a timeline, coupled with Sagan's memorable mantra of "billions and billions" of stars amongst "billions and billions" of galaxies provided powerful arguments for evolution and for the relative insignificance of the human presence.

Rhetorically, the analogy relied on a master rhetorical trope (the metaphor) to drive its point home. The adept viewer may understand that this timeline is not "really" a calendar, that this is only a convenient means to relay a much more complex idea and to underscore the implications of it for our species (though I will discuss later the limitations and misrepresentation that this tactic can construct). The simple analogy implies that, counter to centuries of assumption regarding the primacy and centeredness of the human species in relation to the rest of the universe, we are just another animal—in many ways no more significant or special than any other species, living or extinct, that has walked, crawled, flown, or swum on this planet before us. Even more unsettling is the idea that, given the brevity of our existence here, there is no sign that our ongoing survival is guaranteed. Sagan's message is appropriately startling: We are just as prone to the fickle whims of indifferent nature as the rest of our Terran cousins—worse, we may be the instrument of our own demise. Had Sagan used the technical language of biology or the mathematics of physics to get this point across, it would have been far less dramatic for the general public, reached a smaller segment of the population, and ultimately produced a sterile representation of his primary point: We are *not* the universe. Moreover, Sagan is a successful scientific popularizer because he is a good rhetorician; he understands the dramatic impact of a well-wrought phrase or a gripping metaphor.

In one widely-published photograph called "A Mote of Dust," featuring an almost entirely black field with one white speck inconspicuously nestled in the frame, Sagan explains that the speck is actually the Earth, captured in a photograph by the Voyager I space probe as it was on its way out of our solar system. The accompanying commentary, in typical Saganian spirit, muses about the number of lives lost through the countless wars humankind has inflicted upon itself for territory that, given this reference point, amounts to mere fragments of the "mote of dust" that hangs insignificantly in space. Sagan invokes the trope of the metaphor again as he provides the audience with a view of Earth, not by using its precise position in the solar system, but by comparing it to a fleck of dust that lends perspective to the otherwise narcissistic self-image of human beings. When I showed this photo to one of my freshman English classes, their comments ranged from "That's just stupid. The earth doesn't look like that!" to "This blows my mind! I've never considered how tiny we really are!" In both instances, Sagan has achieved a, albeit slightly altered, goal. In the former response, students are, to adopt a popular therapeutic phrase, "in denial." They can't conceive of the counter-intuitive nature of earthly inconsequentiality. This is in some ways science's power; it is capable of a "magical" transformation from commonsense (a suspicious term, since it can be assigned to rationalize a variety of positions) to mysterious, even surreal demonstrations of reality. The rhetorical impact of science, therefore, is one that challenges our most cherished beliefs, our most unfounded assumptions, and our own self-professed centrality. To drive the mote of dust demonstration home even further, Sagan notes that this photograph represents a perspective of our home within the confines of our own solar system. From even more remote corners of the galaxy (among billions of star systems), the solar system itself resembles a "mote of dust" in a vast cluster of millions of other dust-specks. If one gets far enough away, it does not even register visibly. Leave the galaxy and travel to even further reaches of the universe, and the galaxy looks like just another star, its billions of stars reduced to a single point of light. Considered this way, it is little surprise that some people want to ignore the message; it simply makes our presence too tenuous, too existential, too nihilistic, too unimportant. Dwelling on the implication of nothingness is, perhaps, one of the main reasons we continue to push forward; without an ambitious sense of our own influence on the environment around us, we have absolutely no distinction in the cosmic dust bowl.

Clearly, Sagan was an unapologetic advocate for the sciences, as can be gleaned through the title of one of his later books, *The Demon Haunted World: Science as a Candle in the Dark.* Science, he felt, was the only epistemological savior for humanity, because it was through its "self-correcting" methodology that it would overcome the biases and petty prejudices that plague other ways of knowing, thinking, and speaking (he is especially disdainful of pseudo-science and shamanism, and rightfully so, since these are not, in any meaningful sense of the phrase, "ways of knowing," but rather, ways of blinding). He does not extend appropriate credence, however, to more abstract, theoretical ways of knowing such as philosophy, art, or language. In *Cosmos,* he says that "[t]he essence of science is that it is self-correcting. New experimental results and novel ideas are continually resolving old mysteries" (xiv). He does admit, however, that "science is inseparable from the rest of the human endeavor, [and] it cannot be discussed without making contact, sometimes glancing, sometimes head-on, with a number of social, political, religious, and philosophical issues" (xiv). The phrase "making contact" with these a-scientific human constructs seems to foreshadow the concept, which was something of an obsession with Sagan, of "making contact" with an alien intelligence (*Contact* was the name of his novel, later made into a film starring Jody Foster, which describes humanity's first encounter with aliens). While it may be an accident of phrasing which provokes attention to these similarities, we can nonetheless indulge the idea that "social, political, religious, and philosophical issues" were seen as alien to Sagan, largely because they involve systems that are not scientific (though, interestingly, we often try to make them so), that are distressingly alien in their humanness. Because new results and resolved mysteries and grand discoveries are part of science's domain, he gives the reader the impression that such processes are devoid of the intellectual sloppiness that social networks create and sometimes deliberately perpetuate.

The error here is that Sagan suggests that the influences of these areas of human conduct are incidental to the practice of science. Some may see this as arguing for a "social constructivism" explanation of science, but such an outlook is both misplaced and, actually, misconceived. To insist that social constructivism (however that is defined by those using the term) is somehow foreign to science is thoroughly misguided; science was not "found" as some divine gift but developed, through centuries of trial and error, within a variety of social, political, historical, and philosophical contexts, to produce the system we have now. It is therefore *always* social in the sense that it takes place within these social contexts, amongst social beings, through social processes, and for, ultimately, social purposes. Was it "socially constructed" in the sense that politics, religion, and philosophy somehow got together and "formed" both the means and the ends of science? No, but certainly these intellectual forces contribute to the direction and development of scientific processes. Clearly, the charge of social constructivism is as simplistic as the idea that science is independent of social forces. Sagan, while at least acknowledging such social forces, does so grudgingly, emphasizing instead that science is a self-correcting process, not requiring the extraneous checks and balances needed in other human epistemic systems. It might do us well, however, to question how self-correcting science really is, *especially* when it comes in "contact" with these other human exigencies.

The initiation process into the community of scientists is an intimidating prospect, especially if one wishes to genuinely understand such esoteric theories as Quantum Mechanics. Sagan describes such a process as one where an individual would need to know such "mathematical underpinnings" as Euclidian geometry, basic algebra, differential and integral calculus, differential equations, vector calculus, mathematical physics, matrix algebra, and group theory (249). Most people would find the estimated fifteen-year study an unproductive investment in time, and therefore, few people have any real understanding of what Quantum Mechanics is or is meant to explain. Add to this the thoroughly counterintuitive nature of such a theory, and we can see why the community of scientists who actually possess such knowledge is very small. More importantly, the community then becomes insulated, even self-contained in a way that creates huge restrictions for those who possess both the desire and the ability to enter it. To popularize such a theory in layperson's terms is, according to Sagan, at best difficult, at worst, impossible. In *The Demon-Haunted World,* in a chapter entitled "Antiscience," Sagan gives this explanation of the role of the popularizer of science when it comes to articulating such theories to the general public:

> The job of the popularizer of science, trying to get across some idea of quantum mechanics to a general audience that has not gone through these initiation rites, is daunting. Indeed, there is no successful popularization of quantum mechanics in my opinion—partly for this reason. These mathematical complexities are compounded by the fact that quantum theory is so resolutely counterintuitive. Common sense is almost useless in approaching it. It's no good, Richard Feynman once said, asking why it *is* that way. No one knows why it is that way. That's just the way it is. (249)

One problem here is simple misrepresentation, the forwarding of a theoretical principle as if all the details have been worked out, to suggest that the theory is in fact an unassailable scientific reality. Sagan compares this process to that of the "shaman," especially one who has promised some cure for a disease, malady, or condition. Here, the process looks like this:

> Now suppose we were to approach some obscure religion or New Age doctrine or shamanistic belief system skeptically. We have an open mind; we understand there's something here; we introduce ourselves to the practitioner and ask for an intelligible summary. Instead we are told that it's intrinsically too difficult to be explained simply, that it's replete with "mysteries," but if we are willing to become acolytes for 15 years, at the end of that time we might begin to be prepared to consider the subject seriously. Most of us, I think, would say that we simply don't have the time; and many would suspect the business about 15 years… is evidence that the whole subject is bamboozle. If it's too hard to understand, doesn't it follow that it's too hard for us to criticize knowledgeably? Then bamboozle has free reign. (249–50)

Sagan has, unintentionally I think, put his finger on exactly one reason why science's dominance is virtually unchallengeable by the layperson, but even more importantly, how science can be misapplied, corrupted, and distorted for purposes that, on the surface of them, *appear* to have a scientific face. First, the assumption that most people would approach the prospect skeptically is perhaps an overly-favorable generalization; critical scrutiny of information by the general population has been waning for some time now, largely because such scrutiny requires training and education in areas that are being routinely slashed from or downplayed in high school and college curricula: philosophy, logic, critical discourse, rhetoric, and general training in the rules of argumentation, to name a few. While many people may indeed have a kind of reflexive response to hot-button issues that invoke their culturally conditioned "opinions" for or against something, the majority rarely engage in any sort of practice that can be construed as "skepticism" in the critical sense. Second, because such training in critical discourse is minimal, many people cannot distinguish between bona fide science and pseudo-science, creating an atmosphere where all theories are treated as equally valid (or, in rarer cases, equally suspect). The result is a general willingness to accept what *looks* like science to *be* science, and this is clearly one of Sagan's concerns. "Faith" is perhaps the operative word; we have faith that ideas, methods, or processes that seem scientific are in fact scientifically arrived at, and Sagan's distinction between what passes for science and what does not falls well short of noting the complicated intersection between belief and reason:

> So how is shamanistic or theological or New Age doctrine different from Quantum Mechanics? The answer is that even if we cannot understand it, we can verify that Quantum Mechanics works.[7] We can compare the quantitative predictions of quantum theory with the measured wavelengths of spectral lines of the chemical elements, the behavior of semiconductors and liquid helium, microprocessors, which kinds of molecules form from their constituent atoms, the existence and properties of white dwarf stars, what happens in masers and lasers, and which materials are susceptible to which kinds of magnetism. We don't have to understand the theory to see what it predicts. We don't have to be accomplished physicists to read what the experiments reveal. (250)

The problem with this explanation, of course, is that Sagan makes it sound as though these are all experiments that we can conduct in the privacy and convenience of our own homes. What he means is that *scientists* who are not quantum specialists can use this method to verify the conclusions that the theory predicts; the rest of us have to rely on faith in these scientists that what they tell us is true and accurate. It is no surprise, therefore, that the layperson can no more tell the difference between the reality of quantum theory and New Age shamanism than she can between those news items that are "factual" and those that are cleverly edited. The point here is not that astrology and String Theory should be given equal consideration, weight, and value; on the contrary, it is my basic premise that people are not prepared to make this distinction, and they must therefore rely on an outside scientific agent to make it for them. It is for this reason that we trust scientific popularizers such as Sagan and that we have become increasingly reliant on them, but at a cost: We can't tell the shaman from the scientist because rhetorically they are indistinguishable given the degree of critical attention most people are able to bring to bear on them.

It may therefore also be productive to observe some of the language used by Sagan in his characterization of "bamboozlers." Sagan's free use of this term implies that there are two types of knowing: science and shamanism. He further couples this word with phrases like "New Age" and "theology," suggesting that religious tenets are in the same dubious epistemological category as herbal teas. He refers to these—appropriately, it would seem—as "belief" systems rather than knowledge systems. It is true that the general public today tends to view a "belief" as a valid substitute for a "proof," but in reality, both phrases are overused, and therefore, misused. Many propositions that people set out to "prove" are no more than suggestions—possible ways of looking at a situation, circumstance, event, or idea. For example, I have students routinely say that they will "prove" in their papers that capital punishment is moral. The problem, of course, is that it is not a provable proposition; whether capital punishment is moral is left to a system of ethics that reflects the values and principles of the society that creates it. One might indeed attempt to prove that capital punishment deters crime, but that too is logically misplaced; one cannot prove a causal relationship for events that *didn't* happen, though they may suggest statistical patterns that point to a probability. Even then, people frequently overlook other factors that may contribute to the decline in something like violent crime. "Beliefs" are even more elusive, since they can change at will and require no real substantiation other than the believer's conviction and willingness to act on the belief. Sagan has, therefore, in my mind, made an important distinction.

However, what is troublesome about his language here is that he constructs a false dichotomy again that asserts that either one uses science to gain knowledge or one doesn't really have knowledge (consider the subtitle of the book alone: *Science as a Candle in the Dark*). This is problematic on a number of levels, the first being philosophical. What does it mean to "know" something? What is "real"? What is "true"? Our society's generally simplistic view of what these words mean is not aided by Sagan, who would like his audience to believe that science is the only possible pathway to knowing and certainty. Certainty, after all, may be as deleterious as the search for ultimate Truth because, even if it is available for discovery, what does one do with it if it is found? Certainty and Truth, from a perspective of social efficaciousness, can in fact be both damaging and dangerous, especially if institutions arise around doctrinal declarations of certainty that no one in that society challenges. To cite a perhaps overused example, the Third Reich was unswervingly certain of its place in global and historical events, to the extent that it constructed a hugely intolerant, sadistic, and destructive social order characterized, ultimately, by the suffering and death of millions and the near-annihilation of an entire civilization. Kenneth Burke puts it this way: "Certainty is cheap, [sic] it is the easiest thing of which a man is capable. Deprive him of a meal, or bind his arms, or jockey him out of his job—and convictions spring up like Jacks-in-the-box" (*Counter-Statement* 113). This is, after all, how the Nazi's managed such a swift rise to power: They exploited the basic weaknesses of a vulnerable social structure menaced by depression, joblessness, hunger, and embarrassment and turned those weaknesses into convictions that held no room for other perspectives. And, not insignificantly, they relied heavily on science to do it. Theirs is the best historical lesson available for the unquestioning application of scientific knowledge in the absence of the guiding principles of sound ethical systems (which cannot be reduced scientifically).

Science, undeniably, is unrivaled at discovering physical realities and applying those discoveries in innovative ways. However, science's expertise in matters of human conduct and interaction has a much more dubious track record. Many bad policies have been put into place because science was invoked where science didn't belong, at least not exclusively. Sociological, psychological, and educational "science" (it still baffles—and even angers—me that education majors receive a B.S. degree at many universities) have done at least as much harm as good when trying to scientifically determine how and why human beings behave the way that they do, and most of the harm comes from assuming that because one test group behaves a certain way, all humans will follow suit predictably. Again we see the injurious effects of pseudo-certainty in the misapplication of science to human communal structures because to apply rational criteria to irrational beings is to force a hegemony unto a structure that is independent of the analytical apparatus we bring to bear on it. While it may be appropriate to discuss the physical, chemical, and biological features of nature to better understand its workings (despite its indifference to rationality), we must also concede the limitations of this approach: Science alone can only provide us with *information* about how best to cope with or control natural forces for our own predetermined ends. Our motives for understanding the natural world are independent of the reality of its existence; that is, nature is what it is and does what it does regardless of our attempts to decode it through the lens of science. Science is, in this way, a purely anthropomorphic apparatus, a mechanism of our construction for our purposes. It offers as much or as little in terms of ultimate understanding as the poet who views nature as a grand organism within which we are all intimate components. To then swing this large, often unwieldy instrument into the pathway of human activity tells us as much about what it means to *be* human as studying the mating patterns of a frog tells us what it means to *be* a frog. Moreover, science can tell us some things but not others, and we would do well to keep this in mind when indiscriminately summoning its powers in areas for which it is ill-equipped to supply the answers or to uncritically assume that science is the only valid form of knowledge.

Sagan's language does little to dispel the myth that science is the only proper pathway to knowledge—it, in fact, unapologetically promotes that very position. He has created a logical dichotomy that posits an either/or proposition when it comes to matters of knowledge: Either one employs a scientific method when seeking knowledge or one has participated in shamanism. There seems to be little acknowledgement that many non-scientific ways of knowing are available, and more importantly, used everyday for everything from managing one's finances to raising one's children. What does it mean, for example, to "know" about the Industrial Revolution? Is studying it a matter of scientific experiment, hypothesis, and attempts to disprove one's theories? Or is it more a matter of understanding a historical context, replete with individuals both common and extraordinary, the reasons behind its development, and the interpreter's internalized ability to see the interconnectedness between seemingly disparate conditions, to show a meaningful trajectory from which we can both learn about our past and apply it to our future? While science would certainly figure prominently as a topic of study in such an understanding, using it as a method is of little value to the individual who wishes to understand where, when, and how such an event fits into the narrative of the human past. Sagan's effort is one of educating the masses about the difference between valid scientific argument and proposals that fail the test of scientific scrutiny. Unwittingly, however, he has furthered a blind faith in science itself that misrepresents the appropriate place and function of science and does not allow room for ways of knowing that may in fact aid in the production of wisdom rather than the mere amassing of a body of applicable facts. The difference is at once subtle and far-ranging: Without some acknowledgement that science has its place like any other epistemological system, we will continue to invoke science for issues that really call for an understanding of history, human relationships, philosophical questioning, ethical conviction, social interchange, and rhetorical analysis. The result can be (and often is) an uncompromising certainty about how the world should be and how people should act because this certainty has the weight of science buttressing its conclusions without a corresponding acknowledgement of the limitations of science to temper the partial knowledge that has been created.

The overall impact of Sagan's picture of the scientific project on the viewing and reading public is that it is somehow insulated from the daily realities that the rest of us—in a committee meeting, at the bank, when applying for a zoning permit—must face. Science is not, he naively implies, muddied by the stagnation of bureaucracies or the ambitions of competitive colleagues. It is above all that. It freely corrects its own errors, never holding fast to an untenable idea or hypothesis. It is sublime this way, concerned only with the purity of the process and the pursuit of knowledge for its own sake. And the public likes the idea that there is one arena, at least, that is not contaminated by the cynical pressures of the modern world, not tarnished by the evils of the unscrupulous, the provincial, the ignorant, or the incompetent. Throughout the twentieth century, science and the scientists who practice it have enjoyed a carefully cultivated image, one that gives it an air of dignified authority and benevolent power. And it is through the media vehicles of a modern age that this message has become even more widespread. Certainly, there are those who distrust science and its most transparent manifestation, technology, but most of us readily accept scientific explanations without considering whether they are either scientific or suitable topics for scientific scrutiny. James Burke and Carl Sagan have both undergone the noble enterprise of educating the public about science, its history, and its amazingly brisk progress. Where they have failed, however, is in demonstrating the limitations of an epistemological system by suggesting that it has no limitations. Subtle as this might seem, it perpetuates the very blind faith in a knowledge base that its proponents claim to despise. If, in fact, science is self-correcting, then it needs also to recognize that as a human construct, it should be (and, for all practical purposes, is) self-limiting as well.

Few would deny that science has been, is, and always will be an important—perhaps even central—component of human knowledge. Its method is both sound and valuable, and science itself has created opportunities for progress (whatever that is, exactly) like no other system constructed by human beings. However, this success carries with it a hubris that exposes its corruptibility as a human invention. Both James Burke and Carl Sagan (and any other science popularizer) have a right to crow about the wonders of science and technology. In that self-congratulatory flourish, however, discerning critics will see that the construction of the scientific ethos serves to create a very unscientific complacency that sets the stage for misuse and uncritical dogmatism about the value of science. We have almost completely forgotten that science is a human device and a man-made vehicle for gaining knowledge. In its rational pursuit of precision and intellectual integrity, it seems to be a transcendent form of knowledge, unfettered by the irrational, emotional, and distinctly human forms of abuse. It is at its core a language system, and all language systems are at some point inadequate, failing as they do to completely capture the "reality" of the things that they describe. Mathematics may indeed give us a less ambiguous language with which to describe the physicality of our world and the universe, but even it becomes an interpretive method at the outer reaches of its functionality. Higher mathematical concepts, just like poetry, become an art form—ones that can elicit a variety of interpretations and can be used in a number of different ways to solve the same problem. It is a popular myth that mathematics is certain, unambiguous, and stable, especially at the higher levels. But this too is a component of the scientific ethos: that the methods, the tools, and the practitioners of science are unencumbered by human frailties and misinterpretations and that the figurative tools they use will never malfunction or be used erroneously.

Burke and Sagan serve us as models of the scientific popularizers of our generation. Their project is to educate the masses about the impeccability of science as a tool for human understanding. Where Burke is interested in the historical associationism that makes science a dramatic narrative, bringing it alive for an audience used to being entertained, Sagan is preoccupied with what he sees as an unfortunate trend in American culture to be easily "bamboozled" by alternative belief systems that have not passed the scientific litmus test of validity. Both projects have their value, but in the pursuit of further educating the public, they have offered one and only one option for true knowledge, further perpetuating the mistaken belief that a-scientific pursuits of knowledge have little value or practical applicability. The self-contained and self-correcting nature of science has been traditionally viewed as a truism that is beyond challenge, but I would submit that in the effort to educate others on the value of scientific methodology, the proponents of science have overlooked an important question to ask when receiving scientific information critically: Is science the best possible way to judge a problem or an event? Are there other ways of knowing that might yield results that are more appropriate to the question being asked? By suggesting that science is the only real alternative for solving problems (whether this is what they intended or not), or, for that matter, that solving problems is the only reason we amass knowledge, Burke and Sagan have aided in the misguided public assumption that science is our savior for all issues that publicly and privately affect us. That they use the popular media as the vehicle for their message should suggest that we need to look more carefully at how the message itself is being packaged and dispersed.

4 Scientists Named Steve

From MIT psycholinguist Steven Pinker's website:

> The Big Box o' Steves: "What's more fun than a scientist-writer? A scientist-writer named *Steve!* You'll get Steven Weinberg, Stephen Jay Gould, Stephen Wolfram, Steven Pinker and Stephen Hawking. Hi-ho, Steverino! (Keep Pinker and Gould figures separate to avoid spontaneous combustion.)" (qtd. in Mirsky)

Three other key figures in the popularization of science are a diversely talented set of scientists who have become household names: Stephen Hawking, Steven Pinker, and Stephen Jay Gould. All three men have written prolifically on a wide range of topics and to a wide range of scientific enthusiasts. What is perhaps more impressive than their popularity is that they are also leaders in their respective fields—Stephen Hawking as a theoretical physicist and cosmologist, Steven Pinker as a psycholinguist, and Stephen Jay Gould as a paleontologist.[1] Stephen Hawking is a particular favorite of both the media and the general public, since his is a story of incredible achievement in the face of an extremely incapacitating disability. Hawking, as most people know, suffers from amyotrophic lateral sclerosis, also known as Lou Gehrig's disease, a malady that is characterized by the progressive degeneration of neuro-muscular function. When he was first diagnosed with the disease, his chances for prolonged survival were slim. However, over forty years later, he is considered one of the premier theoretical physicists working today. He is, in this respect, often painted as a hero who overcame overwhelming adversity to become a world-class scientist under circumstances where mere survival was a challenge. His most famous popular book, *A Brief History of Time,* according to the Forward of the 1996 edition, was on the London *Sunday Times* best-seller list for 237 weeks, has been translated into forty languages, and has sold one copy for every 750 people on the planet.

To achieve this level of celebrity is, for a scientist, nothing short of a miracle, and we might say the same for Hawking himself. The circumstances surrounding his rise through the scientific ranks and his triumph over his disability make for high drama, assuring that Hawking was destined to be a media darling. Yet, he is a "real" scientist, likened to some of the greatest minds in intellectual history. With these credentials, as well as his marketable celebrity profile, he is a driving force in educating the public on cosmological matters. Perhaps the most interesting facet of his popularization of topics like the origins of the universe, the nature of time-space, and place of humanity in the cosmos is that these are really expressions of age-old religious questions. *A Brief History of Time* may owe some of its success to its scientific articulation of spiritual inquiries since it helps bridge the apparent gulf between theological and scientific ways of thinking. In truth, however, science and religion have never been as cleanly separated as history and popular opinion tend to hold; in fact, not only are the central questions of science essentially spiritual in nature, science frequently adopts religious and theological terminology when describing its discoveries and objectives.[2]

It is not clear, however, whether Hawking is deliberately capitalizing on the symbolic association between questions of spiritual belonging and questions of our place in the universe or whether he is simply doing what a good theoretical physicist does: ask the big questions. The effect, though, is clear. The reading public responds well to Hawking's instruction because, in the absence of plausible spiritual explanations, his theories help fill a void left by what many see as the retreat of a spiritual consciousness in the American public. While there seems to have been an apparent resurgence of interest in spiritual questions in the last decade or so, it also appears that Christianity may be experiencing a fluctuation in its formal membership. According to *Religioustolerance.org*, church attendance among Protestant institutions declined steadily (around 12 percent fewer people were regularly attending services) during the 1990s. This trend is perhaps a sign of the rationalistic tendencies in American thinking, a pattern that aids science in fulfilling the role once held by religion. I suspect however, that the trend has less to do with overt rationality and more to do with the American propensity toward literalism. Even those who find spiritual solace in scripture seem these days to be increasingly fundamentalist.[3] That is, while there is appears to be an increasing urge to answer questions of a spiritual nature, to push toward a religious dimension in daily life, and to have a pious orientation towards one's moral being, Americans today often propel their spiritual needs with an odd combination of skepticism and anti-intellectualism. For many, in the final analysis, the word is The Word, an unshakable belief in the inerrancy of Biblical Scripture, and people for whom spirituality has been rekindled have a strong impulse to make that spirituality as simple an unadorned as possible. To this end, they read scripture (or, more likely, have it read to them) in only the most superficial, utilitarian way—a practice that guarantees greater certainty and shuts down dialogue about one's faith before it can even begin.

Hawking has enjoyed unparalleled popular achievement with his book not so much because he has bridged the gap between the intellectual and the practical or the religious and the secular, but because he has managed to make the intellectual (in this case, theoretical physics) *seem* practical, and he has made the scientific hold some religious significance. Hawking's success, then, may be attributed to a shift in the public mind-set from one of theological importance to one of cosmological significance. The general public's interest in questions concerning human and cosmological origins might be described as confused, but for those who are suspicious of institutionalized religion, especially those institutions that tend to obscure metaphysical questions, there may be a need to seek answers from sources adjacent to religion itself. Hawking deals comfortably with a variety of esoteric phenomena, ranging from black holes to wormholes to the very nature of time and space, all in under 200 pages. He has made accessible some of the most profound mysteries of the universe, and while he has not, of course, solved every problem, he has certainly given the reader a reference point from which to understand heady concepts and to pursue further inquiry. This is perhaps the most important attribute of the successful scientific popularizer: to bring a readership closer to the complexities of a particular body of knowledge without overburdening it with the intricate complexities that separate the expert from the layperson.

Throughout *A Brief History of Time,* Hawking tempers his theoretical explanations with an effective balance of historical narrative and personal anecdote. This combination is a mainstay of scientific popularizers—Steven Pinker perhaps being the most accomplished rhetorician in this respect. The ability to package highly-specialized technical information in an accessible format is what separates the successful teacher from the isolated researcher. It takes talent, for example, to contextualize the history of discoveries and theories in space/time in a way that is both easy to read and relatively brief while still remaining essentially accurate. That Hawking can also tap into the readership's spiritual motivator means that he was the right guru at the right time. In Chapter 2, titled "Space and Time," Hawking carefully traces the intellectual evolution of the space/time question, summarizing the ideas of all the heavy-hitters: Aristotle, Galileo, Newton, and Einstein. While there are many other important figures in the history of science that address this topic, Hawking has kept the text within easy reach by discussing only the most famous and familiar names. When he does introduce personalities with whom the general reader may not be as familiar, such as nineteenth-century British physicist James Clerk Maxwell, he is careful to show just where they fit into the narrative and what their contribution was to the larger issue. He says of Maxwell, for example, that he "succeeded in unifying the partial theories [of the propagation of light] that up to then had been used to describe the forces of electricity and magnetism," thereby situating Maxwell within the scientific framework that Hawking has vigilantly laid out (19).

While this may seem like a rather obvious stylistic approach to take if one is interested in educating a non-specialist in a highly obscure field of knowledge, it is important to note that only a handful of scientists like Hawking even seem to have any interest in doing so. Of those who attempt it, the most successful are those who, far from condescending to their audience, make readers feel as if they really can embrace a subject matter whose complexity had made understanding it prohibitive in the past. What Hawking applies as a strength is in many ways the weakness of other science-writers. The majority of scientists I've read, when trying to educate the world on their specialty, either a) talk down to the audience in a way that is alienating, or b) dilute the material so extensively that they do more to mythologize erroneous information than they do to clarify remote theoretical ideas. Hawking's project and the success of *A Brief History of Time* are based on the public indication "that there is widespread interest in the big questions like: Where did we come from? And why is the universe the way it is?" (vii). These questions are central to our cosmic identity, and they represent a very old preoccupation with understanding the nature of humanity and with trying to predict our ultimate purpose. The origins of the human species are closely related to our notion of ourselves because we want to know not only where we came from, but also where we're going and how we're getting there. Hawking manages to put or tenuous existence in more humbling terms, and it is this quality that gives his position a more spiritual significance. While we crave to be the center of the universe, we also like to believe that there is something bigger than ourselves. In this way, we have an incentive for moving forward, improving our condition, and envisioning our own future.

Another Steve should be a familiar name to most readers: Stephen Jay Gould. Gould's scientific contribution to both the general public and to the scientific community is prolific by any standard. An abridged list of his books include *Ontogeny and Phylogeny; The Mismeasure of Man* (Penguin 1983), winner of the National Book Critics' Circle Award for 1982; the popular collections of essays *Ever Since Darwin: Reflections in Natural History* (Penguin 1980); *The Panda's Thumb: More Reflections in Natural History* (Penguin 1983), which won the 1981 American Book Award for Science; *Hen's Teeth and Horse's Toes: Further Reflections in Natural History* (Penguin 1984); *The Flamingo's Smile* (Penguin 1987); *Time's Arrow, Time's Cycle* (Penguin 1988); *An Urchin in the Storm* (Penguin 1989); *Wonderful Life* (Penguin 1991), winner of The Science Book Prize for 1990; *Bully for Brontosaurus* (Penguin 1992); *Eight Little Piggies* (Penguin 1994); *The Structure of Evolutionary Theory,* an enormous, 1400+ page tome on the history and implications of the theory of evolution (Harvard UP 2002); and *Rocks of Ages: Science and Religion in the Fullness of Life* (Ballantine 1999). With these sorts of credentials, Gould deserves a prominent place in both the annals of evolutionary biology and in the minds of the public. His project, at least insofar as educating the public is concerned, has been somewhat diverse, but one thread that runs throughout his work is a desire to function as a corrective for misinformation about evolutionary theory and its implications for science and humanity overall. Gould himself, in *The Structure of Evolutionary Theory,* described popular notions of science this way:

> [W]hen distasteful conclusions gain popularity by appealing to supposedly scientific support, and when this 'support' rests upon little more than favored spec-

> ulation in an orthodox mode of increasingly dubious status, then popular misuse can legitimately sharpen a scientist's sense of unhappiness with the flawed theoretical basis behind a particular misuse. (43)

Though hardly an "orthodox mode" in this day and age, Gould's sentiment was particularly true of his efforts to debunk so-called "Social Darwinism," a loosely-woven theoretical branch of Darwinism that suggests that social forces are as much a product of evolutionary processes as biological ones are. The idea is not only dangerous, but untenable; it implies that a basic tenet of Darwin's theory of natural selection is "survival of the fittest" (a phrase popularized by Herbert Spencer, a British economist, upon reading *Origin of Species*) and that this doctrine applies not only to species and their various adaptive offspring, but also to people of particular social groups as well. Literalist proponents of evolution have used Social Darwinism to further arguments of racial, economic, religious, and political superiority, extending the process of natural selection to sociological spheres, usually as a way to rationalize sometimes discriminatory, sometimes outright eugenic, policies. This is not to be confused with "sociobiology" or "evolutionary psychology" which deal more with the role of genes in human behavior (though they are every bit as controversial, and Gould had a long-standing feud with proponents of these ideas—men like Edward O. Wilson, Steven Pinker, and Daniel Dennett); Social Darwinism, conversely, argues that social relationships and institutions "evolve" in the same manner that biological organisms evolve, and that those which flourish have an evolutionary "right" to because it is the order of things for them to use this advantage for any end, no matter how oppressive or immoral. Many proponents unapologetically hold to a "might makes right" philosophy (based erroneously in evolutionary theory), arguing that if a particular segment of society cannot survive without outside aid, they are "destined" to perish and they should be allowed to do just that. Such a stance is certainly distasteful, arguably evil, and has been a target of scientists like Gould.

While the concept of Social Darwinism is considered by many to be archaic, it is clear that the residue of its once-popular status (in the late 19th and first half of the 20th centuries, especially) lingers today, especially among conservative wings. It would do us well, therefore, to examine this phenomenon a bit more if only to see what Gould was objecting to and how he fits into a larger overall scheme of scientific popularization. The Social Darwinistic movement, it should be noted, covers a range of ideas, some more moderate than others. There is also a subtlety present in how it manifests itself, such that certain policies or attitudes that have an underlying Social Darwinistic premise may not be explicitly expressed as such. Take, for example, legislative action that withdraws aid to individuals and families who are living at a level below the arbitrary designation of "poverty" (i.e., "welfare"). The argument for this policy shift might be that it is not the government's responsibility to provide monetary assistance to those who can't otherwise pull themselves up out of their financial predicament, that such people can and must rely solely on their own resources to correct their situation. Taken a step further, those who have received welfare benefits prior to the change in policy must "pay back" the government some or all of the money they received while on the dole, a system known as "workfare." While on the face of it, no one would necessarily view this policy as a product of Social Darwinism, it certainly possesses a "survival of the fittest" mentality, for it suggests that those individuals and families incapable of making their own way in the world are either unwilling or unable to do so. In either case, a society is able to "thin the herd," at least insofar as such people are no longer a burden on that society's tax infrastructure (the burden will certainly manifest itself in other ways: Crime, for example, is likely to rise if this policy remains in effect, but we have a solution for that too: imprisonment.). This, I would argue, is every bit as Social Darwinistic as an active policy of eugenics; it simply isn't as visible as such, in no small part because it is a rare topic on syndicated news broadcasts. Here, then, we see the significance of Gould's crusade. Without a respected scientific figure to challenge the notion of Social Darwinism, a government, with the backing of its main taxpayers, can implement administrative procedures that are clearly designed to maintain inequality, promote opportunities for a privileged few, and widen the gulf between the races and classes in socially toxic ways.

While policies of a Social Darwinistic nature do appear to exist in this country (at least as the nation becomes more conservative in its political orientation), Gould targets a different trend in the academic community with equal vigor: evolutionary psychology. This refers to a theory of psychological development that can be attributed to evolutionary processes just as the development of an opposable thumb can be seen as the product of the same evolutionary adaptations. In other words, evolutionary psychology seeks to answer the question of how it is that humans have developed such hugely sophisticated brains by using the theory of evolution to explain this development. As I have already alluded to, a pressing question in the theory of evolution as it applies to *Homo sapiens* is what adaptive necessities pushed the human intellect to the levels that it currently possesses. Many argue, for example, that the human brain is far more complex than it needs to be for mere survival, a contention that evolutionary psychologists like Steven Pinker, David Buss, Leda Cosmides, John Tooby, Edward O. Wilson, Martin Daly, and others, hope to answer. Pinker, in particular, has done more to popularize this idea than almost any other scientist, and it is his ideas primarily that Gould wishes to challenge (more on Pinker in a moment).

I mention all this not so much because it is necessary to detail the *content* of the ongoing debates about evolutionary theory between scientists like Gould, Pinker, Dennett, and Wilson, but to note that these debates were, relatively speaking, at least, *public* (as opposed to taking place solely in the Ivory Towers of the scientists' respective institutions) and to note how scientific debates, especially debates dealing with so socially provocative a notion as evolution, can easily become *politicized.* This observation challenges the assumption that science and scientists are always objective, concerned only with the discovery of truth, to show how, under certain circumstances, political convictions cannot help but bleed into a scientific examination of a hot topic. This may be no great revelation to science historians, but that's only because they make a living studying such things. For those of us populating the laity, it may be surprising to see just how socially and politically zealous scientists can get, so much so that it can easily compromise the apparent detachment scientists are purported to possess; they often, in fact, use their scientific authority to forward ideas they feel are important, even if that means conceding their own professional disinterestedness in the process. As a politically-minded scientist, Gould is second only to Noam Chomsky in his political convictions,[4] and both are very leftist in their political orientation. The marriage of science and politics, however, can have unintended consequences.

Take, for example, Robert Wright's interesting observation that Gould, despite his scathing indictment against the Kansas Board of Education decision to remove evolution from the biology classroom, has in fact "aided and abetted" the Creationist cause. Wright, no Creationist himself, but a scholar at the University of Pennsylvania and author of *Nonzero: The Logic of Human Destiny,* is interested in showing how unorthodox views of evolution have the effect of giving Creationist ammunition against a well-established theory:

> Over the past three decades, in essays, books, and technical papers, Gould has advanced a distinctive view of evolution. He stresses its flukier aspects—freak environmental catastrophes and the like—and downplays natural selection's power to design complex life forms. In fact, if you really pay attention to what he is saying, and accept it, you might start to wonder how evolution could have created anything as intricate as a human being. (Wright)

Gould's position, Wright explains, furthers the Creationist agenda by casting doubt onto the plausibility of evolutionary theory because it reflects a lack of purpose—the teleological argument—which undermines the theory through its absence of design. What most people misunderstand, because of the human need to impose design and order upon everything we encounter, is that order is itself a human construct. People, especially those of the Creationist ilk, simply cannot reconcile the idea that nature is an indifferent process with what they see everyday—that the "design" apparently inherent in natural structure is an illusion that we have created for cognitive convenience. They cannot live with the prospect that if the universe is allowed enough random readjustments and enough time, the result is something that appears ordered even if the patterns we discern are ones that we have formulated in order for us to adapt more efficiently to our own environments. This may be the most important human attribute, in fact: our ability to find structure in the structureless, to impose design where none otherwise exists. Language is such an ability, and so is technology. Consider our ability to mimic natural processes and "improve" upon them; this can be seen as a definition of technology, the ability to engineer devices, processes, and algorithms that function within the confines of our own environment. An interesting example from Wright illustrates this:

> Consider the bombardier beetle. In one compartment, the beetle carries a harmless chemical mix. In another compartment resides a catalyst. The beetle adds the

> catalyst to the mix to create a scalding substance that he can then spray, through a pliable rear-end nozzle, on tormentors. (This basic idea—making chemicals safe to transport but deadly when deployed—would, long after natural selection invented it, be reinvented by human beings, in the form of binary chemical weapons.) (Wright)

Even more interesting, Wright points out, is the behavior and equipment necessary to deliver such weapons:

> Clearly, a beetle equipped with two munitions tanks and a spray nozzle is more complex than a beetle lacking such accoutrements. And this isn't just any old kind of biological complexity. The beetle's arsenal involves behavioral complexity: aiming and squirting a toxic nozzle. Aiming and squirting—like any impressive behavior—involves information processing, a command-and-control system. In some small measure, then, evolution's promotion of the beetle to bombardier rank involved a growth in intelligence. In other lineages, the evolution of intelligence—of behavioral complexity—has proceeded further. And we have binary chemical weapons, among other things, to show for it. (Wright)

Not only have humans observed the capacity of an insect to use a sophisticated chemical process and imitate it, we have given it a common name which reflects not *its* unusual ability, but *ours*. The bombardier beetle, having possessed the facility to defend itself by combining chemical agents for millions of years—long before humans came on the scene—is referred to by a descriptor that suggests it "bombs," an idiomatic metaphorical reversal implying that we engineered a process that the beetle then copied. This is typical of the solipsistic human tendency to see ourselves as the originators of unique ideas, even if natural selection has a huge jump start on innovative engineering designs. A more accurate arrangement would be to call our binary weapons "beetle weapons," but more to the point, this is an example of how humans observe, mimic, and name things based in what we assume are natural designs—it is then an easy matter of superimposing our design onto it as if the design was always present.[5] Because of this technological and linguistic directionality, we assume order that must have existed prior to the reality of the organism—we assume, in other words, a god that, in his infinite inscrutable wisdom, created such creatures to maintain a balance in the natural order of the world.

But does the beetle's ability really imply "order" in the sense that an omnipotent force drew up the blueprints, imagined a prototype, and put the mechanism into motion? Evolutionary theory says no. Rather, it argues, the billion-year process of natural selection drove the starts and stops, successes and failures of natural elements that randomly coalesced into what we see today. Most people have serious misgivings about such a process because it is so antithetical to our sense of order and certainty. How can a churning mass of compounds without extraneous direction form into something as complex as an orchid? The absence of something (or someone) there to guide the outcome offends our sense of the clockwork universe, our notion of a mechanistic meshing necessary for all things to function;[6] because we manipulate the natural world to do our bidding, so too must the natural world itself operate in this way. God, under this mentality, is created in *our* image; because we are the great manipulators, the great matter-movers of the Earth, so too must God be. It is no coincidence, for example, that theories of natural selection (in some ways an unfortunate descriptor of the evolutionary process, for "selection" implies "will," a criticism Darwin himself had to address) and notions of existentialism share a common historical timeframe. The nineteenth century experienced some of the most profound cognitive shifts in human history, and questions about the ultimate purpose and meaning of life led to what Kenneth Burke refers to in *Permanence and Change* as "the Eternal Enigma": the idea that being and nothingness were equally unthinkable. To the spiritually-minded, the notion of purposelessness and nothingness is simply not an intellectual option, and for them, Gould's description of evolutionary processes only strengthens their resolve against the idea. As Wright points out:

> Gould also performs a more subtle service for creationists. Having bolstered their caricature of Darwinism as implausible, he bolsters their caricature of it as an atheist plot. He depicts evolution as something that can't possibly reflect a higher purpose, and thus can't provide the sort of spiritual consolation most people are after. Even Gould's recent book *Rocks of Ages,*

> which claims to reconcile science and religion, draws this moral from the story of evolution: we live in a universe that is "indifferent to our suffering." (Wright)

The main point of contention between Gould and Wright, moreover (and the detail that gives Creationists the most ammunition), is the notion of evolutionary "progress." Gould argues that, if evolutionary processes and natural selection are random, then there is no guarantee that adaptations will be necessarily advantageous to the organism. That means that an organism is just as likely to mutate in a way that has no benefit for the survival of the species, and may in fact be a detriment. Such adaptations compromise the survival of the organism, and will not be passed on to future generations. Wright, conversely, argues that adaptations almost always move in a direction of greater complexity, and therefore have an affirmative "directionality," which doesn't necessarily mean one that is positive for the organism, but does imply a "progression upward" on the scale of evolutionary complexity. Natural selection, in other words, moves naturally "upward," a metaphorical distinction more than a scientific one. "Up" implies "better," but perhaps a more accurate representation is that adaptation moves neither up nor down, but merely reflects *change.*

This argument has more to do with how the theory is described than it does with the practical outcomes of evolution itself[7]. That is, just as scientists seem to be bogged down in what evolutionary processes *mean,* so too are the Creationists. They are attempting to find significance in the process over a mere description of the process itself. The difference between what something *is* and what it *means* is an important one, for it reveals that the fundamental significance of the question about evolution is what impact it has on us in terms of our *own* speciation. Otherwise, we would have no motive for asking the questions in the first place. This debate, as with all debates, then, boils down to how we talk about it and what people—scientists, fundamentalists, and the general public—can derive from the discussion that provides meaning for their lives. That this is such a hot-button issue illustrates this in no uncertain terms. Scientists have their motives (professional reputation, a desire to know the "truth," the egoistic need to be "right"), fundamentalists have their motives (the impact on "our children," how the issue might compromise their faith, what it means for the centrality and significance of the human species), and the general public has its motives (a need for certainty, a general curiosity, how this impacts education, and hence, how their tax dollars are being spent), and natural selection and the evolutionary processes that drive it continue doing what they have done for billions of years indifferently to our squabbles. It is a rich irony that, long after human beings have ceased their nervous banter about this subject—because they have long ceased being entirely—evolution will show (someone or something) just how truly indifferent it is to us.

And from a practical standpoint, this last notation is perhaps what frightens us most about evolution. We like to believe (in fact, we *must* believe) that the human species is eternal, that it will never become extinct, that it will live on, in some form or another, for all of time. As beings conscious of our own existence—and conscious of the fact that we are conscious about it—we have been given an intellectual curse. We are, so far as we can tell, the only species aware of its own mortality. This single piece of knowledge is, in many profound ways, the foundation upon which all of our institutions are based. We need to lay down a legacy, to let others remember that we were here and that we made some impact on the world. Without this, we are only one more species that has come into and will go out of existence, a prospect that frightens us at our deepest level. This knowledge is, in short, what evolution gives us, and many are not happy about it. Gould himself becomes preoccupied with the implications of mortality, which is one of the reasons he wrote *Rocks of Ages: Science and Religion in the Fullness of Life* in the last years of his own life. The very first paragraph of the Preamble describes his desire to reconcile the apparent conflict between the goals of science and the convictions of religion this way:

> I write this little book to present a blessedly simple and entirely conventional resolution to an issue so laden with emotion and the burden of history that a clear path becomes overgrown by a tangle of contention and confusion. I speak of the supposed conflict between science and religion, a debate that exists only in people's minds and social practices, not in the logic or proper utility of these entirely different, and equally vital, subjects. I present nothing original in stating the basic thesis (while claiming perhaps some inventiveness in choice of illustrations); for my argument follows a strong consensus accepted for decades by leading scientific and religious thinkers alike. (3)

We might ask whether science and religion are, in fact, as "entirely different" (or, for that matter, "equally vital") as Gould claims. It is also interesting to note the subtle invocation of religious language, that Gould sees his solution as "blessedly simple," almost as if the answer to this timeless question had been "revealed" to him. A little further down in the preamble, Gould attempts to clarify the distinction he sees in these two ways of knowing by suggesting that science is concerned with factual explanations while religion, he implies, is the domain of ethics:

> Science tries to document the factual character of the natural world, and to develop theories that coordinate and explains these facts. Religion, on the other hand, operates in the equally important, but utterly different, realm of human purposes, meanings, and values—subjects that the factual domain of science might illuminate, but can never resolve. Similarly, while scientists must operate with ethical principles, some specific to their practice, the validity of these principles can never be inferred from the factual discoveries of science. (4–5)

It is a common misconception (and a central one) that religion is the only area in which moral principles can exist. This idea is dangerous for a number of reasons, and just as Gould aids the Creationist cause through his unorthodox views on evolution, so too does he encourage the notion among the pious that they have a monopoly on ethics and morality. The primary assumption underlying the assertion that morality is the domain of religion is a particularly cynical one: Without external, even supernatural, motivators, this assumption posits, human beings are naturally wicked, prone to behaving in ways that only reflect their own hedonistic urges. God, and his (her/its) requisite promises or threats, is the only force capable of keeping such a basically depraved creation as Man on the path to moral goodness. While this may be a slight oversimplification of Gould's point, we can rest assured that this is how the general readership of his book will respond to his message. I hear this unfortunate "wisdom" all the time, both from my students and from people I speak with every day. To them, religion is the only possible source of moral accountability and meaning. The implications of such a rigid interpretation of morality, of course, are extremely dangerous and can lead to decidedly *immoral* actions, such as the "holy war" currently being conducted by the United States and the insurgency in Iraq (as well as a more generalized "war on terror").

The assumption that a certain religious affiliation (or religion in general) has the final authority on morality is a problem that can be attributed to, according to Karen Armstrong, author of *The Battle for God,* the confusion of *logos* and *mythos.* She explains that fundamentalism, far from a new movement, enjoys a resurgence in times of moral and social uncertainty (though, one might wonder, how often in human history we have enjoyed any real certainty in either of these attributes). What makes fundamentalism a problem is that those who practice it often do not understand the necessary distinction between the rational side of human knowledge (*logos*) and the mythical context which gives it meaning (*mythos*). She, like Gould, posits that there are basic differences between the objectives of science and the goals of religion, but the difference is in knowledge-making, not in relative value. She uses the example of the First Crusade to illustrate what can happen when these two value systems are confused:

> Throughout the long crusading project, it remained true that whenever *logos* was ascendant, the Crusaders prospered. They performed well on the battlefield, created viable colonies in the Middle East, and learned to relate more positively with the local population. When, however, Crusaders started making a mythical or mystical vision the basis of their policies, they were usually defeated and committed terrible atrocities. (xvii)

This statement can be read as an example of what can happen when the zealousness of religious piety encroaches into the realm of the rational, for if one mistakes the religious for the rational, one can easily justify any number of otherwise prohibited activities. This is, according to Armstrong, the problem of fundamentalism writ large, a lesson that America's early legislators took to heart so enthusiastically that they folded it into the very first amendment of the Bill of Rights through the doctrine of The Separation of Church and State. As another by-product of a flimsy educational system, history becomes consigned to the realm of the irrelevant; we would be prudent to read our historical documents more carefully, however, since we are making precisely the mistakes that our ancestors wisely warned against when they framed the Constitution: Religion and politics do not mix well. Radical fundamentalism is the same regardless of religious affiliation, whether it be Muslim, Jew, or Christian, and the US's brand of fundamentalism has made its presence known up to the highest offices in the land, a prospect that has many in this country deeply concerned.

Gould's answer to this problem is what he calls "NOMA"—Non-Overlapping Magisteria. He characterizes this concept as "respectful noninterference—accompanied by intense dialogue between the two distinct subjects, each covering a central facet of human existence" (*Rocks of Ages: Science and Religion in the Fullness of Life* 5). He considers the conflict between science and religion a "false" one, and he, like Armstrong, sees serious difficulties arising when the authority of one branch of knowledge attempts to dictate the ways of knowing in the other. He says that "NOMA represents a principled position on moral and intellectual grounds, not merely a diplomatic solution" and concedes that "NOMA also cuts both ways. If religion can no longer dictate the nature of factual conclusions residing properly within the magisterium of science, then scientists cannot claim higher insight into moral truth from any superior knowledge of the world's empirical constitution" (9). As a basis for this Aristotelian "Golden Mean," Gould asserts—even defends—the apparent human tendency to dichotomize, claiming that this is the natural process of the human mind, to define the world in simple either/or terms. This tendency, he claims, is a holdover from our less complicated past, "when limited consciousness could not transcend 'on or off,' 'yes or no,' 'fight or flee,' 'move or rest'—and the neurology of simpler brains became wired in accordance with such exigencies" (51). This habit of dichotomizing may be, in fact, an indication of our early efforts to categorize, simplistic as they may be. It is certainly true that people today have a habit of oversimplifying issues in this binary way, but how much of this is a matter of socialization or convenience as it is a product of nature? Our brains are obviously capable of more complex operations, but convenience dictates that dichotomizing saves us mental energy. It is easier to say, for example, that you are either a patriot or a terrorist, but it benefits us to question whether this happens because it is our "natural" predilection to do so, or because it is more likely that we simply don't have to challenge our own convictions if we compartmentalize in this superficial way. This raises the question of just how useful it is to conveniently sever religion and science in twain so that they can peacefully coexist. Later on, Gould reminds the reader of his central argument by admitting its simple premise:

> This book rests on a basic, uncomplicated premise that sets my table of contents and order of procedure, and that requires restatement at several points in the logic of my argument: NOMA is a simple, humane, rational, and altogether conventional argument for mutual respect, based on non-overlapping subject matter, between two components of wisdom in a full human life: our drive to understand that factual character of nature (the magisterium of science), and our need to define meaning in our lives and a moral basis for our actions (the magisterium of religion). (175)

Again, the moral question arises, and Gould "conventionally" assigns it to the purview of religion (science, he says, cannot provide moral guidance). Further, he vainly "uncomplicates" what is, at its very foundation, an exceptionally complicated issue. We are left wondering just how valuable such advice is in the face of so large and so vast a body of knowledge represented by these two magisteria. We might wonder, too, just how possible it is to separate them as cleanly as Gould suggests; so tied are these epistemological foundations to the same sorts of questions that it seems impossible to untangle them from the inquiries that propel human curiosity about our world and beyond. Even Gould concedes that "both [religion and science] seem to raise similar questions at the core of our most vital concerns about life and meaning" (51). In the past, Gould points out, religion and science were seen as intellectual systems that could either fight until a single victor emerged or become somehow synthesized. Here, Gould provides a third option, namely, that these areas of magisteria need not encroach upon one another, that they can exist so long as certain "admissible questions" and appropriate procedures are identified for each and left alone by the other. But this is defined in terms of the distinction between the domain of the "factual is" (science) and the domain of the "moral ought" (religion) in a way that fails to avoid the dichotomizing process that Gould apparently resists (55). The proposal, rather, simply creates another binary, one that posits, implicitly at least, that the scientist cannot be moral without the guidance of religion and that the religious cannot be factual without the standards of evidence provided by science. How, then, can we ever keep these domains from overlapping? Taken to its next logical step, we are left with questions like, "where does this compromise leave the secular humanist or the a-religious layperson?" Can, indeed, an atheist *be* moral under this proposal?

"Atheism," interestingly, is viewed by the proponents of the religious straight and narrow not as a belief, but as a lack of one. It does not, therefore, enjoy the same protections as, say, Mormonism. "Atheism" is a dirty word, like "liberal" or "communist," because it literally means "without God"—a condition that most people view as tainted, misguided, even evil. However, atheism can certainly be considered as much of a spiritual conviction as any religion, organized or not. It holds that there is no god (at least not in the sense that we usually envision), and this belief has implications for both how a person views the world and his or her conduct in it. This is not to say that these implications are by definition negative (one is just as likely to encounter an immoral Christian as he is to encounter a highly ethical atheist); an ethical atheist, freed of the fetters of organized religion and the hypocrisy that often accompanies it, may in fact have a moral advantage over his or her pious counterparts. There certainly can be a liberating sense of personal and social fulfillment that occurs when people are not bound by the imperatives of a theological code. For one thing, freed of the doctrinal certainty of a one true god, an atheist is far more likely to suspend judgment of others, to be tolerant of difference, and to see humanity itself as the worthy object of faith. The secular humanist, for example, may hold atheistic beliefs, but is far from amoral. "Humanism," by definition, celebrates the achievements and promise of humanity, not an abstract deity. The humanist is far less chained to an idea of "admissible questions" and is therefore free to explore and inquire, to use the exceptional apparatus endowed upon us through the accidents of evolution. Such a "gift" as the human mind has far more potential if it is seen for what it is and from whence it came; we may value the gift even more by knowing that it was a one-in-a-trillion chance that we should have received it at all, that through the eons of evolutionary development any one fateful shift in the direction of one of our ancestral species might have marked the non-existence of humanity altogether. With this understanding, we can become more tied to the nature from which we arose, appreciating it without trying to dominate it. Perhaps, with this knowledge, we could eventually be a little more grateful for ourselves and our fellow humans, rather than to judge them and kill them when they don't meet our standards of fidelity to a given faith. The atheistic humanist position is far from an amoral stance. In fact, it makes the prudish morality of the religious right seem puerile by comparison, with its preoccupation with sexual conduct and the personal habits of others. The humanist, perhaps better than anyone else, practices the first condition of becoming moral: compassion. Good naturalist that he is, Gould, I suspect, secretly agrees with this, so his effort to give religion its moralistic domain seems disingenuous, a measure taken by a man who is simply tired of hearing all the bickering.

Gould favors a policy of irenics as the basis for NOMA—a promotion of peace in the face of theological differences. "Can't we all just get along," Gould seems to say. Would that it were that easy. He pleads for mutual understanding as one would at the opening of a political rally using language that makes one wonder just how long he has been sequestered in his ivory tower at Harvard. His message to the reader, noble as in may be, is almost insultingly simple, as if he has surrendered to the notion that people are of so limited an intellect that they must be given a grammar school lesson in working and playing well with others:

> I join nearly all people of goodwill in wishing to see two old and cherished institutions, our two rocks of ages—science and religion—coexisting in peace while each works to make a distinctive path for the integrated coat of many colors that will celebrate the distinctions of our lives, yet cloak human nakedness in a seamless covering called wisdom. (209)

Colorful metaphors aside, the message is simple; Gould has come down from his position of academic privilege to let us know how childish we are all being. Like the grand patriarch, he has allowed this squabble to go on far too long and simply must intervene. This condescending stance does little to sway the reader. He attempts to break up the fight with a gentle hand, to allow these two institutions to shake hands and go their separate ways, and to think about how silly they've been for all these centuries. Perhaps this is a harsh reading, but patronizing "solutions" for problems as complex as the one Gould has tackled do not represent a sound rhetorical strategy. Like the media that has frequently given Gould a voice, the assumption regarding the general public is that they are ignorant, self-indulgent fools who can only be mollified using simple dictates from an authority as revered as Gould himself. Such hubris can only lead to more heels digging in, more strident responses from all quarters, and a general sense of exacerbating the problem he hopes resolve. So preoccupied with the method of keeping the issue "simply stated and understood" (176) that he risks alienating the reader—and the public—in the process.

Gould, unfortunately, does not aid us in putting to rest any assumptions about the battle for "truth" (whatever that may in fact be) that has taken place between science and religion for centuries. His book, in fact, is remarkably banal for such an otherwise brilliant man, as if he was not able to muster the intellectual acumen necessary to answer the toughest but most basic questions before he died: What are we, and what is our ultimate fate? He leaves the reader with little to ponder save for the lapses in logic that he guarded so closely as a scientist. He has, in the end, reconciled nothing, and it is telling that *Rocks of Ages* was by far his briefest book. He is out of his element here, a man desperate to record his version of a death-bed confession, but lacking utterly the terministic apparatus with which to do it in any meaningful way. It is a pity that so much of his life was spent dwelling on the minutiae of his profession at the expense of even being able to frame the question he wanted most to influence.

Gould's book is, in the end, thoroughly unsatisfying, devoted as it is to the compromise of the institutional forces of science and religion that represent the conflict rather than the intellectual process of advancing knowledge. Taking the "middle road" of peaceful coexistence is a useful diplomatic strategy, even a desirable one, but it capitulates to the falsities of human social constructs rather than to an understanding of how some bodies of knowledge simply must concede their errors and adjust accordingly in the face of overwhelming evidence those errors pointed out by other bodies of knowledge. In other words, creationists simply have it wrong, and they need to look at science to see why. Absent the unlikely acquiescence from the Creationist camp, however, should we fight tooth and nail over whether it is better to "believe" than to "prove" or vice versa? Of course not. But it is a naïve understanding of the precepts of religion indeed that holds that the pious will suffer quietly in isolation as new discoveries in science chip away at the cornerstone of their beliefs, especially concerning Gould's primary example—the origins of humanity. Gould suggests that religion should simply give up their authority on this question because it is the domain of science, not religion, but the institutional reality on the matter is not that easily compromised. Religious leaders will not loosen their grip on this question because it is at the very heart of the religious project. Regardless of my personal view that science certainly does have the much stronger claim, expecting the devout to acquiesce so easily is simply not a likely solution. As we have recently seen in the creationist/evolution debate, constitutional policies like the separation of institutions such as church and state do not (apparently, will not) stay grounded for very long, while in the face of our increasingly shrinking world and sense of self-importance our significance is challenged by new knowledge, much of it coming from the province of science. Just as church and state seem invariably to overlap, so too, I suspect, will the insistence of the religious reactionary to set the defining borders of inquiry on science never be fully overcome. Persuasion is certainly not likely, and conversion even less so. As Richard Russo so aptly put it in *Straight Man,* "Anyone who observed us would conclude that the purpose of all academic discussion was to provide the grounds for becoming further entrenched in our original positions" (201).

While Gould's project represents a noble, though somewhat impotent, attempt to reconcile this intellectual disparity by not reconciling it, it is, ultimately, a noble cop-out. This is no doubt due to sheer frustration he experienced and by a tendency of representatives of both camps to be so thoroughly entrenched in their positions that there is no rhetorical approach, no matter how brilliantly conceived and executed, that could convert, as it were, the other side to see a different point of view. As tragic as this is, the real failure, as Klinkenborg notes, is on the educational front. Gould is complicit in the very ongoing ignorance that he hopes to quell because he lends credence to an idea that has no basis in anything but the supernatural, a distinctly untestable, unverifiable, and untenable realm indeed. He also suggests that morality, as a system of human conduct, can only be arrived at through faith and that science has no ethical compunction save what religion provides it.[8] This perspective has many serious hazards, for it gives moral authority to certain groups that can and have acted in decidedly immoral ways in the past. As Kenan Malik, author of *Man, Beast, and Zombie,* put it in a review of Gould's book in the February 19th, 2001 edition of the *New Statesman,* religion has "transformed from being the only means to understand reality to being anachronistic dogma," and it is important to remember that the "conflict is between not just science and religion but also science and humanist ethics" ("Inventing Allies in the Sky"). Even if the intent of religious groups proclaiming to posses the moral weight necessary to dictate the behavior of others is sterling, it can so readily get muddled with personal agendas and insecure sanctimony that inhumane, unsympathetic, and intolerant policies and judgments can easily result (consider the current "debate" over same-sex marriage, an "issue" manufactured by the religious right if ever there was one). And science, in the end, should and does have its own ethic, both professional and moral, that it can follow if it demonstrates the integrity of realizing that it has a power over its environment, a capacity for harm that should be a primary concern in all scientific conduct. It should not need, that is, the moral watchdog of religion prescribing when and where it can practice as long as science is as self-monitoring as it claims to be self-correcting.

Yet another Steve, Steven Pinker, represents a contrast to Gould (this is perhaps an understatement—they had been professional nemeses for many years) in that he has not, up until now, attempted to use anything but evolution, genetics, and natural selection to explain both the descent of the human species and the complexities of our language and psychology. His three most influential popular books are *How the Mind Works, Blank Slate,* and *The Language Instinct,* all of which rely on evolutionary psychology to describe the processes that make us uniquely intelligent. In *How the Mind Works,* Pinker claims that there are four prerequisites that make the species *Homo sapiens* unique: the first is that humans (and primates in general) are visual animals, and more specifically, we are visual animals that possess stereoscopic eyesight which allows depth perception; second, we are social animals, and we tend to work more effectively in groups, cooperating toward a common goal and sharing the benefits (this has obvious pitfalls, too, like internal competition and the ability to lie); third, we possess hands—most significantly, hands with opposable thumbs which allow us to grasp things in our environment, an ability that eventually led to tool-making; finally, Pinker says that the need to hunt ushered along the other prerequisites and, through the process of evolution, added a fifth—bipedalism (191–195). Whether one agrees that these are the features of the human species that set us apart from other animals, the thrust of these classifications is that they are all the products of evolution that led to an increased—indeed, oddly disproportionate—level of intelligence compared to our other mammalian, and even primate, counterparts. The common argument against this overview of human capabilities tends to focus, in fact, on other primates. Why, if other apes, monkeys, chimps, baboons, and orangutans possess these attributes (with the exception of strict bipedalism and, in the case of orangutans, a tight social structure) did humans alone evolve the level of brain-power that they currently possess? The answer is, these other animals very well could adapt to our level of intelligence someday, but there is no guarantee that they will; these animal species simply are not at that stage in their evolutionary progression, and may in fact never be. Any or all of these species could become extinct long before they reach that stage in their development, or other environmental factors may push their natural selection process in an entirely different direction. This is perhaps the greatest popular misconception regarding evolution, that given similar conditions, all species would evolve in exactly the same way. That, coupled with the shortcomings of the ladder metaphor—which implies that evolution is a series of discrete steps from one level of complexity to the next (the so-called "missing link" metaphor makes the same erroneous assumption)—conspire to give the public a seriously flawed picture of the actual process of evolution.

Pinker does his best to put to rest many evolutionary myths, but he does underplay the degree of disagreement that does exist among scientists regarding many of the specifics of evolutionary theory, and for good reason. In the past, widespread awareness of scientific disagreements has led to other constituencies opportunistically proclaiming that scientists themselves are unsure about their own theories. In other words, the objection goes, if the details of evolution are under debate within the scientific community, then the entire theory must be flawed. However, Pinker notes that the real problem in determining the origin of human intelligence is not whether evolution was the cause, but what conditions existed that would spawn the need for such large and hungry brains; *this* is one of the areas in which scientists lack consensus, not whether brains were the product of evolution in the first place. Brains, as anatomical organs that require nutrients, are fuel hogs. It has been estimated that the human brain requires more energy from the body than all the other major organs combined, and this, from an evolutionary standpoint, is puzzling. Adaptations of any sort usually involve a kind of crude, incomplete cost-benefit analysis. That is, the adaptive advantage of having a large brain must, in some calculus of natural selection, outweigh the enormous cost of feeding it, otherwise humans would not have survived. On the other hand, as Pinker points out, natural selection does not necessarily produce outcomes that work in favor of the organisms at all times, and it certainly does not rely on what we would consider rational processes. Evolution does not, say, "observe" an environment and systematically extrapolate and deploy the best adaptation for survival within it. Mutations are rare, random, and not necessarily a direct reaction to the environment in which the organism lives. As Pinker puts it, "Natural selection is not a guardian angel that hovers over us making sure that our behavior [or our natural attributes] always maximizes biological fitness" (41). Natural selection is a trial and error process, and this is perhaps the best argument against "Intelligent Design," for it does not require us to account for all the myriad trajectories and unusual series of steps that must occur for one organism to reach the complexity of another; it is enough to say that the "accidents" of nature, along with the time necessary to correct the design flaws of a given organism or see it perish in the process, will account not only for an instrument of complexity as astounding as the human brain, but also for the staggering variety of life on this planet. Pinker notes that in many cases

> [t]he organisms reach an optimum and stay put, often for hundreds of millions of years. And those that do become more complex don't always become smarter. They become bigger, or faster, or more poisonous, or more fecund, or more sensitive to smells and sounds, or able to fly higher and farther, or better at building nests or dams—whatever works for them. Evolution is about ends, not means; becoming smart is just one option. (153)

This range of adaptive attributes—and many, many more—help explain how evolutionary processes, through ancient, relentless, and infinitely varied trial and error attempts, use whatever adaptations allow them to survive. This is the only prerequisite for evolution: to advance the evolutionary pool. It is difficult to look at a praying mantis that almost identically resembles an orchid and not conclude that this is a specialized adaptation that must have been generated through natural selection and the evolutionary transformation that allowed an insect to resemble a plant in its environment for predatory efficacy. Pinker is a compelling rhetorician in this respect, because he provides entertaining narratives of the ingenious paths that evolutionary directionality takes to aid organisms in survival in all kinds of environments.

The irony, of course, is that perhaps the most impressive product of evolution, the human brain, is at once the mechanism that discovered the process of evolution and the one that questions it. We are, so far as we can discern from studying other animals, the only species with a foreknowledge of our own mortality, but we have this pesky emotional side to our brains that finds the idea of the finality of death unpleasant.[9] "Emotion" is an odd sort of evolutionary adaptation in its own right, but there are those who hold that the entire range of possible human emotional responses can be condensed to a few basic ones: fear, anger, pleasure, and sadness (Pinker adds "disgust" and "surprise," but these classifications are debatable). Emotions are, according to Pinker and others, in fact, not bestial leftovers from an animalistic past, but a very key component to our survival and propagation. Without the basic emotions that all humans experience, we would not be able to function as anything more than vulnerable automatons. A computer does not "care" whether it exists or not because it has no emotional point of reference to make it care nor any glandular or chemical dispensers to provoke emotional reactions; it simply does what it's programmed to do, and (for those who have been frustrated enough by computer "stubbornness" to throw the machine out of a three-story window when it doesn't "cooperate") it is unconcerned whether it "survives" or not.[10] Therefore, emotions are an important component to survival: Fear alerts us of danger (perceived or real); anger provides an appropriate (though in civilized society, it is often "inappropriate") response to a fighting situation; pleasure can aid in a sense of being in the right place, reproduction, and social cohesion; and sadness gives social animals like humans a feeling of empathy that may help maintain mutually beneficial interactions with members of the same (or even other) species.

This is all a very clinical (and even reductionistic) view on the emotions so central to the human experience, but Pinker is interested not in furthering a Romantic outlook on the centrality of emotions, but in forwarding an evolutionary one. He notes, for example, that we have a habit of describing cultures foreign to us as lacking the emotional depth that we ourselves have. This is a myth, says Pinker, because all cultures that have been catalogued by anthropologists have demonstrated the same basic emotions that we all share as a species—they may simply describe them, act upon them, or channel them in different ways. He cites Darwin himself as claiming that "[t]he same state of mind is expressed throughout the world with remarkable uniformity, and this fact is in itself interesting as evidence of the close similarity in bodily structure and mental disposition of all the races of mankind" (qtd. in Pinker 365). He further claims that the relatively recent tendency to separate the intellect from the emotions is technically correct from a biological standpoint, but it is a mistake to assume that emotions are an unsavory holdover from our cruder hominid history; they are, rather, part and parcel of the survival mechanism of human beings:

> The problem with the emotions is not that they are untamed forces or vestiges of our animal past; it is that they were designed to propagate copies of the genes that built them rather than to promote happiness, wisdom, or moral values. We often call an act "emotional" when it is harmful to the social group, damaging to the actor's happiness in the long run, uncontrollable and impervious to persuasion, or a product of self-delusion. Sad to say, these outcomes are not malfunctions but precisely what we would expect from well-engineered emotions. (370)

The mistake from my point of view is to make the implicit assumption, as Pinker seems to do, that separates the animal from the human in the first place, as if these were distinct categories of life. Perhaps we would have less difficulty accepting the role of the emotions if we viewed ourselves properly as animals—separated, in Kenneth Burke's words, "from our natural condition through instruments of our own making" (*Rhetoric of Religion* 40). We like to congratulate ourselves on our transcendence above such animalistic behaviors as reproduction and a might-makes-right mentality, but these are our basic emotional drives, and we both act on them and rationalize their propriety all the time. Dress them up in the attire of civilization, and we still have the same functional motives that drove our ancestors and every other living thing on this planet. We have simply superimposed a rationalistic order onto the emotive chaos.

Pinker's point is that the human brain, while an amazing apparatus for problem solving, is equally adept at problem creating. Our evolutionary legacy, and burden, is to be so self-aware that we not only anticipate our own deaths, but construct elaborate institutions to deal with that very eventuality. More than that, we are aware of our capacity to use and misuse our biological endowment, and have the unusual ability to think of ourselves in terms that are outside our own sphere of experience; we can, in short, imagine things that have never happened and project our visions into the future, making them a reality. This sets us apart from our fellow creatures which, while able to predict certain outcomes on a very rudimentary level,[11] do not understand their own role in the reality within which they live. When humans do not like their surroundings, they have the ability to change it on both large and small scales. But where Pinker's error lies is in the way he underplays the role of language when assembling our reality. It is a naïve assumption common among scientists (and the general public) that language is only a convenient tool we use to get things done. Pinker uses a specious argument from his explanation of universal emotions to drive home a point that needs serious reconsideration. He says,

> The common remark that a language does or doesn't have a word for an emotion means little. In *The Language Instinct* I argued that the influence of language on thought has been exaggerated, and that is all the more true for the influence of language on feeling. . . . A language accumulates a large vocabulary, including words for emotions, when it has had influential wordsmiths, contact with other languages, rules for forming new words out of old ones, and widespread literacy, which allows new coinages to become epidemic. When language has not had these stimulants, people describe how they feel with circumlocutions, metaphors, metonyms, and synecdoches [sic]. (367)

As a psycholinguist, one would think Pinker would know better than to make such a sweeping claim. The dismissal of the rhetorical tropes he lists is part of the problem; the cavalier use of the word "thought" is another. According to Pinker, the thought precedes the language used to describe it, a move he feels he must make in order to argue for the precognitive hardwiring for language he posits is part of human evolutionary anatomy—like opposable thumbs and bipedalism. But there is no problem intrinsic to accepting the idea that humans learn language so quickly and use it so incessantly because it is part of our evolutionary make-up to do so; there is, however, a problem with the idea that language is only a template superimposed onto preexisting thought, that we use language in a strictly utilitarian way to simply describe what we see and categorize how we act, as if the activities of thought and speech were independent of one another.

The reason I have such difficulty accepting the idea that thought supersedes language is relatively simple but has far-reaching implications. Recall your first memory (or at least a very early one). What do you remember? Did it involve strictly images, or were there words involved (either receiving them or speaking them or both)? For me, and for anyone I have ever asked, images do play a part in the memory, but the pictures in the mind are almost always accompanied by words. Prior to the acquisition of language, in other words, we have no memories. This suggests to me that, even from a strictly biological, language-acquisition standpoint, language is key to developing memory. Without memory, we have no thought. When someone is "thinking"—of a complicated idea, of how to manage one's bank account, of what to have for dinner—does he or she have a mental picture first that is then draped in language to give it verbalistic form, or do we tend to think in words? For me, and I suspect for most people, the latter is true; we tend to solve problems by speaking them through in our mind, and while these words may have many images as company, we don't truly "know" something until we can articulate it. We may have impressions, impulses, even premonitions that we have difficulty describing, and these might be considered "thought," but it is not until we able to give the thought substance that we truly consider it a thought proper. But this is only the tip of the iceberg. Pinker seems to assume a definition of "thought" that is exasperatingly basic, a mere urge sparked by some sort of stimulus. But what about how we frame ideas, issues, assumptions, even biases and prejudgments? Don't these have a direct correlation to the way we are accustomed to talking about them? When you ask the person on the street why we are fighting in Iraq, the stock response is "freedom"—this single word holds so much currency for us that it casts the manner in which we think of the conflict.

And as soon as we begin heaping symbols on top of words on top of language—leading to the institutions and social structures that are central to our identity—we have a complicated network of linguistic predetermination that forces us to think of things in the terms that the words have assigned. It takes great effort to remove oneself from the terministic frame we construct to see things from an entirely new linguistic vantage point, and without an alternative vocabulary with which to do it, we are stuck in a terminology rut from which it is nearly impossible to climb out. Language's influence has been "exaggerated"? If anything, the impact of language on the way we think has been underemphasized, sometimes with grave consequences.

Pinker's overconfident declarations about the relationship between language and thought manifest themselves in the very oversights he has regarding this issue. His dismissal of rhetorical tropes is, put bluntly, ill-informed, particularly considering he is apparently familiar with the research done that shows the obvious centrality of these devices (he cites Lakoff and Johnson and their book, *Metaphors We Live By* on several occasions, for instance[12]). Pinker also appears to contradict himself as he explains the significance of the particular metaphors that George Lakoff and Mark Johnson say are central to our cognitive understanding of the world around us:

> Once you begin to notice this pedestrian poetry [i.e. the use of metaphors like "argument is war"], you find it everywhere. Ideas are not only food but buildings, people, plants, products, commodities, money, tools, and fashions. Love is a force, madness, magic, and war. The visual field is a container, self-esteem is a brittle object, time is money, life is a game of chance. (358)

The role that the master tropes play in coordinating that which we experience and think and how we articulate it symbolically through the use of rhetorical expression like the metaphor is fundamental to understanding our symbolic use of words, and Lakoff and Johnson are key to this realization about the construction of our language. Pinker, however, seems to downplay the importance of this elsewhere and has therefore missed the entire point underscoring *Metaphors We Live By:* metaphor, and its related tropes like metonymy, synecdoche, and simile, are not merely "device[s] of the poetic imagination and rhetorical flourish—a matter of extraordinary rather than ordinary language," but a mechanism "pervasive in everyday life, not just in language but in thought and action" (3). Further, what we tend to think of as our "ordinary conceptual system, in terms of how we both think and act, is fundamentally metaphorical in nature" (3). The preceding passages are from the very first paragraph of *Metaphors We live By,* and the thesis is unmistakable (and one which Pinker ostensibly agrees with, even though he later claims that the impact of language on thought is "exaggerated"): metaphors not only contribute to but *define* how we *think* and how we *act.* If a culture observes the metaphorical connection between "knowledge" and "light," for example, this association will not only manifest itself through the language it uses but in the actions it takes to improve the acquisition of knowledge. The related associations of "light" with other features of the word "reflect" the quality of light: "She is a very *bright* person." "The claims made by that presenter were *clear.*" And, conversely, when knowledge or intelligence is lacking, the same metaphorical description is employed: "He is a *dull* person." "The central claims are *murky.*" Pinker cannot deny that these orientations have a tremendous impact on the way we view our world both physically and abstractly. But scientists generally prefer to underemphasize the role of such "unfocused" uses of language; for them, rhetorical tropes create ambiguity. They are right. But they are not beyond them nor are they capable of avoiding them themselves. All language, by its very nature, is ambiguous, including Pinker's.

Pinker himself uses the metaphor "thinking machines" to describe the human mind; he dedicates an entire chapter to describing the human mind as a "processor." He lists the discussion of the computer model for the brain in his index under "computational metaphor." He says explicitly:

> Artificial computer programs, from Macintosh user interface to simulations of the weather to programs that recognize speech and answer questions in English, give us a hint of the finesse and power of which computation is capable. Human thought and behavior, no matter how subtle and flexible, could be the product of a very complicated program, and that program may have been our endowment from natural selection. (27)

Now, using the computer metaphor to describe the human mind is hardly new, but much like the camera metaphor used to describe the human eye or the bomb metaphor used to describe the beetle, the relationship is offered as an inversion of reality. The human mind is not "like" a computer any more than the eye is "like" a camera. Being the imitative species we are, we took from nature a model and then extended it to our own technology. Therefore, we would rightfully say that a computer is a human attempt to artificially imitate the human brain, just as a camera is an attempt to replicate the recording functions of the human eye. It is interesting to me that this relationship is so frequently inverted, as if through the hubris of our own technological accomplishments, we feel it necessary to give ourselves credit for the paradigm that we, in fact, were copying from. And it is equally important to note that, in our typical reliance on the metaphors we construct, we mistakenly set our standards to the model we think is most efficient and productive. That is why we have moved toward the horribly damaging habit of training people to learn like a computer, to see like a camera, or remember like a tape recorder. In acting on our metaphors, we create ideas that make those metaphorical associations the norm and the reality, and thus, if the mind is "like" a computer, then it should consequently function like one. Hence, rote memorization, the downplaying of the role of imagination, discouraging independent thought or fresh ideas, and the expectation that people can recall at whim that which they have learned seeps into areas like education and the workplace. For Pinker to suggest that the language we use and the tropes we erect have little to do with the way we think and act in the world is tragically naïve and, in my opinion, humanistically irresponsible. Worse, he has reinforced this erroneous idea to the public, for whom books like *How the Mind Works* was written.

But given his evident certainty about both how the mind works and how language contributes to its operation, we should turn to his other ambitious book, *The Language Instinct,* to get a fuller sense of his take on the nature of language. In this study, Pinker argues that language is an evolutionary attribute to the human species in the same way that intricate camouflage schemes are intrinsic adaptations to a cuttlefish. For him, language is an "instinct" that has evolved to a complexity in humans unlike any other animal, and it is this capability that has set us apart from our animal cousins and from our less-articulate ancestors. His disciplinary basis is eclectic, stemming as it does from "cognitive science," a branch of biology that "combines tools from psychology, computer science, linguistics, philosophy, and neurobiology to explain the working of human intelligence" (3–4). In an effort to drive this point home, Pinker simplifies the process this way: "In nature's talent show we are simply a species of primate with our own act, a knack for communicating who did what to whom by modulating the sounds we make when we exhale" (5). While the physical/anatomical description is, of course, essentially correct, the depiction of human communication as merely a matter of expressing "who did what to whom" is far too reductionistic to be of practical use when attempting to comprehend the degree to which language constructs our reality. But this, as one might expect from a scientist, is not Pinker's goal; rather, he is interested in documenting the evolutionary, biological, and paleontological evidence that establishes his thesis that language is simply another adaptation, albeit an adaptation to a complex set of historical, environmental, and biological contingencies, that has allowed *Homo sapiens* to dominate this planet.

He further claims that "[o]nce you begin to look at language not as the ineffable essence of human uniqueness but as a biological adaptation to communicate information, it is no longer tempting to see language as an insidious shaper of thought, and, as we shall see, it is not" (5). This confident proclamation, that language is not a "shaper of thought" but merely an "adaptation to communicate" flies in the face of a huge corpus of language theory that all but proves the contrary, an historical fact that alone should verify just how invalid his position on the connection between language and thought really is. That is, if language were nothing but a tool of information, language *theory* (which is itself an application of language)—and the subsequent influence that thinkers ranging from Samuel Coleridge to Michel Foucault have had on literary criticism, philosophy of language, rhetorical studies, and social theory—would not exist at all. The theories adopted by such critics do far more than "communicate information"; they influence thought, and consequently, action. He also constructs a specious dichotomy by implying that either language is used to communicate information or it is used insidiously to program other users of language. The staggering lack of awareness regarding the subtlety of the vast range of language applications in-between these extremes seems equally strange for a man who is using language himself in an attempt to shape thought—namely, how we *think* about language and its place in the human world. It is doubly vexing that he should miss this inherent contradiction, but on the other hand, a man who thinks of language in only these most basic terms might very likely overlook the inconsistencies in his own statement. It is as if Pinker feels that, in order to validate the likelihood of his own premise of language as an "organ" like any other specialized endowment, he cannot also concede that its complexity is far more significant for our perception of ourselves and our environment because this might undermine the assertion that language is, in fact, biological.

There is no reason to conclude that the idea that language is an "instinct" and that it is also used for far more than the conveyance of information are mutually exclusive concepts. While it may, in fact, probably *is,* true that language evolved like anything else that marks our species as distinct from other species, it is also true that we have taken this basic attribute and mastered all its myriad applications, such that language becomes, in the word of Kenneth Burke, "symbolic action." The clothing of civilization needs language in order to exist; our mastery of language is why we are the only species to possess great cathedrals, stunning works of art that stem purely from the imagination, literature that captures the very quintessence of humanity, education, politics, technology, music, fashion, business, or any of the other achievements that are the identifiers of a civilized society. Is all that was required to reach these summits the "communication of information"? Apparently so. In an attempt to be fair, however, I should point out that Pinker makes a distinction between language that is "naturally" learned (i.e., by a child discovering the wonders of speech, using his natural human instinct to do so) and between literacy (in the sense of reading closely and writing critically), which he views as an activity that is learned through the same artificial means we would use to learn how to play the trombone. The former is instinctual because our brains are pre-wired to accept language, regardless of its form or origin. The latter is a much more conscious attempt to learn something that we do not have a biological predisposition to do. This is why learning a second language is so much more difficult; we have bypassed the original neural formation process and moved on to "training" ourselves in a new way. However, children raised in a bilingual environment have a much easier time learning both languages because their brains are still in the process of shaping the neural pathways that will be impressed there for the rest of their lives.

If we acknowledge this distinction, it is difficult to see how one can make the claim that language and thought have little bearing on one another. To expound on the cathedral example, it is one thing to convey information about the structure, dimensions, architectural style, materials, etc. of the project, and it is quite another to motivate an entire community to build the thing in the first place. Whereas information is obviously needed to consider the logistics of erecting the building, it does little to explain how language was used to inspire people to such a grand project. And much of what is included in a cathedral has no practical value in helping the structure stand up; if it were constructed strictly for utilitarian purposes, a large box would suffice. Rather, it is a testament to an entire faith, the spiritual grandeur reflecting the mightiness of a deity that makes a cathedral what it is. While flying buttresses have an architectural value, they are also aesthetically absorbed by the form. While a gargoyle may disperse water, a simple spout would do the job just as well (perhaps better). Stained glass not only allows light to enter, but also manipulates it in beautiful ways and provides a spiritual narrative of some of the most important events and people of early Christianity. These and many other artistic flourishes that complement the actual design of the building aid the person who enters the building in experiencing the splendor of the Christian god. The cathedral is therefore not only the product of an idea inspired by the symbolism of a culture's faith; it is itself a text. It is, as a matter of fact, a text designed to tell the story of Catholicism to the illiterate, the condition afflicting 98 percent of the population at the time that most Medieval cathedrals were built. This is a far more inclusive definition of language than Pinker allows, and we must also consider the language that gave birth to the idea before the first stone was ever put into place. The cathedral is the culmination of centuries of rich cultural history, representing the doctrines and faith of a powerful religious institution. Those who hatched the notion of building it and those who commissioned its construction were educated using the language of Catholicism. The language that informed their orientation was highly symbolic, very abstract, and carefully ritualized. The cathedral not only represented the Catholic faith, but also the power of The Church. It was viewed as an appropriate conduit for receiving and transmitting the spiritual messages that took place between God and Man. The mere transference of information cannot accomplish all this. To realize this symbolic feat, language must be viewed as a means of symbolic action, words carefully orchestrated to express to an entire civilization the centrality of The Church in the lives of its followers. Like the Coliseum of Rome, the cathedral was a testament to the enormity of this civilized juggernaut, a manifestation of the might of The Church and the god it sponsored. And, it was a product of language, for without language, the materials used and the technology applied would lay fallow and mean nothing.

Pinker contradicts himself again in the chapter entitled "Talking Heads" when he asserts that

> [h]uman communication is not just a transfer of information like two fax machines connected with a wire; it is a series of alternating displays of behavior by sensitive, scheming, second-guessing, social animals. When we put words into people's ears we are impinging on them and revealing our own intentions, honorable or not, just as surely as if we were touching them. (230)

The reader is never sure where Pinker stands on the issue of the breadth and scope of language as it dictates our thinking and behavior, but what he seems to be arguing is that, since language is an evolutionary instinct, it cannot also be a "social construction." Again, the simplistic binary of either nurture or nature rears its head (though Pinker elsewhere complains that this very question is overemphasized and admonishes those who place too much stock in the distinction) as he suggests that human language use is strictly a product of evolutionary forces. He proclaims, for example, that his way of viewing language is of course the only valid one, and that his innovative thinking "inverts popular wisdom, especially as it has been passed down in the canon of the humanities and social sciences. Language is no more a cultural invention than is upright posture" (5). I know of no humanistic scholar who makes the blunt claim that humanity "invented" language (the social sciences may, however). Most claims about language forwarded by literary theorists, philosophers, or rhetoricians do not involve any declaration that it is a "cultural invention" (this tiresome phrase again presents itself as some overarching definition of humanistic language studies—a typical misrepresentation encouraged by scientists and positivists) but argue, rather, through a variety of demonstrative methods, that the ways we *use* language have a great impact on how we move, act, react, and comprehend in our social environment. Pinker, moreover, I suspect, is less certain about his own claims than he would lead the reader to believe, but he does, ostensibly, rely on science to stake out his position. The problem is that he oversteps his bounds; while science may in fact be capable of establishing an evolutionary link between human dominance and our abilities to acquire language, when it comes to rhetorical analysis of the vast applications of those abilities, he simply does not base his claims in the appropriate epistemological domains. Science, in short, isn't likely to help him very much since it does not address the symbolic function of language—only the biological, anatomical, and structural relationships between species and speech act. He assures the readers of scientific advancement in the understanding of human cognition, but he overstates the certainty that science has about how humans use language except on the most rudimentary level. He says, for example, that

> there is no scientific evidence that languages dramatically shape their speaker's ways of thinking. But I want to do more than review the unintentionally comical history of attempts to prove that they do. The idea that language shapes thinking seems plausible when scientists were in the dark about how thinking works or even how to study it. Now that cognitive scientists know how to think about thinking, there is less of a temptation to equate it with language just because words are more palpable than thoughts. By understanding *why* linguistic determinism is wrong, we will be in a better position to understand how language itself works . . .. (48)

So science, according to Pinker, has already figured all this out. They *know* "how the mind works" (at least Pinker thinks he does, hence the overly-confident title of his other book: *How the Mind Works*) and they understand how the mind uses language, and they know what thought "is." Therefore, science, a la Pinker, has determined that the evidence for a language/thought relationship is non-existent or, at least, highly exaggerated. The hubris of these claims is astounding. There is virtually no consensus, at least in any of the scientific material I've read, regarding either how the mind "really" works or how language influences thought. The consensus about whether science is the best mechanism for answering these questions is even more remote.

Pinker, like so many dubiously intentioned popularizers of science, has misrepresented an agreement within the scientific community where none exists, but because he has a Harvard degree and is a professor at MIT, his ethos with the general population is enough to cement his premature proclamations in the minds of the public for whom this book was written. But when all is said and done, he is using the wrong instrument to examine the impact language has on social orientations. Science may be able to tell us the what of language acquisition, but in the realm of language meaning, it is ill-equipped to supply answers to the why and how, largely because the methodology is a mismatch for what it is attempting to measure. How, one might ask, does one "measure" meaning? Why would we want to? What would it tell us if we were able to?

While Pinker has some very valid and convincing arguments regarding the origins of the "language instinct," his willingness to degrade any way of knowing besides the scientific and to misrepresent scientific consensus is simply irresponsible. However, this is an unfortunate commonality among those scientists who, either for the public good or for fame or for some of both, lower themselves to a layperson's level to explain to us how things really are. The scientists named Steve all do this with varying degrees of success, but they all have one thing in common—they are Ivy League or English elite scientists who are appreciated by the public for their willingness to emerge from their Ivory Towers long enough to educate us in their respective fields. While we should be thankful for the lowering of professional standards that must be involved in so altruistic an endeavor, we should be equally mindful of the motives and intentions of scientists like these. In the tradition of Alfred Kinsey,[13] the Steve Scientists are benefiting not only the public and themselves, but also their educational institutions and any foundations or other funding sources that have bankrolled their various projects. As cynical as this may sound, it is an important and revealing consideration. Without such popularizers of science, many of our most pressing scientific issues would go un-debated in public forums, yet it is important to understand what, exactly, the role is that they play and to determine under what conditions they aid in public understanding of science and under what conditions they may confuse important ideas even further.

So far, we have been concerned with how and in what forms scientific authority has developed recently in this country and to establish the nature of the interface between the public and the scientific community, especially as it has been projected through the lens of the popular media. We have seen a number of popular scientists—reputable and responsible in their own respective fields—who have actually been unintentionally complicit in the misrepresentation of scientific ideas, theories, and activities. If this has resulted in troublesome ripple-effects such as an increased chasm among the participants in the evolution/creationism debate or in forwarding the erroneous notion that the workings of the mind are fully understood, we can forgive these fine scientists for making the attempt in the first place. The result of their efforts is not that they have failed so much as it is that we as a society have short-changed the populous in the area of scientific literacy (and cultural literacy generally) through faulty, uninspired, and indifferent educational practices. These scientists have attempted in good faith to improve this condition, but, like a governmental administration who wants to help the poor but has no experiential point of reference from which to understand the reality of the situation, they are often powerless to bridge the gap in a way that makes any real difference. Under the best of circumstances, they can produce good, quotable nuggets of wisdom; under the worst, they help foster the very ignorance and apathy that they have charged themselves with correcting. This is truly unfortunate, because it is clearly not through simple acts of arrogance or grandiosity that they hope to improve public/scientific relations for personal or professional gain, but in a genuine desire to make people understand what they do.

The Steve Scientists contribute to but also distort scientific questions from often unscientific perspectives. The three Steves represent a range of scientific inquiry: Steven Hawking is the "hard" scientist, even though theoretical physics is as much an exercise in exotic equations as it is an empirical science like practical physics (hence the contentiousness between these two branches of physics); Stephen Jay Gould represents the biologist/paleontologist, an area of science that relies heavily on the interpretive skills of the field scientist—someone who seeks out, uncovers, and reads fragmentary texts in an effort to put together a cogent narrative. But Gould violates his own directive in the act of explaining it. He is not prepared to forward an argument that is clearly outside his comfort zone. Steven Pinker is the social scientist, representing the softest of the scientific disciplines and, not surprisingly, making some of the most dubious claims about the nature of language and science's ability to make sense of it. What is central to all these men is the power lent them through a strong ethos, a quality that will be dealt with in the next chapter. Armed with ethos, scientists and their representatives are able to command respect from their audience and authoritativeness in their claims. This quality will be key to understanding how it is that scientists have become the sages of our generation.

5 Scientific Ethos

Professional rhetoricians have written widely on *ethos.* It, along with the two other main Aristotlean appeals, *logos* and *pathos,* have figured prominently in the tradition of rhetorical studies, but the scholarship has been especially prolific in the last ten years. Craig R. Smith, for example, has written on the issue of credibility using a hermeneutic reading of Aristotle's *On Rhetoric;* Robert Wade Kennedy has examined the relationship between ethos, truth, and metaphor in his "Truth as Metaphor"; Barbara Warnick sees ethos as a central appeal in the critical practice of rhetorical criticism; John Poulakos has speculated on the intersection between ethos and questions of beauty; Eric King Watts has explored the centrality of ethos for determining a racial aesthetic; R.D. Cherry has written on the impact of ethos on self-perception in "Ethos and Persona: Self-Representation in Written Discourse"; Martin Medhurst uses ethos as a primary sounding board for understanding the religious rhetoric of our contemporary democracy; Carole Blair and Neil Michael have discussed the role of ethos in American national identity; and Carolyn Miller has attempted to tease apart human-computer interactions and the role of ethos in determining expertise and agency. And these are just contemporary rhetoricians discussing specific modern aspects of ethos. Jeanne Fahnstock has written widely on figures, tropes, and appeals as they relate to science specifically, as in her articles in *Written Communication,* "Accommodating Science: The Rhetorical Life of Scientific Facts," and "Preserving the Figure: Consistency in the Presentation of Scientific Arguments," or her book *Rhetorical Figures in Science.* H.P. Peters has written on the relationship between journalists and scientists and the influence of scientific ethos in *Media, Culture, and Society* in "The Interaction of Journalists and Scientific Experts: Co-operation and Conflict Between Two Professional Cultures." There are scores of other rhetorician doing similar work, most of which is written for professional rhetoricians or other language experts. This chapter provides a brief overview of the importance of scientific ethos by laying out several key features of this important appeal that contribute to scientific credibility when addressing a public audience.

The constant shifting, realignment, and transformation of language does not mean we are destined to operate with incomplete "data" or are doomed to be deceived by those who control the discourse that we rely on to make important democratic contributions to our society. Perhaps one of the overarching impulses that has led us to these problems of desperate confusion in the first place is an obsessive quest for certainty in Western civilization. In our pursuit of an absolute truth that is "out there" somewhere, we forget how fallible language is and how utterly unreliable the people who use it are. We seek an established bedrock of discursive reality when language is by its very definition incapable of supplying it. And even for those who realize this fundamental linguistic quality, the search then becomes one that relegates language to a subservient position in the hierarchy of knowledge (as we saw in the last chapter with Steven Pinker). This is one reason rhetoric has taken such a thrashing in recent centuries; it is seen as language at its "worst," when what it really champions is an attempt to embrace language on its own intrinsic terms, to treat it as the reality of human expression on a basis that recognizes the inherent contradictions, ambiguity, and imprecision that is necessarily part of language use and misuse. Rhetoric, then, provides us with both a means and a technique, allowing for a useful intersection between practice and theory and a realization that language is a supple mechanism for achieving specific discursive purposes and a hermeneutic for analyzing the language that we encounter.

Several adjustments in our own attitudes may be necessary to come to terms with the paradox of knowledge and reality and its relationship to language, and none of them are easy given the magnitude of our own cultural inheritance about the nature of truth and the reliability of "clear" language. First, if we understand that language is a symbolic approximation, not a direct correlation between object and word, we have considerably more freedom to treat words and ideas with some interpretive flexibility. Rather than creating a world of relativity destined to confound all reliable evaluation of texts and discourse, both formal and informal, this knowledge embraces the richness of our symbolic constructs and helps reveal their associative nature. We understand how ideas interconnect through rhetorical study, and we applaud a disciplinary anti-essentialism that ultimately gives us keener insight into how we relate to one another rhetorically. Gideon Burton defines rhetoric beyond its classical role as a study of the varied means of persuasion to include the very important issue of *meaning:* "Indeed, a basic premise for rhetoric is the indivisibility of means from meaning; *how* one says something conveys meaning as much as *what* one says. Rhetoric studies the effectiveness of language comprehensively, including its emotional impact (pathos), as much as its propositional content (logos)."

What Burton leaves out, however, is the even more significant attribute of ethos, which is traditionally defined as "character" but is perhaps more accurately described as the cumulative effect of a speaker's (or writer's or communicator's) personality, dress, manner, status, authority, and presumed or explicit level of expertise. The classical, Aristotlean reading of "ethos," taken from Aristotle's *On Rhetoric,* has almost always involved the original sense of "character," particularly that of a moral sort, though Aristotle himself, as translator George Kennedy clarifies, used the term "ethos" "to refer to qualities, such as an innate sense of justice or a quickness of temper, with which individuals may be naturally endowed and which dispose them to certain kinds of action" (163). "Ethos," in this sense, retains the inherent qualities of personality that a speaker (for Aristotle, rhetoric almost always indicated formal speech, as opposed to writing or general communication) may possess (whether moral or not), but it seems also to point to a speaker's propensity for taking decisive action. Even Aristotle's own text at 1.2.4 of *On Rhetoric* complicates a straight-forward usage of ethos as just "character" by suggesting that ethos, as a figure desirable for any speaker, has as much to do with "credence" or "a natural proclivity for projecting confidence" as it does with "moral conviction." I reproduce the Kennedy translation of that section below because more recent, "readable" translations that have been made available seem to sacrifice much of the subtlety of Kennedy's reading for a brisker prose:

> [There is persuasion] through character whenever the speech is spoken in such a way as to make the speaker worthy of credence; for we believe fair-minded people to a greater extent and more quickly [than we do others] on all subjects in general and completely so in cases where there is not exact knowledge but room

> for doubt. And this should result from the speech, not from a previous opinion that the speaker is a certain kind of person, for it is not the case, as some of the technical writers propose in their treatment of the art, that fair-mindedness [*epieikeia*] on the part of the speaker makes no contribution to persuasiveness. (38)

Aristotle's cautionary stricture that all judgments about a speaker's ethos should come "from the speech" and not from "previous opinion" about the speaker is a prudent warning, but unlikely nonetheless. Audiences frequently have preconceived notions of a speaker's or writer's character, and this does color a recipient's judgment about the content of a speech or text. Scientists, especially well-known ones, come to the podium with an advantage in this respect, since their status as "scientist" is often enough to cement an audience's willingness to judge their ethos favorably.

Other academics go even further with the "character" reading by equating ethos with something as apparently reductionistic as "source credibility." James McCroskey and Thomas Young, in a 1981 essay for the *Central States Speech Journal,* are early examples of rhetoricians interested in challenging this limiting view of ethos which had been largely established during the mid twentieth-century as a "credibility construct" that dealt only with the areas of "expertness, trustworthiness, and intention toward the receiver" (25). This, they rightly point out, is not inclusive of the domain that so important an "artistic mode" (to use Aristotle's classification) as ethos occupies. I mention these sources mainly as a matter of demonstrating that the centrality of ethos has often been eclipsed (at least partially) by the more commonly-discussed pathos and logos (which can be translated loosely as "emotion" and "reason," respectively), such that when ethos is treated in rhetorical analysis, it tends to be subordinate to these other, more familiar, figures. This illustrates again a preference for the binary distinction between rationality and emotion that has so dominated Western thinking; ethos, however, is a more malleable concept in that it not only helps illuminate what is being said, but how and by whom. William Haskins defines ethos as "the perceived degree of character or credibility that a person believes exists in another person or object," but this too reduces ethos to a simple impression of a speaker's authority to say what he or she professes to be qualified to say. While this is certainly a major quality of ethos, a more complete picture of this mode must include some recognition of the cultural currency that lends the discursive agent his or her initial authority. Michael Hyde, for example, has recently published a collected series of articles that, he feels, returns to a more elemental meaning of the term, such that we can come to

> understand the phrase "the ethos of rhetoric" to relate to the way discourse is used to transform space and time into "dwelling places" where people can deliberate about and collectively understand some matter of interest. Such dwelling places define the grounds, abodes, and habitats where a person's ethics and moral character take form and develop. Together the contributors define ethical discourse and describe what its practice looks like in particular communities. ("The *Ethos* of Rhetoric")

Beyond ethics, however, ethos also encapsulates an iconographic image of the speaker, writer, or communicator. While the etymology of the ethical derivation can be seen in synonyms like "reputation" and "character"—where the speaker's perceived integrity and sound value system can be projected to an audience—it is also important to recognize that the ethics of ethos are as much about proximity to an issue and timing within a historical context as they are about any innate moral makeup a communicator may have. For the scientist, a good deal of the ethical center of his or her ethos is attributed to the knowledge base he or she is able to draw upon and is able to successfully articulate to the audience. This is true of many professions, but science—with its emphasis on reputation and expertise—seems especially prone to the needs of a well-established ethos.

We productively reflect on the key element of ethos as a way of compensating for competing ideas within particular fields of science. Bruno Latour, in his book *Science in Action,* describes the increase of controversy surrounding an especially heated scientific topic as a process of creating more, not less, "noise." This noise serves to obscure rather than clarify issues for the listening public, and even for the professionals involved in the argument. This noise is caused by more furious debate, certainly, but the tenor of the debate is escalated by an increase in the technicality of the discourse that is used. The means of escalation is another interesting spatial phenomenon, according to

Latour; to improve the leverage that one side has over another, new data, information, numbers, and arguments are brought to bear from different times and places: "People start using texts, files, documents, articles to force others to transform what was at first opinion into fact. If the discussion continues then the contenders in an *oral* dispute become the *readers* of technical texts or reports" (30). This is a process that is not, of course, unique to science, but it may be more contentious in scientific fields because it seems as if the stakes are higher and more significant than, say, whether Hamlet was "really" insane or just feigning madness in order to pursue his revenge. From the point of view of ethos, the collection of "facts" in the form of documents and authorities from other texts written at other times, often for a different audience and relying on different conclusions, the scientific community can be easily caught up in the quagmire of debate on a given issue for some time.

The difference between what takes place in the scientific community and what the general public audience "hears," however, is an important distinction. Any examination of popular science will indicate that, as in any other venue, the person who gets heard is the one who has the opportunity to *be* heard. The sequestered biologist who has been living her professional life in the south wing of a university laboratory for the last twenty years with virtually no contact with anyone outside of her own immediate scientific community is going to find it difficult to convince non-specialists and specialists alike that Stephen Jay Gould was mistaken in his assessment of Punctuated Equilibrium, no matter how right she is about it (I am assuming for the sake of simplicity that this is an issue that the public would view as important and already have formed opinions about based on the work of Gould, which is not likely to be the case). Gould has been out in the public world spreading his word, whereas the sequestered biologist has been in a lab somewhere with no such contact. From the point of view of science, the sequestered biologists has been doing exactly what a scientist should be doing, not attempting to establish a hegemony with the laity or allowing her work to be rhetorically contaminated by public opinion. But who is more convincing? The sequestered biologist cannot simply let her findings "speak for themselves"; beyond compensating for the sheer lack of training to judge scientific findings that would be a determining factor in convincing the public, she must also contend with Gould's reputation, his visibility, his rhetorical acumen, his perceived as well as his real expertise, not to mention his minions and disciples, all of which coalesce to create a formidable opponent indeed.[1] On this last point, Latour says that allegiances are always an important component of any scientific enterprise:

> The adjective "scientific" is not attributed to *isolated* texts that are able to oppose the opinion of the multitude by virtue of some mysterious faculty. A document becomes scientific when its claims stop being isolated and when the number of people engaged in publishing it are many and explicitly indicated in the text. When reading it, it is on the contrary the reader who becomes *isolated*. The careful marking of the allies' presence is the first sign that the controversy is now heated enough to generate technical documents. (33)

Ethos, in this sense, is more than a matter of "mere" integrity and character; it is an iconographic transformation of the individual making an argument into something that is greater than the sum of his expertise and findings. Ethos involves a massive inculcation of all the characteristics of what science is as it comes to bear on the audience, and it is almost always symbolic of more than one person. While we may have an iconographic image of the "face" of science (Albert Einstein leaps to mind), that face is emblematic of an entire regiment of scientific attributes, authorities, and ideas. This is the broadest demonstration of ethos: the presence of a representative figure that is larger than life, all that is respected in science in one personified package. But ethos may have more to do with persuasion and identification than just a symbolic face or presence. It may also reflect ideas that are well-known but often poorly understood or even mythologized for public intake.[2] In this sense, Latour's discussion of how texts get appropriated for certain purposes is especially useful. He emphasizes that the use of extraneous authorities has a cumulative effect wherein the challenger (or just the general reader) must account for the accumulation of material and the preponderance of ideas in order to make any valid objection to the claims under consideration. This has an impact on ethos in the sense that the related articles in a scholarly text are tantamount to gathering together a group of "friends" who are allies in defense of a particular claim, finding, or argument. References of this sort are persuasive as a body of knowledge, but they also create an insular agent in the ethos of the person invoking them:

> The effect of references on persuasion is not limited to that of "prestige" or "bluff." Again, it is a question of *numbers.* A paper that does not have references is like a child without an escort walking at night in a big city it does not know; isolated, lost, anything may happen to it. On the contrary, attacking a paper heavy with footnotes means that the dissenter has to weaken each of the other papers, or will at least be threatened with having to do so. [...] The difference at this point between technical and non-technical literature is not that one is about fact and the other about fiction, but that the latter gathers only a few resources at hand and the former a lot of resources, even from far away in time and space. (33)

The ethos of the scientist is enhanced by the collective ethos of the references offered as patrols to guard a particular claim. Here we see a symbolic accumulation of "friends" in the form of many supporting texts, and these textual collaborators represent a well-established corpus of knowledge that is intimidating and well-defended. This "intertextuality," as Charles Bazerman calls it, creates a scientific mass of discourse that is internalized by individuals and the scientific community. Texts become a staple of any professional conversation, a "sea" of words: "We create our texts out of the sea of former texts that surround us, the sea of language we live in. And we understand the texts of others within that same sea" (83–84).

But the more generalized audience rarely encounters such technical texts in their casual pursuit of new knowledge. Instead, the ethos of allegiance in popular science is made more accessible through the informalization of the footnoting method seen in professional journals and books. Rather than overwhelming the reader with a long list of footnotes, popular scientists adopt a more familiar approach when citing other scientists and colleagues. Steven Pinker is very adept at this. His method is to refer to the work of others in an explicitly intimate way, as if he is talking about people whose work he not only knows, but who themselves appear to be old acquaintances. One even gets the impression that he has had many social discussions with the people he cites (which, in fact, is very likely), such that the attire of sterile professionalism is stripped away and replaced with an air of personability. The representation of alternative arguments, especially, is conversational, even chatty. The following quote from Pinker's *The Language Instinct* will give the reader some idea of what this technique looks like and the effect it has on the reader: "How plausible is it that an ancestor to language first appeared after the branch leading to humans split off from the branch leading to chimps? Not very, says Philip Lieberman, one of the scientists who believe that the vocal tract anatomy and speech control are the only things that were modified by evolution . . ." (359). Note the "rhetorical question" followed by an informal paraphrase of Lieberman's position, a combination of devices that has the familiar effect that Pinker has had this conversation before, perhaps with Lieberman himself. A technique such as this creates a climate of friendly exchange on the one hand, but also the sense of a professional who is well-fortified on the other. Pinker implies subtly that he is so well-acquainted with the people he cites (in both favorable and challenging ways) that he can adopt this air of familiarity as a peer would one of his associates. While this is in fact the case with someone as recognized as Pinker, it is interesting to note that he makes no attempt to conceal his status with such prestigious men and women. He is a peer letting the reader in on his personal/professional relationship with people who think seriously about such matters. It is entirely likely that Pinker—and others who use this method—is not fully conscious of his enlistment of this rhetorical mechanism, but it does nevertheless create a desirable effect. Pinker's ethos is strengthened through his regular contact with such people, and while he doesn't always agree with them, he is at the very least in a position to challenge them on their claims. The general reader, of course, is not, and Pinker's public relations package is reinforced as a result.

Latour's observation that texts become increasingly technical as the claims they make become more radical (by which I mean deviating from the established paradigm) is an important consideration for both a scientific audience and a more public one, though in obviously different ways; in either case, however, the intent is to further brace the ethos of the author. For the scientific audience, Latour suggests that external references become an accumulation of technical elements that create a layered effect in the prose that is generated: "The mobilization of these new elements [i.e., the introduction of further references to buttress a particular argument] transforms deeply the manner in which texts are written: they become more technical and, to make a metaphor, **stratified**" (45). If technical prose can be considered a means to increase the layers of complexity in a text, thereby making it more difficult to challenge, it can be downright intimidating and even prohibitive to understanding for the general reader. The reader feels that if an author/scientist can master a level of complexity that requires advanced degrees to decode, then there must be some validity to it. Rhetorically, this is a sophisticated maneuver; it is certainly the case that difficult problems require specialized knowledge and restricted discourse to address, but it is equally true that just because a text is complicated doesn't mean there is a proportionate level of truth value to the claims. An accomplished popular scientist will attempt to keep the specialized "jargon"[3] to a minimum in order to achieve another important element of ethos: to let the reader in on the "secrets" of his or her profession. Such knowledge is the very definition of "esoteric"—that knowledge which is guarded, protected, and privileged—and that scientists like Pinker are willing to let us into the inner sanctum of such knowledge only aids in the construction of his ethos, suggesting that he has a prophetic quality in the wisdom that he brings down to the people.

To take an example from theoretical physics, we can turn to a relatively newly-minted popularizer of science in the figure of Brian Greene, author of the surprisingly fashionable book, *The Elegant Universe.* Greene is so typical of the ethos-oozing scientist that his credentials can almost be guessed at: By the age of 12, he was considered a mathematical prodigy, and was tutored by Columbia professors throughout his teenage years because he had so far surpassed the high school math offerings. He holds an undergraduate degree from Harvard, a doctorate from Oxford (where he was a Rhodes Scholar), and he was a *full* professor at Cornell before taking a job at Columbia as a professor of physics and mathematics in 1996. He has written a number of books and scores of articles for both a general and a professional audience, and he had reached all these achievements by the age of 33. His ethos is unassailable from a standpoint of credibility. Of the book itself, it has been described as on par with Hawking's *A Brief History of Time* and as "luring lay people into learning cutting-edge physics with its engaging prose." It is "remarkable," "a classic," "rewarding," "poetic," and even "beautiful" (front notes). String Theory, as most people probably know, is an area of theoretical physics that suggests that the universe is really comprised of vibrating particles known as "strings" which creates a "malleable" universe that has multiple dimensions and a space-time fabric that can actually "rip" (x).

So successful has Greene's book been that it was made into a PBS series that reached an even wider audience. Few scientists can claim this level of popular support, and Greene is following a path forged by Carl Sagan. In order to illustrate how Greene achieved this level of ethos, some samples from the text will aid us in seeing his method and his language for educating the lay public about cosmological matters. Part of the challenge of educating a non-specialist on the nature of exotic physical phenomena is that, like so much of what goes on in science, one must be able to reorder habitual ways of thinking in a manner that allows for comprehension of a counter-intuitive physical event in terms that are more familiar. That is, colloquial vocabularies must be called upon in order to describe ideas that really are well beyond the capacity of common language to describe.

Take, for example, the misnomer "black hole." The term "black hole" refers to a phenomenon that is at once an object and a physical principle; a black hole is both a thing and an agent for distorting space. It operates under the same basic principles of gravity as any other star or planet, but to an extreme. While the general enthusiast knows that objects in the universe have an effect on the space around them that we call gravity, this characteristic is eclipsed in our minds by the physical presence of the object itself. A garden variety star or planet "warps" the space around it, and the physical effect can be observed on other objects that pass within a certain vicinity, but the gravitational pull an object exerts is overshadowed by the existence of the object as a primary feature. The non-specialist does not think of a star as a "gravity maker," but is aware of this effect on surrounding space. A black hole, by contrast, literally "hides" its primary characteristic as an object because its other property—the extreme gravitational field it emits—is far more noticeable than the existence of the object creating the gravitational field. Here is another example of how a colloquial term has distorted the reality of the phenomenon: A black hole is not really a "hole" at all, but a "compressed star" that has the mass of a sun (in the most impressive cases, millions or even billions of suns) compacted into a sphere only several kilometers in diameter—as Greene describes it, "the whole of the sun squeezed to fit comfortably within upper Manhattan" (80). The density of such a star takes traditional physics to such a theoretical extreme that many physicists doubted that Newtonian physical explanations could account for it (or even Einsteinian physics, though, according to Greene, general relativity can account through the study of x-ray emissions for the *existence* of black holes [81], suggesting that even the reality of such objects is still uncertain). It was not clear, for example, whether the atomic structure of particles could actually withstand such density, and this has often been described in terms of what a baseball-sized amount of the collapsed star's material would "weigh on earth"—usually millions of tons. Greene himself takes a slight variation on this comparison, saying that "a teaspoonful of such a compressed sun would weigh about as much as Mount Everest" (80). It is hard to imagine an object of such minute size weighing so much, but even more interesting is the physical impossibility of the hypothetical demonstration. It would be impossible to mine material from the black hole using any conventional technology that currently exists, and even if it were possible, it would never make sense to attempt to transport such an object to earth to test its weight. Such descriptions and measurements are mathematical shorthand, and for the convenience of the public, scientists describe the amazing density of black hole material in terms that emphasize the unusual properties of such dramatic celestial objects, but they must do so only within the confines of familiar reference points that are accessible to most human beings—teaspoons, baseballs, Manhattan, Mt. Everest. Such comparisons have the effect of demonstrating how outside the realm of common physical experience a black hole really is; when placed in juxtaposition to these familiar places, objects, weights, and sizes, black hole physics seems monstrous indeed.

The "black hole" moniker, Greene tells us, was not the first scientific choice for a common name for this phenomenon. The "black" property of black holes captures the notion that the star's compression is so great, and its gravitational force so extreme, that nothing passing near its "event horizon" can escape its pull, including light. This concept in itself is counter-intuitive for the non-physicist since light is not something we usually think of as possessing mass, which it must have in some minuscule proportion in order to be affected by gravity. Since light is captured in this way, no light is emitted from the event horizon inward toward the compressed star. Hence, they were initially referred to by astronomers as "dark" or "frozen" stars until eminent physicist John Wheeler coined what Greene calls the "catchy" term "black hole": "black because they cannot emit light, holes because anything getting too close falls into them, never to return. The name stuck" (79). However, this term, while simply a descriptive phrase of convenience for the physicist who understands the limitations of such a descriptor, is taken far more literally by the general public. A black hole is often viewed as a great abyss and a kind of gateway to another dimension (a misconception we can thank Disney for), even though there is no necessary suggestion of this in the term "black hole" itself. Interestingly, "hole" refers to the extreme warping of space that occurs around a compressed star, not to some corridor to another place or time. The truth is, because of the radical nature of such objects, physical examination of them is limited to what we can see from our remote instruments on and around earth and what astronomers and physicists can postulate about the physical implications of what they observe. Direct experimentation is impossible, at least insofar as coming into physical contact with the object is concerned, and so much of what is known about black holes is necessarily based upon visual and radio examination of the object and upon the mathematics and theoretical physics used to make sense of the data that is received from these sources.

Greene uses some other innovative rhetorical techniques in an effort to clarify for the general reader the unusual properties of this mysterious cosmic occurrence. In the chapter of *The Elegant Universe* entitled "Of Warps and Ripples," Greene gives us a graphic depiction of the effects a black hole would have on normal matter, in this case, a human body:

> For example, if you dropped feet first through the event horizon, as you approached the black hole's center you would find yourself getting increasingly uncomfortable. The gravitational force of the black hole would increase so dramatically that its pull on your feet would be much greater than its pull on your head (since in a feet-first fall your feet are always a bit closer than your head to the black hole's center); so much stronger, in fact, that you would be stretched with a force that would quickly tear your body to shreds. (80)

This scenario is at once fantastic and effective, since any examination of a black hole would never involve dropping anyone into the event horizon, for obvious reasons. Yet the bodily demonstration, no matter how unlikely, has an intriguing rhetorical effect on the readers: It impresses upon us the degree to which matter is transformed when within the vicinity of a black hole using a sample matter that is most immediate to our experience, our own bodies. The effect is pronounced, since it impresses upon the reader both the profound danger of such an object to our ordinary understanding of celestial objects and the degree to which direct physical examination of it is prohibited. A black hole is mysterious (as the term also suggests—"black" is typically associated with the unknown and the feared) and fascinating, but it is also off limits so far as achieving any intimate understanding of it is concerned. Black holes are typically located at the center of a galaxy, and some astrophysicists even suggest that a black hole is a necessary condition for the formation of galaxies, meaning that every organized galaxy must have one. Its centrality to a galaxy has the added effect of fostering a notion of great secrets laying in wait there, centeredness being a standard spatial metaphor for the "most significant" nuclear source of anything possessing power.

Greene's esoteric knowledge of such objects adds to his ethos as one of the elite, the select few scientists who preserve the privilege of studying and understanding the nature of one of the most mysterious features of the universe. Certainly, as a scientific discovery, black holes are well-situated within the realm of scientific inquiry, and this alone makes any knowledge shared with the general public regarding black holes seem impenetrable to those unaccustomed to thinking in such radical physical terms. Greene manages to strike an effective balance between scientific accuracy and analogical embellishment, not because he needs to enhance the already amazing properties of black holes, but because he needs to impress upon the reader the alien physical properties as they operate within the proximity of such objects. In order to accomplish this, he must exercise both precision and drama, a combination of rhetorical requirements that can be at odds with one another. By using the reader's own body, Greene resituates the physical strangeness of the black hole by allowing the reader to hypothetically and vicariously participate in the discovery process.

Despite George Johnson's declaration that there is a "great tradition of physicists writing for the masses" (front cover of *The Elegant Uni-*

verse), one is hard-pressed to count them on more than one hand. Certainly Hawking makes the list, and Greene himself, but the only other widely-known physicist is Albert Einstein, whose theories of relativity are known but (to my knowledge) were never offered in a comprehensive volume for *public* consumption. Certain other physicists have recognition value—Robert Oppenheimer, Freeman Dyson, and Enrico Fermi, for example, all because of their involvement with the Manhattan Project—and would occasionally address the public (especially Oppenheimer)—but their efforts in educating the public can hardly be called expansive, much less a "tradition." The secrets of theoretical physics have, rather, been "traditionally" kept out of public hands by virtue of the obscure nature of professional scientific research. As I have noted, the primary source of science for the interested public has been periodicals expressly designed for that purpose, publications like *Scientific American, Science,* and *Popular Science,* not professional journals since, as Latour reminds us, such publications are so densely layered that they prohibit public comprehension even if an interested party has ready access to them.[4]

The public impression of an author's ethos is further confused by the next most available domain of information, the Internet and the proliferation of online sources, perhaps the most common place for a general science enthusiast to go to seek whatever range of materials he or she is interested in learning more about. Barbara Warnick has offered some compelling discussions of the nature of online ethos for the general public in her article "Online Ethos: Source Credibility in an 'Authorless' Environment," which appeared in the October 2004 issue of *The American Behavioral Scientist,* by suggesting that the criteria used by people to discern the credibility of online sources is less than scientific. Credibility judgments by readers of online sources (of all types, apparently), according to Warnick, "are driven by social and normative factors that have to do with the nature of the Web environment and by values and priorities attaching context and community values" (259) rather than by the reputation of the author or the site sponsor or the quality of the content, insofar as this can be determined by the amateur. One Princeton study she cites involved phone interviews with 1500 Internet users over the age of 18 who were asked about their online habits in 2001 and 2002. The summary of the results yields some interesting information about what general users view as valid online sources:

> [A]lthough respondents reported that they generally have a low level of trust in commercial Web cites (only 29% said they trusted these sites, and only 33% said they trusted their advice), the respondents use the sites anyway. Respondents reported that they wanted the sites to provide clear information on who runs the site, how to reach those people, the site's privacy policy, and other factors related to site authorship and sponsorship. The study also concludes, however, that "credibility and trust online are the product of many factors, including each person's overall view of the world and the level of trust of people in general." (259)

However, I am hesitant to draw any definitive conclusions from the Princeton study, mainly because it involved a survey of general Internet usage (as opposed to users specifically seeking scientific research) and because vague phrases such as "use" of the Internet ("used" how—as a source of general information, as a place to locate research documents, to determine a pattern in sites that may have nothing to do with their actual credibility?) and "overall view of the world" (how does this affect credibility criteria? What *is* an overall view of the world such that it would contribute to the generation of specific criteria or a lack thereof?), and the suggestion that the "trust level" of "people in general" may skew results.

Also, what Warnick and the Princeton study apparently find statistically insignificant seems decidedly significant to me; 29 percent of users *do* evidently trust commercial sites and 33 percent trust their advice (this distinction is a bit unclear), which means a full one-third of the survey participants will use dubious commercial sites as information and consultation sources. This means that one out of every three people visiting commercial sites trust their content, regardless of more substantial factors that might be used to determine credibility and ethos such as sponsorship by a non-profit organization or educational site or the reputation of the author—assuming this can be determined (one new wrinkle in the reliance on Internet sources is that individual authorship is frequently obscured or absent altogether). More importantly, perhaps, is the finding that site packaging has more to do with the ethos of the site's author or the credibility of the sources than almost anything else. Warnick reports "that people rarely used the rigorous criteria that Princeton Survey Research Associates

(2002) had indicated were most important" (260). Rather, according to the report itself, "the average consumer paid far more attention to the superficial aspects of a site, such as visual cues, than to its content" (qtd. in Warnick 260). This may account, in the case of online science sources, for the popularity of online versions of magazines like *Popular Science,* which is visually very appealing, relatively easy to navigate, and packed with brief articles that allow the reader quick access and even quicker processing time for information about science and technology. This layout is in marked contrast to *Scientific American,* which, while also very professionally done and utilizing the full range of computer technology, tends to adopt a more traditional layout for its feature articles. In a culture that seems accustomed to a rapid succession of images (any attempt to dwell futilely on an image in most television commercials and nearly all music videos should give us idea of the truth of this statement) means that the Internet becomes the ideal venue for packaging visually stunning graphics—often at the expense of more substantial content—such that today's audience has become conditioned to receive its information this way. What suffers, then, is any critical scrutiny of the information we receive in favor of the aesthetically pleasing or graphically flashy, perhaps because these qualities suggest funding and professionalism, factors that the public has been conditioned to respond favorably towards; many Internet users have been culturally trained to respond to the cosmetic elements over the substantial ones, leading to a blurring of the valid and credible source with the merely well-presented one.

But the question of why we find some sources credible and others not must have a more deeply rational origin since we have also been conditioned to value "fact" over mere "opinion." It is interesting to note that these two states are frequently divided in our description of an idea's validity, as if opinions are never based on anything but whim or founded on facts or data. Judy Segal and Alan Richardson offer a two-pronged explanation for scientific credibility that has circulated among scientific theorists: that we hear of science through scientists and that the credibility of an individual scientist is enhanced by his or her reputation or status as a scientist (139). But scientists are also "linked to the larger society by virtue of shared values instantiated in the character of science" (138). In other words, reputation is for scientists the ability to represent the ideology of science to other scientists. While reputation is, as I have already noted, central to the success or failure of a scientist, it means something very different for those laypersons merely interested in scientific developments. Reputation for the general public must necessarily have as much to do with the packaging as with the content since there is little in way of specific knowledge about areas in which the scientist claims expertise. Reputation for the general public, then, would involve criteria like credentials (a PhD from Harvard in Behavioral Psychology), past awards (a Pulitzer Prize), past works (has written five books for a popular audience), and, often, some indication that the scientist has enjoyed a popular reception from the public through mainstream media sources (has appeared on the "Oprah Winfrey" and "Today" shows). A popular reputation might also include more intangible credentials like personability, charisma, and accessibility.

One other intangible might be added to this list, perhaps the most important one of all for the purposes of a strong scientific ethos: a shared sense of ideological purpose. While it is certainly true that scientists unleash counter-intuitive theories and unpleasant realities on the general population—an important and necessary role—many of the most successful popular scientists are able to combine innovative (or, at least, apparently innovative) ideas, discoveries, and even advice with the value structures we already hold. This is clearly true of television personalities like Dr. Phil, but Dr. Phil is not a scientist or even a good psychologist—he just plays one on TV. Phil parrots our own biases and moralistic values, translates it into pseudo-clinical language or bumper-sticker aphorisms, and sells it back to us as something new (or, at least, newly validated). However, Carl Sagan, Steven Pinker, Stephen Jay Gould, and Steven Hawking *are* scientists—"real" ones—yet they have, to retread an overused phrase, "captured our imagination" in a way that renders their ability to tap into some of our most deeply held emotional and inquisitive responses. Sagan and Hawking, in particular, represent not only the power of the human mind, but the quest for adventure and exploration that is intrinsic to the human soul. They deal in the unknown, and as banal and predictable as our own lives become, they are able to take us to other worlds, to fulfill the basic human wanderlust in a vicarious and imaginative journey. They show us the possibilities of human exploration, intellectually, scientifically, and emotionally. They mirror our curiosity, even if that curiosity has been retarded by years of attention to the mundane and typical. They help us release an innate characteristic of inquiry that has been buried under the constant need to direct our mental energies to the everyday. In the war of attrition that modern life exacts on our psyche, figures like Sagan and Hawking project not only an ethos of competence and knowledge, but of the potentials of the human spirit, each in slightly different ways. Sagan lets us into the imagination of the planetary scientist keen on the idea of space exploration for the sake of gaining knew knowledge, not because it may be economically profitable or may help support an ideological war against a different way of governing. Hawking reminds us that our physical bodies are but a very small part of who we are, that they are, in fact, incidental to our greatest human attribute—the mind. Hawking's story of overcoming Lou Gehrig's disease not by conclusively beating the disease itself, but by developing his mind to an astounding level, is the very picture of the triumph of mind over matter. I do not wish to belittle the pain and suffering that Hawking has surely experienced as a result of his illness, but rather to note that there are many ways to overcome a physical disability, to show that when one door closes, another opens. This part of Hawking's character, more than any other, is what gives him his strong ethos with the public. That he can do this while also maintaining one of the most respected reputations in science makes his accomplishments even more remarkable.

Ethos, then, is a combination of reputation (especially between scientist and scientist), credentials, personality, opportunity, and values. I use the last word hesitantly because, while it does certainly imply values of an ethical nature, it also encompasses values that we hold about our own identity, place, and purpose in the world. "Values" also manifest themselves in epistemological ideologies, and this is another important feature of the scientific ethos. The positivistic strain of empiricism and analytical rationality that informs the scientific methodology is a large part of its mystical allure, because while the rest of us struggle with the irrational, emotional, and discriminatory (in the most anti-intellectual sense), science remains aloof, transcendent of the more primal impulses that taint other forms of discourse. This description of science is not entirely mythic; it has managed to achieve more objectivity than most other epistemological methods (like logic or philosophy), but it is important to remind ourselves that the total objectivity of the scientific process is an illusion. It cannot be emphasized strongly enough that this is not the same as saying that objective realities are also illusions, despite the popular but specious representation of alternate epistemologies among scientistic detractors, especially those ways of knowing that have been casually and loosely assigned the postmodernism moniker. I shall repeat for emphasis the unjust representation that postmodernists have received: I know of no postmodern thinker who is so simplistic in her thinking that she denies, for example, the existence of gravity or the reality of physical objects. However, how we deal with these realities, how we describe them, what we do with our knowledge, where we apply the language that describes them, and what form that language takes are certainly subject to human interaction with those realities. It is unavoidable as agents both part of and separate from the realities we encounter that they will be observed through the murky lenses of human understanding, and in this sense, they are "cultural constructions" because we must pragmatically put a human perspective on the realities we observe, describe, catalog, and manipulate.

I bring this up again because it is a central feature of the scientific ethos to obscure the complexity of human knowledge systems by, ironically, claiming that such knowledge is much simpler than it actually is. Scientists like Alan Sokal and his like-minded crusaders (Steven Weinberg, Kurt Gottfried, Norman Levitt, Paul Gross, and Jean Bricmont, to name a few) in the so-called "culture wars" construct polemics designed to close down inquiry about the nature of knowledge and to dismantle any dialogue that might pose questions about the nature and self-regulatory accountability of science, and they boldly use the primacy of the scientific epistemology as a basis for their own ethos and the ethos of science itself (Sokal will be discussed in more detail in Chapter 10). It is necessary to examine their discursive process, then, if only to see why they would object so stridently to other (and not necessarily counter) discourses to those that dominate the sciences. This is a complex issue, and much has been written on it, so my purpose is far from attempting to put the debate to rest, but rather, to describe why it has become so heated, what is at stake, and how it contributes to or detracts from the maintenance of the scientific ethos. Michael Truscello has written a fine survey of the literature that has surrounded the Sokal affair, not just that which came about as a result of the hoax perpetrated on the journal *Social Text,* but also the subsequent books that were written as a direct result of this increasingly heated discussion. In the *Rhetoric Review* article "The Clothing of the American Mind:

The Construction of Scientific Ethos in the Science Wars," Truscello characterizes the tenor of the debate as, in his words, a

> contest over the cultural hegemony [that] has been framed by the scientists as a war against superstitious, antidemocratic, and anti-American ideologues who threaten to 'clothe' the American mind with radical propaganda simply out of spite for the failure of counter-cultural activism in the 1960s . . .. When scientists are involved in the rhetorical construction of ethos, especially in response to a problematization of their professional interests, they are engaged in the rhetorical negotiation of demarcation criteria, the definition of what is and is not 'science.' (329)

Two areas of interest in these statements are fundamental to understanding the culture wars debate and its effect on scientific ethos: that "postmodernist" discourse is an indignant reaction by the faux left to failed political restructuring in the counter-culture movements of the 1960s and 1970s, and that many scientists, when defending their limited (and limiting) definition of science, cast the debate not in terms of inclusion of other possible perspectives on the nature of science as a cultural force, but in terms of a binary attempt to prevent radical proletariat intellectual factions from storming the Bastille of scientific truth. The ethos such defenses erect for the scientists who want to patrol the perimeter of intellectual purism is very strong, because scientists who adopt this position (and not all of them do, to be sure) are employing the most effective of all possible dichotomies: the battle against ignorance by the forces of truth. Truscello sees the efforts of scholars outside of science as dealing with an attitude well-entrenched in many scientific circles, that science is insulated by "extrasystematic" analysis because its truths are obvious and its authority is unassailable. (It is unfortunate that so many cultural dialogues [to put it favorably] have been framed by militaristic metaphors, such that a "culture war" implies that this issue is a holy battle that can only leave one victor standing once the smoke has cleared.) Nevertheless, this is the orientation we are confronting, and providing alternate views to the public on the nature, impact, and relevance of science meets with a reductionistic "either/or" position that is akin to the Liberal/Conservative political wars that we find ourselves in today:

> The shift in modern ethos toward a social construction of ethos predicated on cultural sites of discursive instantiations 'shifts its implications of responsibility from the individual to a negotiation or mediation between rhetor and the community' (Reynolds 328). This social negotiation of ethos is especially problematic for scientists who conceive of their ethos as self-evident, as something defined from within the boundaries of 'science' as it is broadly conceived. (330)

In other words, any discussion of the social formation or validation of scientific ideas is usurped by the self-appointed authority that drives the scientific ethos. Sokal et al. are not interested in discerning distinctive ways of looking at science that might alter its purist reality because to be so would also be to relinquish the hold that scientific power has on its guardianship of "real" or "actual" knowledge. This could, in turn, weaken an otherwise powerful ethos that such scientists would lead us to believe has been achieved through a hard-won campaign against the forces of superstition and ignorance.

The problem with such an outlook is that it insulates science from extraneous scrutiny and it reinforces a public perception that science is intellectually untouchable. The political ramifications are significant; the suggestion that the "academic Left" has waged total combat on science is a deliberate misrepresentation of the real purpose of science studies. Andrew Ross, in his book, *Science Wars* offers another list of possible motives for those scholars wishing to illuminate the activities and influences of science on society. For him, the scholarly motives driving science studies fall into four possible categories, which I will paraphrase as:

1. A desire to provide "an accurate, scientific description of empirical scientific practice";
2. To help science reclaim its ideals that have suffered at the hands of internal abuse and external contamination;
3. To help scientists recognize the social origins of their research and history and to protect it from "technoscientific development" that can be exploited by external agents;
4. To create alternate scientific methods (or, perhaps, priorities) that address the social needs and interests of a segment of the population that is not aligned with the "managerial elites in

> business, government, and the military." (qtd. in Truscello 331)

The purpose of the discussion here, I should note, is a variation on these categories, and this is by no means an exhaustive list of what those of us in science studies might hope to accomplish. Truscello sees a statement like Ross's as a direct reaction to the Sokal affair and as "one in which both sides have opined on the affair as the litmus test of the efficacy of scientific and postmodern discourse" (331). If I were forced to align myself with any of these motives, it would most closely resemble number four, though with a caveat: I don't have a problem with the scientific method per se, which, in the words of Kenneth Burke, "are mainly converted into menaces by the inadequacies of present political institutions" (*Permanence and Change*). Rather, my concern is the degree to which science is subservient to the power structures that benefit most from the products of its labor. Science, far from being the autonomous intellectual institution that Sokal, Levitt, or Weinberg claim, is really the servant of the deep-pocketed institutions that fund it. It is through the starkly essentialist assumptions of scientists like Sokal that a climate of intellectual intolerance is fostered ("tolerance" is not an accurate word, for it suggests that science should "put up with" science studies that cast it in anything but a reverential light). Sokal's ethos is predicated on the authoritarianism he assumes is the birthright of science. To challenge it is a heresy, an impiety that he cannot forgive.

Which brings me to the last point I would like to make about scientific ethos, the somewhat nebulous issue of trust. Ethos, in any manifestation, is achieved through a commitment to trust on the part of the audience. A powerful ethos commands trust, because without it, statements, findings, claims, and advice ring hollow. The concept—much less the practice—of trust has become increasingly elusive in American society. We distrust our leadership because we have been routinely lied to; we distrust our institutions because they have regularly failed us; we distrust many of our traditional values because they seem inadequate for guiding us through our moral problems; and, increasingly, we distrust science as its ethos erodes in the face of a cynical response to the problems that modern living has brought. When science fails to engender trust in a popular audience it is because, according to Carolyn Miller, it has overemphasized logos at the expense of ethos. Scientists have, in other words, mistakenly assumed that because science values logic, rationality, and sound argumentation as the definitive factors in persuasion that its audience intuitively shares in these principles. She concludes that "we might therefore have predicted that an *ethos* that allies itself too closely to logos, like the *ethos* of expert systems, will fail to persuade" ("Expertise and Agency" 207). Strict logical analysis will almost always fail to persuade a general audience, largely because the expression of logical strictures is often mistaken for arrogance or, ironically, an attempt to deliberately obscure an argument in voluminous prose or loquacious speech. Too much logos, it seems, erodes trust, even though rationality should dictate that it would have the opposite effect. People do not respond well to carefully and logically constructed arguments mainly because they have not been properly educated in what they look like. I have students who, having just read an academic article that by my standards of argumentation was flawlessly arranged and logically articulated, say, "he could have said in two pages what took him fifteen." They have failed to see the subtleties, subtexts, and subordinate premises (and sometimes the major ones), and they have often failed to understand the conclusion as well, but that doesn't matter; those are mere "details" that are not important to the "main point" (which they also frequently miss, or misinterpret, or reduce to the point of absurdity). This is perhaps another indication of the literalist, materialist, anti-intellectual tenor of American society today, a vague feeling that anything worth saying is dispatched with alacrity. A scientist must defend his ethos by capitulating to these expectations, at least to a degree. If he doesn't, he risks estranging an audience for whom complicated analysis or careful scrutiny is tantamount to disingenuousness.

Logos, then, occupies the strange position of being venerated in principle but disdained in practice by a public who has little experience with the properties of sound argumentation. For much of the public, arguments are "democratized" in the sense of majority rule—that argument which most people agree with is the best. Since people don't trust complexities, they don't trust those who rely on them in order to make elaborate arguments. Even our educational system supports this idea, often forcing students to articulate a proposal, write a paper, or answer a question in a very finite space, as if this will magically eradicate the "BS" problem. The ethos of the modern scientist must embrace a reduced attention to the details of a claim or an idea, generating in the audience a sense of faith in his or her knowledge of these details that are simply not important to the central claim they are considering. People no longer respond with deference to science just because it's science; rather, the contemporary scientific ethos dictates that some attention to the expectations of the modern audience be observed, and these expectations have to do with character, expertise, brevity, lucidity, directness, and even belief.[5]

How, then, are science and belief related, especially when considering the importance of ethos in the production of audience identification and the issue of trust? Martin Medhurst, in "Religious Rhetoric and the *Ethos* of Democracy," sees the indelible mark of a common value system on our form of government "as a signal of an emerging consensus over the nature, purpose, and identity in America." This common value system, he says, is "a sign of something deeper—an *ethos,* an identity, a character, a spirit that binds Americans together rather than drives them apart" (116). While politically this often manifests itself in the form of religion, as in G.W. Bush's declaration that schools "must cultivate conscience" and "must not oppose religion" or in the move to adopt "faith-based initiatives" (121), the apparent commonality of our values would seem misplaced in the secular pursuit of science. Certainly, many scientists, such as Gould and his notion of Non-Overlapping Magisteria, endorse the idea that religion and science should be kept separate; in practice this is rarely the case. So many scientific issues have religious implications that it is naïve to believe that they can be kept as sequestered, independent entities that have their own unbreachable boundaries. The origin of humanity is, as I have mentioned, the most significant area of overlap, but other ethical issues like stem-cell research, abortion, genetics research and cloning, euthanasia (or, conversely, the prolonging of one's life artificially), vivisectionism, and a host of other relatively minor issues demonstrate just how closely tied our value system is to everything that touches our world. Perhaps one of the areas where trust is sacrificed in scientific portrayals of these problems is in the claim that it is not science's domain to treat them with moral concern in the first place. Science just does what it does, some argue, and we'll leave the moral untidiness to the religious folks who are apparently responsible for such things.

While scientists like Gould might explain their position as one that recognizes the limitations of science and scientists, most people would probably not see it that way. Rather, the scientific ethos is compromised through what might be seen as a casual dismissal of the significance of moral questions, especially as they affect the process and outcomes of science. Frank H.T. Rhodes, president emeritus of Cornell University, sees science as a public activity that should reflect public values and address public needs:

> Science is public knowledge, discovered with public financial support, disseminated for public use; its practice is a public trust. The health of our society requires not only a robust scientific enterprise, but also a basic public understanding of science: without these there will be no firm foundation for our public health or our environmental safety nor essential support for our civil infrastructure, or our military security, no source for our technological inventiveness and no secure basis for our industrial competitiveness or long-term economic prosperity. ("Science as a Public Trust" 1)

The emphasis on such key areas as "public health," "environmental safety," "support for the civil infrastructure," "military security," etc., are all statements of value; these are the areas that science should serve, Rhodes seems to say, and a particular emphasis should be placed on "industrial competitiveness" and "economic prosperity." This expression of value mirrors the areas of concern that most Americans see as important, even imperative. But science, according to Rhodes, is enduring too many setbacks—financially, educationally, politically, and socially—and it is therefore necessary to revise the autonomous privilege that science has enjoyed in the past. This is because the scientist has a unique social burden: "They carry a particular public trust that involves not only the responsible practice of science itself but also the promotion of public understanding for the scientific endeavor" (1).

John Stuart Mill said in *On Liberty* that the higher one's stature in society, the greater his social obligation. Because of that, he argued for greater taxes to be levied against the rich and for higher penalties to be imposed on the powerful who betray the public trust. But this obligation was also positive; it was important, he felt, that the more influence a body or an individual had in society, the more important it was to tailor objectives toward a public end. For good or ill, we rely on science in modern society because it is so inextricably tied to the infrastructure we inhabit, maintain, manipulate, and penetrate. The technology, the medicine, the environment, the very rational make-up of our citizenry is dependent on science for both its practical and intellectual well-being. The ethos that science and scientists command is a powerful one, but it has been damaged in recent times because of a failure of science to recognize its close connection to society. As a result, trust in science has eroded. Whereas science once occupied a central position in the hierarchy of social power, it has found it increasingly difficult to identify with the community in its charge. I have argued that science is dominant, that it has overtaken our worldview in such a way that it is difficult to think in anything but scientistic terms or to pursue knowledge by any standard other than a scientific one. It is perhaps because of this that the issue of trust in the ethos of the scientist becomes so important; American society is increasingly disillusioned with all of its important leadership agencies. In politics, we have been routinely lied to and dismissed as children who don't understand the "way the world works"; in religion, we feel that faith may not be enough to address the daunting problems that confront us personally, socially, or globally; in education, we fear that our children are not learning what they need to survive in an increasingly hostile and complicated world; in science, we wonder whether the competence and will to help us through the quagmire of environmental, social, medical, and technological problems that science itself has helped create exists anymore.

From a rhetorical point of view, science may simply become one more voice in the wilderness shouting over the din of competing discourses. Without the instruments to hear what is really being said, by whom, and for what purpose, trust in science may continue to disintegrate. And as a nation conditioned to the principles of scientistic thinking, when that voice no longer resonates, we will be one step closer to a rational and emotional blankness that may leave us groping for another locus of meaning. The ethos of science, strong and commanding as it has been in post-war America, may not continue to dominate if the trust level maintains an irretrievable drop. Perhaps such a challenge to science will lead to a positive synthesis that values not just the rationalistic application of apparently objective criteria that can have the effect of treating social problems as a mechanistic failure of some grand machine, but will allow us to embrace all areas of knowledge in the pursuit of the most compassionate and human solution to a given problem as opposed to simply the most efficient one. Understanding how scientific discourse is used and what purpose it is designed to serve may aid us in the pursuit of a happy unity, one where educated people have a broad base of awareness in all fields of knowing and the ability to see where and when particular epistemological choices should be made. Our situation today is so pluralistic that it only makes sense that our knowledge should be too. Science can help itself by recognizing this and adapting to it so that it becomes the friend of the civic body it serves and allows for alternative ways of thinking about the problems that beset us most in these complicated times. In this way, the ethos science once enjoyed (and still enjoys on several levels) might be recaptured, strengthened, and deserved.

6 The Sound of Punditry

One aspect of the scientific ethos that scientists enjoy is a collective reputation as noble, ethical, and highly principled professionals–professionals who would never falsify or deliberately misconstrue data. Falsification or selectivity is against what every ethical scientist believes about the scientific methodology, namely, that the core of successful science is reliant upon accurate and truthful data that can be verified by independent experimentation. To purposely sabotage this hallowed method of science is counterproductive to the profession, because it doesn't yield precise results that can be reproduced elsewhere and used for the furtherance of knowledge. But like any other vocation, the professional scientist is vulnerable to the whims and follies of other economic, political, and rhetorical pressures. Given this reality of contemporary society, scientists are sometimes forced to compromise this ethic in order to maintain good stead with their employers, to further personal ambitions, or simply to survive professionally. There are many examples: Expert witnesses who sell their credentials to a plaintiff or defendant to win a lawsuit or even a criminal case; research scientists who fudge data to continue funding for a stalled project; scientific representatives for health associations who indicate that cures for diseases are far closer than they actually are in order to secure donations. This is not to say that scientists are more prone to such activities than any other white-collar group; their role as knowledge-makers, and the often esoteric nature of the knowledge, however, allows them a latitude that other professions may not enjoy. The only people who can check on the precision of scientific findings are other scientists, and this is, ostensibly at least, built into the method. However, by the time some scientific projects are investigated by independent scientific parties (if they are at all), the slanted results may have been released (or hinted at or implied) to other pundits in this rhetorical process who spread the word about a new drug, a new treatment, a new gadget, or a new technique to those with the power, money, and economic motive to make such findings public domain. The result is a sometimes confusing quagmire of contradictory information that the general population is not prepared to sort out and evaluate shrewdly. Such phenomena are yet another by-product of a consumer culture of wish-fulfillment and denial that turns to science to answer its most difficult questions and solve its most troubling problems.

In a standard dismissal of those professional academics who might critique science from an intellectual standpoint, Robert L. Park, in his essay, "Voodoo Science," maintains that:

> Unwilling or unable to comprehend the technology on which they depend, [people] are deeply distrustful of the science behind it, and reject the Western scientific tradition that created it. It is a romantic rebellion, led not by the semi-literate yahoos of fundamentalist religion, who are the traditional foes of science, but by serious academics who regard themselves as intellectuals. They range from hysterical environmentalist Jeremy Rifkin, who sees disaster lurking behind all industrial progress, to a University of Delaware philosophy professor, Sandra Harding, who seems to believe that the laws of physics were constructed to maintain white male dominance. (150)

By "standard dismissal" I mean that this broad description paints the cultural study of scientific dominance in only the most stark, extreme, and absolute terms, thus leading a public who tends to hold steadfastly to the epistemological dominance of science to believe that challenges to scientific authority are perpetuated only by the lunatic fringe, a group of professionally jealous academics who do not, cannot, and will not understand the science that governs their lives and therefore react as startled savages to the sheer might and power of the scientific juggernaut. Such a declamatory tenor, unfortunately, does little to educate, and functions rather to establish a climate of patriarchal condescension. The notion, for example, that the "laws of physics" have anything to do with a conspiratorial, patriarchal, Anglo-Saxon scheme to keep the great unwashed in its place is ridiculous, and it is as equally thorough in its misrepresentation of Harding's position as are the distressingly reductionistic claims that postmodernism is a kooky relativistic trend where "anything goes."

Park's glib dispatching of Harding functions to illustrate what amounts to intellectual sabotage of alternative perspectives and possible ways of knowing. Sandra Harding is a feminist, and as such, she is an easy target for devout followers of scientism because her philosophical and political perspective often leads critics to a premature judgment that has little relevance to what she is actually arguing. In other words, her arguments are caricatured and then summarily discharged on the basis of that caricature (The Straw Man). But she is no marginalized academic weirdo who has independently hatched bizarre schemes to challenge the White Patriarchs and their Laws of Physics for some mystical, sprout- and herb-laden form of alternative reality; she is an established and respected professional philosopher with half a dozen books to her credit. In the preface to *Whose Science? Whose Knowledge?*, Harding describes her project this way:

> [T]here has been an outpouring of critical examination of Western science, technology, and epistemology from the peace and ecology movements, the left, philosophers, historians, and sociologists of science, and Third World critics, as well as from Western feminists. Consequently, the feminist critiques are not isolated voices crying in the wilderness (if they ever were) but are linked thematically and historically to a rising tide of historical analysis of the mental life and social relations of the modern, androcentric, imperial, bourgeois West, including its sciences and notions of knowledge. The present book is a study about this moment in the feminist criticisms of science, technology, and epistemology. It is *a* study, not *the* study; other participants in these debates would focus on issues other than the ones I have chosen. It a book about Western sciences, technologies, and claims to knowledge from feminist perspectives, and in the later chapters it begins to be about appropriate focuses on such issues from the perspective of "global feminisms"—feminisms capable of speaking out of particular historical concerns other than the local Western ones that continue to distort so much of Western intellectual life. (viii)

This description of her project neither says nor implies that physics is a fiction designed to keep white men in power. She does, however, challenge the dominant notion that physics should, by default, be considered the model on which all other science is based. She explains in the same book cited above that

> the model for good science should be research programs explicitly directed by liberatory political goals—by interests in identifying and eliminating from our understanding of nature and social relations the partialities and distortions that have been created by socially coercive projects. It does not *ensure* good empirical results to select scientific problematics, concepts, hypotheses, and research designs with these goals in mind; democratic sciences must be able to distinguish between how people want the world to be and how it is. But better science is likely to result if *all the causes* of scientific conclusions are thought to be equally reasonable objects of scientific analysis. Since sexism, racism, imperialism, and bourgeois beliefs have been among the most powerful influences on the production of false scientific belief, critical examination of these causes, too, of the "results of research" should be considered to be *inside* the natural sciences. We could say that the natural sciences should be considered to be embedded in the social sciences because everything scientists do or think is part of the social world. (98–99)

The idea that science can be driven by discriminatory practices and is a socially or economically motivated enterprise is a far cry from saying that the make-up of the earth's atmosphere is the way it is because white men say so. One might question the intellectual honesty (or the interpretive acumen) of scientists who, like Sokal, Kimball, Bricmont et al., have either read what they wanted to in texts like Harding's or haven't read them at all. Either way, it deals damage to the ethos of a scientist who cannot be bothered to discover the authentic message of those who challenge scientific epistemology but only "interpret" such a message in a way that forwards their own intellectual agenda. This alone helps demonstrate Harding's point; we cannot have blind faith in the scientific method or a scientistic epistemological model when they are applied by people in a position of authority who are highly motivated to maintain that authority (as well as being motivated by all other possible human impulses and appetites). This is what detractors disparagingly call the "social constructivist" view of science—the idea that science, like anything else, is practiced by people within social contexts. This is far from a radical view. It is at least as old and as accepted as Thomas Kuhn, but could historically be traced back even further, to any era in the past—the Civil War, the Industrial Revolution, the First World War, for example—when the impact of science on society was questioned and, consequently, so was the social element of science itself. Such glib dismissals as Park's lead to a troubling question: Do such scientists practice more objectivity in their science than they do in their textual analysis?[1]

The message Park forwards is strangely paradoxical: on the one hand, he suggests that claims against science are the result of mere ignorance, a simple inability to understand what science is and how it is conducted; on the other, he does not seem to be advocating further education which, ironically, might give one the instrument with which to examine science as a social force. Park's words imply that one must be a scientist in order to critique the outcomes of science, and any other intellectual vantage point is irrelevant, invalid, and wrongheaded (or just a matter of professional sour grapes). This is tantamount to saying that one must be a dentist in order to know that one has a toothache. Park's message is that the effects that science has on people have no bearing on whether people can then interpret and, if warranted, critique those effects. The suggestion that to scrutinize the effects of science on society is somehow to throw out the scientific baby with the intellectual bathwater is silly: one does not need to know thermonuclear physics to know that the hydrogen bomb is a dangerous thing and, more importantly, to see that its development and potential use has far-reaching effects on social and political policy and the general tenor of popular consciousness. As a matter of fact, science is probably not the best institution to critically self-correct in these matters—we require a perspective based on judgment and knowledge of human agencies and language. One has the right, even the obligation, to study the effects of science without necessarily knowing all the technical minutiae that brought about those effects.

But Park's polemic stance is important in maintaining scientific dominance, and one that is typical of powerful cultural paradigms. To ridicule detractors is a common rhetorical move to discredit those who might question certain aspects of the hegemonic elite, and to over-simplify counterpoints so that they can be easily destroyed is to commit the fallacy of the straw man. The fact that science stands at the gate to discourage "outsider" criticism speaks to its privileged status. It is also important to clarify the point that Park is not entirely mistaken; there are those who foolishly object to science on grounds that are less than compelling, positing bizarre alternatives that are, to adopt a phrase used by both Sagan and Park, "anti-science." The reality, social or otherwise, is that the circumstances and the methodologies propelling the cultural study of science are far more complicated (and real) than Park paints them. His examples of anti-science, which range from a Harvard-educated particle theorist who claimed to be able to reduce violent crime through the use of transcendental meditation to the Unabomber (153), are trite dismissals that no one, scientist or layperson, should take seriously. That he uses them as examples of potential detractors, effectively placing Sandra Harding in the same category as Theodore Kaczynski, demonstrates that he himself has a woeful (or, perhaps, deliberate) misunderstanding of the cultural and rhetorical studies behind authentic and valid criticisms of science.

So while Park is not entirely off-base, he does, like the very critics he hopes to expose, oversimplify the issue. His book (of the same name as the article cited above), *Voodoo Science: The Road from Foolishness to Fraud,* is an anecdotal account of the failure to recognize "junk science" from the genuine article. The effort to expose bad science is, of course, a valuable pursuit; it is important for conscientious scientists to identify when, how, and why pseudo-science is being perpetuated for anti-scientific motives. Park is to scientific purism what Michael Moore is to political expose—he provides a valuable service by showing possible interpretive alternatives, and loudly declares (when the need arises) that the emperor has no clothes. Like Moore, he hopes to represent the "truth," but also like Moore, he has a position to defend, and this requires a rhetorical stance. He is selective in his representation because to narratively reveal "the whole truth" would surely weaken his argument. For example, in the introduction to *Voodoo Science,* Park contends that, "[o]f the major problems confronting society—problems involving the environment, national security, health, and the economy—there are few that can be sensibly addressed without input from science" (viii). He excludes, conveniently, any discussion of science's complicity in the *creation* of these problems. While it would be unfair to suggest that science is to "blame" for the pollution problems endemic to American society, it would be equally unfair to suggest that science has in no way contributed to them. Technology, modern industry, and corporate science have all propelled pollution levels because much of what practical science produces is geared toward widespread industrial use. That is one of its chief appeals. But just as guilty are those of us who misuse or overuse the technology at our disposal. We have only recently begun, as a nation, to understand the truly finite nature of our most coveted resources. We are just starting to realize that the environment which we have taken for granted is far more fragile that we once believed.

But also figured into this equation is the culture of waste that is encouraged in order to be good consumers. We closely guard our "right" to convenience, and disposability is a key element in protecting the freedom this provides. Capitalistic, consumer culture is based primarily two things: waste and greed. Put more euphemistically by the corporate super-structure, we might describe these as "choice" and "competition." The semantic distinction, however, results in the same thing: The U.S. consumes well more of the world's natural resources than the rest of the globe taken together. "Shopping" is, by many families, considered a kind of bourgeois art form; it is a demonstration of buying power, it shows one's sense of "taste," it protects the security of one's class stratum, and it is seen as a patriotic duty.[2] We throw away as much as we keep, and this cycle creates an ongoing strain that has been aided by the capitalistic doctrine of greed that suggests the more one has, the better one is, personally, socially, nationalistically, and morally. We cannot place the responsibility for this cavalier, egoistic attitude on the doorstep of science, even if it has helped make all of it possible. It is a co-complicity between industry, science, technology, corporate competition, and the general public. Howard Gardner, Hobbs Professor of Cognition and Education at the Harvard Graduate School of Education, reminds us that "the ways in which Americans evaluate individuals whom they know from a distance and even those whom they know personally occur increasingly in market terms" (99). We all have something to gain, but we also have something to lose. In the end, we feel we must turn to science in order to define and solve the problems we have made together.

Park is also correct when he points out that "[. . .] many people choose scientific beliefs the same way they choose to be Methodists, or Democrats, or Chicago Cubs fans. They judge science by how well it agrees with the way they want the world to be" (ix). Perhaps one of the most overlooked aspects of the public's affinity for science is the notion that it is considered a "belief" like a religion or a political stance. The very foundation of science, as a methodology, however, is anti-belief: The data or evidence exists to support a conclusion so that we don't *have* to believe. I am often asked (especially by those of a religious orientation) whether I "believe" in evolution. My usual reply is that the beauty of evolution is that I don't have to believe in it; it is sufficiently established as a natural phenomenon that it usurps faith. Most of the time, this response elicits either bewilderment or indignation. There is the sense that any conviction worth holding must involve some degree of blind acceptance—belief—in order to be valid (a strange notion, since the "validity" of a belief system is thus based on a *lack* of evidence). The questioner requires that science hold itself up to the standards of religion, not the other way around. It is much easier to maintain a kind of intellectual sanctimony as long as one embraces a faith because having faith admits to something beyond human understanding. In this respect, not only is the scientist atheistic if she "believes" in evolution, but also she is amoral as well. She cannot possibly have a set of moral convictions, the argument goes, if she relies on her own mind and her own sense of innate goodness to arrive at her moral position. This strange paradox, one where we live in a practical society that at the same time expects anyone of character to "believe" in something, is one of the reasons that science has adopted the unusual façade that Park laments.

Park rightfully asks, "Did we set people up for this? In our eagerness to share the excitement of discovery, have scientists conveyed the message that the universe is so strange that anything is possible?" (ix). To those for whom science is a mysterious oddity, an esoteric brand of magic practiced by people with multiple PhD's who often condescend to the public or ignore them altogether, the answer is an emphatic "yes." Science is mysterious to many because its findings are frequently so counter-intuitive. Common sense, ironically, has little place in most of the great scientific discoveries. One need only attempt to imagine the age of the earth (estimated at some five billion years) or the distance of the *nearest* galaxy (some 2.5 million light-years) or the speed of light needed to measure such distances (186,000 miles per *second*) to see how it is entirely possible that the uninitiated would view science as a magical, even divine, process. It goes against the human experience to visualize such an expanse of space or time to truly consider such numbers as "facts." How can the light I'm seeing through the telescope from the Ring-Tail Galaxy *really* be 70 million years old? That means that many of the stars that contribute to the glow from that galaxy no longer exist! It means I'm looking at something that is no longer there! It means that I'm looking backward in time! Such ideas are beyond immediate sensory observation. One cannot "see" something move at the speed of light in the way that one sees a train speed past while standing next to the tracks. One cannot experience enough time to understand how *long* a billion years is. There is simply no point of reference in human sensory experience or in the human capacity to experience space and time to have the concept make empirical sense. Scientists try to put these concepts into graspable terms, but what they are really asking people to do, at least insofar as the general population interprets the process, *is* to believe. Believe me when I say that the earth used to be populated by alien-like creatures living on continents that are entirely different from what they are now. Believe me when I tell you that microscopic organic molecules contain the "blueprint" for who and what you are. Believe me when I interpret rocks and can tell you how old they really are and, more than that, can extrapolate from that the possible composition of the universe. It is no surprise to me, when considered in this way, that people ask me whether I "believe" in evolution. Creationism, by contrast, is perhaps a better expression of Occam's Razor, even if it isn't true.

Some examples from Park's own work will help illustrate how people are easily taken in by claims that have a scientific exterior but not a scientific basis. It is important to mention here how scientific pundits, who often lack scientific credentials, are given an audience for their claims through the conduit of the mass media. Park himself provides clues for why the media phenomenon of sensationalized science happens, yet he does not go far enough in explaining the reasons behind the issue of reporting bad science. Take, for example, the case of Joe Newman's "invention" of the first "true" perpetual motion machine. The perpetual motion machine, long a pursuit of inventors and second only to alchemy in its mythic impossibility, is a device that actually delivers more energy than it uses. Such a device, as Park points out, runs thoroughly counter the first law of thermodynamics: a mechanical system cannot conserve more energy than it uses (6). Any device that requires energy to operate (and a machine, by definition, requires energy) will lose more energy than it produces. The appeal of a perpetual motion machine is that it would, if physically possible, mark the end of the energy crisis since it could theoretically run forever without needing additional fuel. The applications would be unlimited. Apparently, Joe Newman had solved the problem of the first law of thermodynamics (insofar as a "law" can be considered a problem to be solved) and was more than happy to share his findings with the public via demonstrations, media interviews, and other high-profile stunts. In demonstrations that hearken back to the magic elixir and miracle tonic days of the nineteenth century, Newman staged elaborate exhibitions of his new device to the awe and wonder of large audiences.

What is interesting about Joe Newman is that he was not, in any formal sense of the word, a scientist. He had little more than a high school diploma, and in the words of Park, "[t]he prospect of someone with little education and no scientific record of accomplishment overturning the most basic laws of physics, laws that have withstood every challenge, seemed much too unlikely to bother with" (8). If the scientific community chooses not to "bother" with correcting such errors and misapplications of science, it should not then be surprised to find that the general public, which has no theoretical understanding of thermodynamics and no elite bias about the academic track record of scientists, should be duped by the efforts of a sincere, if misguided (and self-appointed) representative of science. Park wonders aloud why scientists themselves don't do more to challenge voodoo science. He says:

> The reluctance of scientists to publicly confront voodoo science is vexing. While forever bemoaning general scientific illiteracy, scientists suddenly turn shy when given an opportunity to help educate the public by exposing some preposterous claim. If they comment at all, their words are often so burdened with qualifiers that it appears that nothing can ever be known for sure. This timidity stems in part from an understandable fear of being seen as intolerant to new ideas. *It also comes from a feeling that public airing of scientific disputes somehow reflects badly on science.* The result is

> the public is denied a look at the process by which new scientific ideas gain acceptance. (27, emphasis added)

Park's admission that disputes reflect badly on science (presumably a fear of scientists themselves) is of particular interest here. In its quest for certainty, the American public might very well see scientific debates as weakening science as a whole. It is a failure of scientific education to project the impression that science is always certain, that there are no detractors of any given idea in the scientific community, or that science's main objective is to find and present an unshakable Objective Ultimate and Immutable Truth. Such a stance feeds into the notion that science is infallible, and scientists themselves must enjoy that reputation, no matter how erroneous, or they would feel more secure about sharing with the masses the reality of scientific disagreement. The public desire for certainty, as part of the scientific ethos that fuels the public adulation of science and its practitioners, propels an odd anxiety among scientists that suggests that if people really knew how much disparity flourished in the scientific community, they might lose trust ("faith," if you like) in science altogether.

Interestingly, both Newman and genuine scientists conspire, as it were, to create the atmosphere of reverence that science has on the general public. Though Park is correct in his assessment that bad science is almost as dangerous (perhaps more so) than no science, he neglects the implicit message that the public—through its devotion to the quacks as well as the Nobel Prize Winners—is sending about its interest in (and its veneration of) science in all its myriad forms. While scientific dabblers like Newman do little to emphasize the tedium and disappointment that is a large part of conducting science, they do manage to capture the awe and admiration that is innate in public questions about science and its promise for the future of humanity. From a rhetorical point of view, people like Newman do help maintain the primacy that science has enjoyed because he is able to guide the imagination of the public in a way that implies all science is important, even if it is not "real" or "true" science. He may in fact be riding in on the coattails of genuine science, but he is also functioning as an ideological custodian, reminding the public that science can be exciting, promising, important, ground-breaking, and accessible. Is he irresponsible for making claims that have not been validated by the scientific community proper? Only if he knows that he's wrong and foists bad science on the public anyway. There is no indication, in the narrative provided by Park, that this is the case. The failure of recognition is perhaps more the fault of the scientific community itself than it is an insidious intent on the part of Newman. And as Park himself admits, even "real" scientists fall prey to the temptations of hopeful new, though theoretically unlikely, discoveries like cold fusion because they crave the excitement and celebrity (not to mention the implications for knowledge) that such discoveries promise (27).

So even though Park himself asks the question, perhaps his conceptual error is his underestimation of the idea that science itself has contributed greatly to the very anti-science that he objects to. In order for a population to be bamboozled (a favorite word of science purists) by *bad* science, they must have a faith in the capacities of "real" science first. Park, however, sidesteps this by claiming that people, in general, are "greatly distrustful of science" (*Voodoo Science* 150), while failing to acknowledge that, on the contrary, the vast majority of the American population has great conviction in the ability of science to address any issue that needs to be addressed. It is, perhaps, a symptom of America's general intellectual laziness that affords science its privileged status in the first place. Without an agency willing to step in and fill the mental void, original thinking might very well come to a standstill. This is due in part to the lifestyle that Americans have become accustomed to; what is valued is material, and the way materialism is satisfied, as I discussed in Chapter One, is to create opportunities for making money. The cycle, one where we work to pay for the conveniences, status symbols, and contributing means to further wealth, has a number of interesting, if upsetting, side-effects. Despite the "ease" of living in the postmodern age—that is, the products of our labor are generally no longer governed by the need to merely survive—we seem to have little time to actually enjoy the fruits of what we do acquire or to intellectually improve ourselves in any significant way.

Gilbert T. Sewell, in his inappropriately titled essay, "The Postmodern Schoolhouse," offers some possible explanations for the problem of how to educate our children in science or anything else by declaring most public schools as forwarding " . . . academic programs [that] are flaccid, where moral education is a hodgepodge of relativism and radical individualism" (57–58). He also notes that schools require little work from their students, grades become inflated, and "[. . .] a laissez-faire attitude toward dress and courtesy is expected" (58). Sewell, yet another example of someone thoroughly misguided in his use of the term "postmodern," does have some important points, the first of which is the conclusion that school curricula have become thin on content and academic preparation and heavy on a variety of politically motivated teachings that have less to do with historical and scientific "fact" and more to do with power redistribution—that is, concentrating on the peoples and events that have been underrepresented in the past so that an increasingly "fair" and "complete" picture of history and its implications can be painted. Laudable as this goal may seem, it has also been vulnerable to abuses, especially from those who would use the doctrines of diversity and revisionism for their own professional or political ends. Science can and does get caught up in the politics of teaching, as we see in the evolution/creationism debate, but aptitude in academic areas can also suffer in general from de-emphasis on the merits of "thinking critically" and of logic and of the basic rules of argument. One might reasonably ask how it is that these skills suffer in a climate of tolerance and diversity. The answer is, they needn't, but they often do because (though not a postmodern doctrine by any means) the idea that every opinion is as valuable as every other is the decades-long paradox behind the so-called "self-esteem" movement that many feel has been the bane of education since at least the 1970s. My main point here, however, is not whether self-esteem, diversity, and tolerance are the "cause" of "lower standards," but whether we are in the midst of an educational antithesis that is having difficulty making meaningful change when it comes to the educational needs of the country. Certainly, students are not receiving the same rigorous instruction in many of the basic criteria that once defined critical discourse in the past, and this may contribute to an overall reduction in the standards that once existed, but are these standards still appropriate, and if so, how can we balance the needs of the present with the conservative expectations of the past?

And what, if anything, does Sewell add to the debate besides more division, more suspicion, more intolerance, and an air of righteous indignation? Is the old standard that has been challenged above reproach? Does education stagnate in favor of a "traditional" reading of history that never reexamines its own mythic precepts? Sewell, speaking of new academic standards proposed by the National Alliance of Black School Educators, claims that the proposal

> [. . .] reinvented the European discovery of the New World, changing a once triumphal Columbian con-

> quest into a three-way "encounter" of Europeans, Africans, and Native Americans. From the beginning, disease-carrying Europeans encounter and enslave innocent people of color. Older paradigms of federalism, industrialism, and expansionism were minimized, along with heroic figures and their achievements. Hamilton and Jefferson, the Erie Canal, Gettysburg, and Promontory Point did not exactly vanish, but they were not much savored either. Teachers and students inherited a solemn, often bitter chronicle of unfulfilled national promise. Historical sufferers and victim groups receive belated recognition and redress. Participation in history becomes an empathetic act. By sharing the pain of exploited groups and learning the gloomy "truth" of the U.S. past, students presumably learn to become more virtuous and sensitive. (59)

The word "reinvented" suggests that the proposal is based on a mere whim, as if the authors simply manufactured an historical narrative out of thin air, based on nothing beyond a political agenda. In fact, the best "revisionist" history is that which is grounded in repressed texts, and what makes revisionism a necessary component of scholarly correction is the fact that many texts—either suppressed because they did not match the glorified fantasy of traditional discovery narratives or because they were ignored altogether—add significant details to the corpus of historical documentation to give us a fuller picture of the events of the past. Well-executed revisionism is more honest, more in line with the truth of the past than history written to deify the imperialistic conquerors of the Renaissance. Furthermore, Sewell's contemptuous mockery of learning history as "an empathetic act" is especially inflammatory here. How does one really understand history without an act of empathy, without the imaginative ability to put oneself in the time and place of others, to really try to transport oneself to another time, to "be there" to feel what it must have been like? No wonder history is often viewed by students as nothing more than a stupefying stream of meaningless dates and names. No wonder they see it as irrelevant, or worse, a colossal waste of time. No wonder they are turned off and are unable to see the correspondence of historical events to the conditions in which they themselves live. No wonder they are doomed, like their fathers, to repeat the mistakes of the past.

If we look beyond Sewell's abject crankiness, we further discover that he is simply wrong about the trajectory of history education. According to James W. Loewen in *Lies My Teacher Told Me,* not only are traditional historical narratives alive and well in high school history textbooks, but they are generously salted with jingoistic, patriotic, and optimistic ideologies about the grandness of the American experience. According to Loewen:

> [High school history textbooks] portray the past as a simple-minded morality play. "Be a good citizen" is the message that textbooks extract from the past. "You have a proud heritage. Be all that you can be. After all, look at what the United States has accomplished." While there is nothing wrong with optimism, it can become something of a burden for students of color, children of working-class parents, girls who notice the dearth of female historical figures, or members of any group that has not achieved socioeconomic success. The optimistic approach prevents any understanding of failure other than blaming the victim. No wonder children of color are alienated. Even for male children from affluent white families, bland optimism gets pretty boring after eight hundred pages. (3)

Contrary to Sewell's contention that Columbus has been vilified by modern historical educators, according to Loewen's analysis of the twelve most popular high school history textbooks of the 1990s, the glorification of Columbus's "discovery" is consistent throughout all of them, sometimes even recycling the same passages in different texts. The accounts all center on the "courage," "daring," "spirit of adventure," and "eye for opportunities" that Columbus apparently possessed, and the level of scholarship informing these extravagant claims is woefully thin. In fact, many claims are simply unverifiable, as in the case of one textbook that describes the climate of Europe prior to the Age of Exploration as "stirring with new ideas" and "burning with curiosity" (31). Rather than assess this apparent trend in any scholarly way, the textbook writers, says Loewen,

> provide no real causal explanations for the age of European conquest. Instead, they argue for Europe's greatness in transparently psychological terms—"people

> grew more curious." Such arguments make sociologists smile: we know that nobody measured the curiosity level in Spain in 1492 or can with authority compare it to the curiosity level in, say, Norway or Iceland in 1005. (31)

If we think about the basis of Sewell's objection in contrast to Loewen's findings, an especially interesting flaw in Sewell's argument bubbles to the surface: His denouncement of modern educational practices involving history, which one would assume is based on problems of "scientific" verifiability, methodology, or compromised "facts," is in fact based on something far more subjective, namely, Sewell's own nationalistic pride in the myth of Columbus and his indignation about the impiety of revisionists who might dare to challenge the superficial reading of history represented in textbooks. In other words, he attacks the "revisionist" for doing the very thing that a good scientist should do: verify that the data is accurate and correct that which cannot be empirically (or even textually) authenticated. His willingness to cast aside a detached scientific stance in favor of a mythologized nationalism demonstrates that scientists are hardly above the subjective impulses that affect the rest of us. He relies, moreover, on his inherited ethos as a scientist to establish credibility instead of on his methodological acumen. He can persuade, that is, by virtue of *being* a scientist instead of actually *doing* science.

The language Sewell uses to describe efforts to correct this oversight is dismissive and insulting, but it does hearken back to the notion that truth and certainty are static conditions that we must seek and embrace once we ostensibly "find" them. Is he denying that Europeans did bring diseases to the "New World" (a very old world for those already indigenous to the Americas) that nearly wiped out entire civilizations? Is he saying that the doctrines of federalism, industrialism, and expansionism are goods in themselves that need not be critiqued? "Expansionism" is, incidentally, merely a euphemism for less savory synonyms like "conquest," "colonialism," and "imperialism"—jealously guarded Western ideals that many see as an American birthright. He, in fact, unapologetically calls the "Columbian" invasion a "conquest." Is he ignoring the flaws of our forefathers, the flaws that made them, indeed, more human? Can he seriously have difficulty believing that certain historical details were suppressed in favor of more complimentary accounts of our ancestors' deeds? Sewell's representation of proposals for curricular change give one the impression that, despite the shortcomings of the proposal or the over-zealousness of its advocates, such reexaminations were in fact needed. But have revisionists gone too far? Perhaps. Were they political motivated? Often. Were they important for reawakening the complacent slumber marked by centuries of denial? Certainly. But the impact of new "standards" and "preparation" did have some unwanted effects, at least in the short term, and perhaps the biggest casualty wasn't content as much as it was methodology. Knowing that schools employ real people with real ideas and their own wills, it is clear that an inappropriately ambitious teacher might begin overstepping his or her bounds when teaching. However, this can be countered with a greater emphasis on the *process* of learning over the mere "mastery" of material. Hence the intersection between the lack of scientific education in particular as it relates to the lack of inquiry-based education in general. Sewell is bemoaning the details of historical content when he should be concerned with the methods students are using to learn in the first place. Are all "traditional" (read "conservative") accounts accurate and all "revisionist" (read "liberal") accounts false? How can we tell unless we have the critical apparatus necessary to distinguish between the valid and the embellished? Likewise, how can we ever hope that our students will be able to tell the difference if we don't impress upon them the need for careful analysis, a weighing of the existing evidence, and a sense of humanistic judgment? Moreover, what Sewell seems to be arguing for is a literalist interpretation of the past, the strange notion that history can and should be practiced as some form of social science. At the same time, however, he undermines the very methodology that would help makes this elusive task possible: the questioning of received knowledge and the reassessment of old data. The stridency of his objections echoes those of Park, as if "revisionists" were denying some natural law rather than a set of fragmentary, remote, and selective texts that have been brought together and interpreted by a handful of historians. He argues for an absolutist version of history, much as one would argue for the immutability of a physical reality, and he assumes that the existing data are incontrovertible. The problem, of course, is that history is not a science, not even a social one. The reason is that it deals with texts, and dealing with texts means dealing with language, which means dealing with rhetoric. Texts are by their nature isolated, shards of another person's interpretation of events in another time. Revisionists are simply trying to provide more complete pictures, to include texts and interpretation that have been, for whatever reason, repressed, undervalued, unavailable, or ignored. Moreover, the irony in Sewell's position is that, as someone arguing for a scientistic methodology for providing conclusive readings of historical "fact," he is ignoring some pretty compelling evidence that suggests not everything was as clear cut and wonderful as we were taught in high school. In other words, as a scientist, he should be the *first* to readjust his hypothesis in light of new evidence.

Other critics of the contemporary educational and social situation, however, have some very real, very valid points, conservative or not. In the preface to Ken Kalfus's editorial, "Last Night at the Planetarium," editors Katherine Washburn and John F. Thornton note the poignancy of the image of a planetarium in disrepair, seeing it as a symbol of our lost sense of wonder, of our willingness to allow media forces, not independent learning and personal inquiry, to have a monopoly on our imagination, of a generation that is willing not only to let others think for it, but also to dream for it. These criticisms are valid because they point to the same disturbing anti-intellectual trend that views education as a means to a vocational end and a generation that is jaded by older principles like honesty and integrity. To Washburn and Thornton,

> [w]e are forcibly exposed [in Kalfus's essay] to more than the dimming down of the cosmos, to the rupture of a long discourse with scientific knowledge, and to the by-now routine boutiquification of a great museum. The substitution of science fiction for science, instant fun for deep pleasure, is a phenomenon which becomes a symptom of a wider malaise. We have replaced the childlike with the childish, we renovate without first carefully restoring and maintaining, and perhaps, as we extend multiculturalism into dreams of extraterrestrial life, we'll even discover space travel is narrowing. (139)

These nostalgic musings, the symbolism of the planetarium itself, and the observation that we have become a disposable society all have a ring of truth for those of us who did our growing up even as recently as twenty-five years ago. To give a personal example, my neighbor, a fairly young man who ostensibly possesses a college degree, came over to my yard to do some star-gazing through my modest, kit-built 6" Newtonian reflector telescope. I started the session with some simple objects, the moon and Jupiter. While the usual questions surfaced during the session—how great was the magnification of my telescope, where was the Big Dipper, etc.—I was flabbergasted when he asked me, with all seriousness, which of these objects, the moon or Jupiter, was closer to the earth. Could a man who actually had some education really not know the answer to that question? Could he really believe that it was possible for a planet as massive as Jupiter to actually be *closer* to us than the moon? Did he not understand the relative composition and orbital distinctions between these two celestial objects, information that would be covered in any course on the natural sciences, certainly in any course in astronomy? It seemed impossible, but he didn't know. Kalfus himself frames the matter this way:

> When Mark Twain's Connecticut Yankee, a character possibly based on [*Scientific American* founder Rufus] Porter, found himself in King Arthur's court, his practical knowledge of science made him a wizard even greater than Merlin. With native ingenuity and an awareness of scientific principles, he reinvented (or pre-invented) the telegraph, the revolver, and other nineteenth-century artifacts. But if he had been transported instead to the late twentieth century, would his wizardry have been any less remarkable? If a twentieth-century Connecticut suburbanite found himself in the Dark Ages, would he be able to reinvent anything more complicated than a gin and tonic? Less hyperbolically, can a conventionally educated adult American of the twentieth century claim to know more about science than a conventionally educated adult American in the nineteenth century? (143)

(For the record, my neighbor's favorite drink is gin and tonic). Kalfus asks an important question here because he is implying that the modern American adult has been educated into a regressive state, that a late-twentieth century education, despite our technological advances and greater overall knowledge, is actually less impressive, less complete, less broadly applicable, than an education of 150 years ago. Is this true? If so, how can it be?

Part of the explanation is the way popular images of space, space travel, and extraterrestrial life are portrayed in movies, on television, and by the popularization of space exploration. Movies and TV, of course, provide the most unreliable picture of humanity's eventual confrontation with space. From *Star Trek* on, the notion of routine contact and interaction with alien life has completely transformed the motives behind why people are interested in space and space exploration. The "Are we alone?" question seems to be a mass preoccupation with the American public (thanks largely to television programs like *The X-Files* and movies like *Contact*), and the quest for an answer to this question is even validated through legitimate scientific groups like the Search for Extraterrestrial Intelligence Institute (SETI), an organization of scientists who have been scouring the cosmos with radio telescopes since 1984 and whose self-professed purpose is to "explore, understand and explain the origin, nature and prevalence of life in the universe." Oddly, the segment of space that gets the most attention in regard to its promise for extraterrestrial life is Mars, which, if it ever hosted life of any sort was probably of the most rudimentary kind. There are many sound political and professional reasons for the interest in Mars; it has been a target of NASA for a number of years, and this creates a political motivation for painting Mars as a likely candidate for life, even if the existing evidence supporting this contention is slim to none.

NASA, like any other governmental agency, must deal with its public relations image in order to garner support from its public constituency. While traditional wisdom dictates that public opinion does not have a direct effect on NASA, certainly the men and women in office and in a position to grant funding for space exploration must consider public priorities when allocating money to unpopular projects. Much of the nation feels that space exploration is, in fact, a poor use of tax revenue, citing more pressing domestic concerns as a higher priority. Writing in 2006, David Livingston, science journalist, host of *The Space Show* and professor of Space Studies at the University of North Dakota, describes the difficulty of making an argument for space exploration in his Apogee Space Books Series article, "Space as a Popular National Goal." Typically, he argues, the reasons given to the general public for exploring space are far too lofty to be taken seriously by a literalist material culture. Going to Mars to "save humanity" is not, in other words, nearly as compelling an argument as going to Mars to exploit its resources. Yet, interestingly, a rhetorically shrewd advocate for space exploration can have it both ways. A guest on Livingston's radio talk show (he does not provide the name of the guest, though he was apparently a member of the space exploration community) argued in a listener Q&A session that

> Mars may very well prove to be a productive place for the discovery of new medicines and pharmaceuticals. He [the guest on the show] said that it might be possible to use what is unique on Mars to significantly improve our health and well-being on Earth. [. . .] This guest then logically added that sickness and disease reduce economic productivity. For a fraction of what is cumulatively spent on these conditions on Earth, if we as a nation would invest some of our wealth in going to Mars, we might well create a national and even global economic boost, let alone realize significant benefits for all mankind. (85)

While academic questions regarding the fate of humanity can be left to philosophers and clerics, it seems, a gain in economic productivity is the strongest motivator for the American citizen to devote some tax dollars to space exploration. Accordingly, Livingston concludes, the real obstacles to dedicating resources to space exploration are not technological or logistical, but rhetorical. The right arguments (i.e., those that foster widespread public support and have a tangible return on a material investment) are what speak to the 21st Century American taxpayer, not abstract trivialities like the future of the human species. Given this atmosphere, NASA must tap into the imaginative possibilities that rest in the symbolic power of planetary neighbors like the Red Planet. More than that, they must convey an image of being on the cutting edge of science and technology. Consider, for example, the following list of proposals to NASA for the 2005 Phase 1 awards and the people who are presenting them:

1. A Contamination-Free Ultrahigh-Precision Formation Flight Method Based on Intracavity Photon Thrusters and Tethers

(Principle Investigator: Dr. Young K. Bae, Bae Institute, Tustin, Calif.)

2. Extraction of Antiparticles Concentrated in Planetary Magnetic Fields (PI: Mr. James Bickford, Draper Laboratory, Cambridge, Mass.)
3. Customizable, Reprogrammable, Food Preparation, Production and Invention System (PI: Dr. Eric Bonabeau, Icosystem Corporation, Cambridge, Mass.)
4. Scalable Flat-Panel Nano-Particle MEMS/NEMS Propulsion Technology for Space Exploration in the 21st Century (PI: Dr. Brian Gilchrist, University of Michigan, Ann Arbor, Mich.)
5. Antimatter Harvesting in Space (PI: Dr. Gerald P. Jackson, Hbar Technologies, LLC, Chicago)
6. Multi-MICE: A Network of Interactive Nuclear Cryoprobes to Explore Ice Sheets on Mars and Europa (PI: Dr. George Maise, Plus Ultra Technologies, Inc., Stony Brook, N.Y.)
7. Artificial Neural Membrane Flapping Wing (PI: Dr. Pamela A. Menges, Aerospace Research Systems, Inc., Cincinnati)
8. Lorentz-Actuated Orbits: Electrodynamic Propulsion without a Tether (PI: Dr. Mason Peck, Cornell University College of Engineering, Ithaca, N.Y.)
9. Magnetically Inflated Cable System for Space Applications (PI: Dr. James Powell, Plus Ultra Technologies, Stony Brook, N.Y.)
10. Ultra-High Resolution Fourier Transform X-ray Interferometer (PI: Dr. Herbert Schnopper, Smithsonian Astrophysical Observatory, Cambridge, Mass.)
11. Positron-Propelled and Powered Space Transport Vehicle for Planetary Missions (PI: Dr. Gerald A. Smith, Positronics Research LLC, Santa Fe, N.M.)
12. Modular Spacecraft with Integrated Structural Electrodynamic Propulsion (PI: Mr. Nestor Voronka, Tethers Unlimited, Bothell, Wash.). (National Aeronautics and Space Administration)

These proposals read like something out of science fiction, for most people don't realize that physical phenomena like "anti-matter" are real except as they are mentioned as part of the theoretical explanation for warp drive in *Star Trek* episodes. Also, following such mission disasters as Challenger and Columbia, it is important for NASA to project an image of both competence and innovation. These proposals lend a degree of credibility to the administration while concomitantly showing the ultra-high tech nature of the space exploration game. The "Positron-Propelled and Powered Space Transport Vehicle for Planetary Missions," for instance, if accepted, would be a direct contributor to future projects like a mission to Mars. Research and development for such a transport, however, would take years and cost billions. Space exploration is not only high-tech, it's high-finance, and since space programs are still largely the purview of the federal government (as opposed to private companies), without the full support of the American public, politicians in a position to grant such huge sums of money are likely to be very cautious. Such is the nature of a democracy; tax-payers do have a say, no matter how indirectly, in the ways that their money is spent, and this is largely accomplished through voting into office those who will guard a constituency's interests (in theory, at least).

Writing in 1980, in his book, *Cosmos,* Carl Sagan outlined a number of interstellar spacecraft designs that were apparently on the drawing table at the time. Sagan, as perhaps the most recognizable figure in the popularization of science, made travel between the stars seem not only possible, but likely. His wishful thinking may reveal his professional bias as a planetary astronomer, but the effect of his prediction is hard to underestimate. In the chapter titled "Travels in Space and Time," Sagan promises deep space exploration through the use of three main starship designs: the Orion Nuclear Starship, the Daedalus Nuclear Fusion Starship, and the Bussard Fusion Ramjet (204–205). While there have certainly been new technological developments since the time Sagan wrote about these ship designs in 1980, the important result is the renewed interest Sagan's predictions generated in the public. Here, a well-respected scientist in a relevant field was endorsing—and providing details—about the likelihood of interplanetary and even interstellar space travel. His book, coupled with the PBS television program of the same name, helped to validate space travel as a project that was both feasible and desirable. Nearly everyone has at least a passing interest in the topic of space since it represents the uncharted territory of the "final frontier" and it may hold the answer to questions about our relevance in the universe and our purpose on this planet. Sagan was able to provide hope for the future of interstellar travel while maintaining the signature cautionary guidance characteristic of the scientist. His message, in short, was, yes, interstellar travel is theoretically possible, but there are some serious obstacles to overcome. The most pesky of these problems is the Theory of Relativity, which posits the interconnectedness between time and space, a complication which presents practical problems for interstellar travel from a research perspective. Space travelers voyaging to distant stars are restricted in both travel time and communication by the speed of light. This means that while the life span of the crew would proceed normally as it would by standard time-keeping methods, the time passage back on Earth could be millions of years. Sagan provides this example:

> A nearby mission objective, a sun that may have planets, is Barnard's Star, about six light-years away. It could be reached in about eight years as measured by clocks aboard the ship; the center of the Milky Way, in twenty-one years; M31, the Andromeda Galaxy, in twenty-eight years. Of course, people left behind on Earth would see things differently. Instead of twenty-one years to the center of the galaxy, they would measure an elapsed time of 30,000 years. When we got home, few of our friends would be left to greet us. In principle, such a journey, mounting the decimal points ever closer to the speed of light, would even permit us to circumnavigate the known universe in some fifty-six years ship time. We would return tens of billions of years in our future—to find the Earth a charred cinder and the Sun dead. Relativistic space flight makes the universe accessible to advanced civilizations, but only to those who go on the journey. There seems to be no way for information to travel back to those left behind any faster than the speed of light. (207)

What Sagan has managed to do here rhetorically is to suggest that while interstellar and intergalactic space travel has technical plausibility, it is practically prohibitive. This implies that we should set our sights on objectives that have research value, on destinations for which we can actually transmit home useful scientific information. He has primed the public imagination with the wonders of our own technological know-how, but then made it clear that the very reasons for traveling to distant star systems would be contraindicated by theoretical physics: we could gather information but we could not report on what we find. The rationale breaks down because there would be no return on our investment; we could not relay the information that would be the fruit of our efforts. Beyond that, however, he has given his readers a dose of the breadth and depth of the scientific challenges that confront such projects as interstellar space travel in stark and uncompromising terms. This contributes to the ethos of the scientist, oddly enough, by suggesting that dealing with the obstacles of interstellar travel are routine ones for the serious scientist. Technology, as always, is only one factor in the equation; the rest involves the details of the imaginative researchers to sort out, problems that a good scientist would welcome as part of the great scientific puzzle.

Given such seemingly insurmountable problems, the solution a la Sagan is much more modest, however; we must readjust our goals to something more scientifically justifiable. But what are some good reasons for, say, a mission to Mars? People in general, as I have already alluded to, want a more practical outcome than something as abstract as "scientific discovery"—there must be a payoff for such a huge venture. In the 1960s, this was accomplished through an ideological means. "Beating the Russians" to the moon was, in the aftermath of the Sputnik scare, a top nationalistic priority, and one that presidents like John F. Kennedy and Lyndon B. Johnson used to their advantage. Not only did the moon mission serve to unify the whole country, it gave our politicians (and by extension, the scientists they hired) economic and procedural *carte blanche* to do research and development that would have taken decades without it. Nationalistic identity is a strong rhetorical motivator, and given the pride most Americans felt about their expeditionary past, the political climate was ripe for pushing forward—in record time—the research, design, and experimentation necessary for such a daunting objective. But without the ideological other in the form of Soviet Russia, a mission to Mars must be rationalized in other ways. Kalfus offers this explanation as he describes

> NASA and aerospace companies that have a stake in the future exploration of Mars. They have led the campaign to drive public policy in favor of such exploration, frequently publishing articles in popular science magazines that exaggerate the possibilities of past, current, or future life on the planet. One of the space-propaganda market's high-tech catchwords is the verb

> "terraform," meaning to modify another planet's environment to make it more like earth's, and it is usually employed in connection with making Mars more livable for human pioneers (of course, humans have shown greater promise of success in modifying *Earth's* environment to make it more like the environment of, say, smog-shrouded Venus). (145)

That there is an entire "market" for "space-propaganda" is itself an interesting claim, since it runs thoroughly counter to the notion of science as an objective, rational process that does not allow institutional interests to deliberately distort information. Here is another example of how we have exited the realm of science proper and entered the realm of rhetoric, a domain that uses language—and sometimes realigns information—to achieve a desired end, in this case, public support for space exploration. Are scientists or their PR spokespersons "lying" per se? Not exactly, but they are embellishing the likelihood of life on Mars because they recognize that this, along with the potential material benefits of Mars missions, is an important motivational button to push in the American public. Between the constant flow of "documentaries" exploring the alien question (including lending credence to the idea of regular alien "abductions"), the media juggernaut that seizes on the drama of contacting new life, and the blitz of movies and television programs that showcase the exciting universe of extraterrestrial multiculturalism, it is not difficult for NASA to use this public preoccupation to its own rhetorical advantage. Hint that we might actually discover "extraterrestrial life" (even though this would likely take the form of single-celled organisms—a probability that is downplayed because it would hardly qualify as a "first contact" that carries the necessary drama) and suddenly people start paying attention. The truly interesting facet of this rhetorical strategy is that the discovery of any life outside the earth would be of huge scientific significance, single-celled or not. It would confirm the hypothesis that, given the vastness of the universe, life on other worlds must exist, and if life exists, it is possible to imagine that intelligent life exists. But this, of course, is too subtle to be of rhetorical value with a population who is conditioned to expect the fictitious extraterrestrial beings that they see in science fiction, beings which are really only composites of ourselves, humanoids with good prosthetic makeup.[3]

But let's examine the likelihood of life elsewhere in the universe. It is estimated that there are 200 billion stars in the Milky Way galaxy alone, and that of those stars, a significant number have planets or planetary bodies orbiting them. Given the magnitude of this number, it would seem unlikely that ours is the only planet that has had the good fortune to develop life. In a recent Imax documentary film, "Aliens of the Deep," the producers postulate that life is not nearly as fragile or choosey about its environment as we once thought. They draw this conclusion based on deep sea expeditions, using submersibles and probes that have plunged to the deepest trenches of the Atlantic and Pacific oceans, sometimes reaching underwater points as abysmal as 3500 meters. Even at such crushing depths, life not only exists—it *thrives.* Elsewhere in the ocean are thermal vents that pulsate forth jets of water that are heated by the pressure of subterranean physics to 750 degrees Fahrenheit. In water this hot, one could boil just about anything, yet swimming nearby (and often into the jets themselves) are shrimp—*thousands* of them. From this, the scientists exploring these regions speculate that, as long as certain conditions are met (the most important of which is liquid water), life almost invariably results. Wherever organic compounds can get a foothold, it seems, they will, and the diversity of life forms that we observe on a planet like Earth is the inevitable result. There is, in fact, nowhere on Earth that doesn't support some form of life, from the frozen reaches of Antarctica to the sweltering deserts of Australia to the deepest trenches of the Pacific. The question is, then, not so much how many "Earth-like" planets exist, but how many planets with the minimal conditions necessary for life exist. An unexpected candidate in our own solar system is not, technically, a planet at all, but a moon: Jupiter's Europa. Here is a frozen iceball of a world, thoroughly unlike Earth in almost every respect, which at first glance is seemingly antithetical to life of any sort. Yet, many scientists believe that underneath miles of frozen surface, Europa may have a liquid ocean—an ocean with far more water than even the Earth itself. This hypothesis is based upon the extensive fissures in the moon's icy surface, a condition which indicates that there is significant movement beneath the ice which could be attributable to the water movement of a global, subterranean sea. If this is the case, it is entirely possible, even probable, that life exists there. To find out whether this is in fact the case, however, NASA would have to design and dispatch an automated probe capable of landing on Europa, tunneling through the ice, and swimming through the ocean while relaying its findings back to Earth. This is no small technical and engineering feat, and the prospects for success are contingent on further research and development for such a probe, a prospect that relies, of course, on an ongoing financial commitment.

Elsewhere in the galaxy we have far less information on which to base even remote speculations. It has only been in the last 15 years or so that we were even certain of the existence of planets around stars other than our own. When such planets were in fact discovered, they were always gas giants like Jupiter that had the gravitational pull necessary to affect their host stars. These planets were discovered indirectly, in other words, and the first ones were almost always found very close to the parent star, indicating that the existence of rocky, Earth-like planets was unlikely. In our own solar system, for example, Jupiter plays a key role, for it is the gravitational protection of this gaseous behemoth that prevents the rocky inner planets like Mars and Earth from being pulverized by passing asteroids, meteors, and comets. It is far more difficult to assess whether a star system contains a Jupiter-like planet further out since the gravitational pull on its star is much weaker and more difficult to detect, especially with terran-based telescopes. However, recently, astronomers have detected just such planets, and many estimate that where such planets exist, if our own solar system is a consistent model, then rocky inner planets are likely to exist as well. Even in our own solar system, one must remember, only one in four rocky planets are (currently) capable of supporting life (so far as we know), a pattern of probability that further reduces the number of systems that are life-friendly. We begin to see an equation emerge, though, and some scientists have constructed just such equations, the most famous of which is Drake's Equation developed by radio astronomer Frank Drake. Interestingly, this equation is not designed to merely predict the number of possible planets capable of sustaining life, but of sustaining *intelligent* life. The equation is explained below:

> [Drake's Equation is a] proposed equation to estimate the number of technological civilizations that might exist in our galaxy. The Drake Equation was first presented by Drake in 1961 and identifies specific factors thought to play a role in the development of such civilizations. Although there is no unique solution to

this equation, it is a generally accepted tool used by the scientific community to examine these factors.

$$N = R_* \times f_p \times n_e \times f_l \times f_i \times f_c \times L$$

N = The number of communicative civilizations

The number of civilizations in the Milky Way Galaxy whose radio emissions are detectable.

R_* = The rate of formation of "suitable" stars

The rate of formation of stars with a large enough "habitable zone" and long enough lifetime to be suitable for the development of intelligent life.

f_p = The fraction of those stars with planets

The fraction of Sun-like stars with planets is currently unknown, but evidence indicates that planetary systems may be common for stars like the Sun.

n_e = The number of "earths" per planetary system

All stars have a habitable zone where a planet would be able to maintain a temperature that would allow liquid water. A planet in the habitable zone could have the basic conditions for life as we know it.

f_l = The fraction of those planets where life develops

Although a planet orbits in the habitable zone of a suitable star, other factors are necessary for life to arise. Thus, only a fraction of suitable planets will actually develop life.

f_i = The fraction life sites where intelligence develops

Life on Earth began over 3.5 billion years ago. Intelligence took a long time to develop. On other life-bearing planets it may happen faster, it may take longer, or it may not develop at all.

f_c = The fraction of planets where technology develops

> The fraction of planets with intelligent life that develop technological civilizations, i.e., technology that releases detectable signs of their existence into space.
>
> **L = The "Lifetime" of communicating civilizations**
>
> The length of time such civilizations release detectable signals into space. ("Drake's Equation for Extraterrestrial Life")

Based on Drake's own estimate of the values of the places in this equation, he suggests that there are at least 10,000 such "civilizations" in our own galaxy alone. The problem, however, is that the values of nearly all the variables in this equation are unknown. If one changes the value of one or all of the variables, that is, the corresponding number for "N" can be anything from 1 to 1 million. (We know that there is at least one planet capable of producing intelligent life that can communicate via radio: our own). The value that is often overlooked by those who imagine so many civilizations is "L," since, while there may in fact be many such interstellar societies capable of radio communication, it is difficult to imagine that they would all arise at the same time. The implications of this are profound, since while radio waves may float around the universe in perpetuity, it does not mean that the civilization that created them is even in existence any longer. Ergo, while communication waves detected from a civilization that has been extinct for five million years does in fact prove the existence of extraterrestrial intelligence, it has no practical value in terms of making contact with it. One can imagine, after all, that if the estimated age of the universe is even remotely accurate (some twelve billion years), there have been thousands—perhaps millions—of civilizations that have come into and out of being throughout the eons. Considering that *Homo sapiens* is a species that, as nearly as we can estimate, has been in existence for less than 150,000 years and has possessed the technology necessary for radio communication for a paltry century, one gets a better picture of the temporal constraints of making "first contact."

This is also an example of the rhetorical component of a scientific equation. For those who might question whether mathematics in fact has any rhetorical influence, one need only look at Drake's Equation to see that is retains the features of both persuasion and identification. The variables suggest that, once the appropriate and accurate numbers are "plugged in," we will have a conclusive idea of the number of civilizations in the universe, a very compelling piece of evidence for continued financial and intellectual support in the search for extra-terrestrial life. It is persuasive in the sense that it engenders hope for "first contact," and it makes a similar type of contact with an audience excited about the prospect of discovering new species, intelligent life, and civilizations. The problem of the value of the variables is somewhat underplayed, however, giving the audience an illusory sense of the certitude such an equation produces. People often overlook the fact that while mathematics as a closed system is usually a "black and white" prospect—i.e., there is a "right" answer and a "wrong" answer—Drake's equation identifies how a formula is only as good in practice as the numbers that one can assign to the variables. From a pragmatic perspective, the equation gives us hope but no certainty since the numbers will, for all practical purposes, always be changing as new information comes to light and new values are placed in the appropriate slots as they are made available. "First Contact" becomes contingent on the fluctuations of the numbers, and this is far from a black and white prospect. The equation relies, therefore, on the supportive response that it can generate in the audience, and this too will vary depending on the level of enticement that the possibility of discovering interstellar societies generates in the person considering the equation.

Drake's Equation is rhetorically compelling because "First Contact" is, in fact, a popular concept with the general public. Science fiction is preoccupied with the notion, and the mainstream media is quick to pick up on this. Indeed, it is rare to hear in the news of space exploration or of missions around the solar system without some reference to the science fiction that informs the general public's perception of what space travel is about. On July 3, 2005, NASA scientists from the Jet Propulsion Laboratory in Pasadena, California saw the fruits of their mission to impact a comet with a space probe. On July 4 the Associated Press introduced the story by noting that, "it sounded like science fiction-NASA scientists used a space probe to chase down a speeding comet 83 million miles away and slammed it into the frozen ball of dirty ice and debris in a mission to learn how the solar system was formed." So entrenched is our only point of reference to the unknown wonders of space that we tend to frame actual missions in terms of our fictional contact with the concepts that agencies like NASA are trying to make a reality. In this instance, it seems, life imitates art. Steve Lin, a physician from Honolulu who had the privilege of witnessing the impact on a giant movie screen in Hawaii commented, "It's almost like one of those science fiction movies." The article goes on to explain the mission this way: "The cosmic smash-up did not significantly alter the comet's orbit around the sun and NASA said the experiment does not pose any danger to Earth—unlike the scary comet headed for Earth in the 1998 movie, *Deep Impact*" ("Deep Impact Scores Bull's-Eye"). NASA mission engineers even went so far as to name the operation *after* the movie, calling it the "Deep Impact" mission. Strange as it seems, media influences, particularly fictional ones, so far outstrip other modes of information that, in an attempt to keep the public informed, mission designers are compelled label their activities in terms that are accessible by means of movies rather than more scientific modes. While details of the mission were provided in the AP article, they came toward the end, suggesting that the journalists writing the story deemed it necessary to reveal the nature of the operation through popular culture, not scientific data.[4]

Returning to the public relations campaign surrounding Mars, I should also note that the media imagery of modern broadcast techniques by no means marks the first time we have gazed expectantly at the Red Planet. An oft-cited source for earlier interest in Mars is, of course, H.G. Wells and his famous 1897 novel, *War of the Worlds,* which suggested that intelligent life from Mars does not necessarily mean benevolent life. The Orson Welles radio broadcast in 1938, a movie version of the novel in 1953, and then a remake of the same film in 2005 suggest that our interest in Mars has been an on-going one.[5] Edgar Rice Burroughs wrote an entire series of "John Carter of Mars" adventures that were second in popularity only to his *Tarzan* books. Countless other books and movies have been made about the "little green men" who hail from this planet, so any attempt to reacquaint the public with our nearest planetary neighbor is made easier by this significant cultural history. The reports of "canals" of seemingly intelligent design on Mars written by Percival Lowell added some scientific credence to the fantasy notions we already harbored about the planet, making it an even more attractive subject of inquiry for the professional and layperson alike (these "canals" turned out to be natural grooves in the planet's surface, and while they do not suggest intelligent life, they do point to the existence of dried-up river beds, evidence that Mars once had liquid water). Moreover, many professional scientists, including Robert Hutchings Goddard, a pioneer in rocket research, and Carl Sagan himself, were inspired by these narratives. It seems that the distinction between fantasy and reality is not as clear-cut as we often assume. Much of the fantasy play engaged in by scientists when they were children propelled a search to make such ideas a reality. Sagan explains in *Cosmos:*

> Even if Lowell's conclusions about Mars, including the existence of the fabled canals, turned out to be bankrupt, his depiction of the planet had at least this virtue: it aroused generations of eight-year-olds, myself among them, to consider the exploration of the planets as a real possibility, to wonder if we ourselves might one day voyage to Mars. John Carter got there by standing in an open field, spreading his hands and wishing. I can remember spending many an hour in my boyhood, arms resolutely outstretched in an empty field, imploring what I believed to be Mars to transport me there. It never worked. There had to be some other way. (111)

The overlapping of science and fiction is a major contributing factor to the efforts already put forth toward the exploration of Mars, even, it appears, among scientists themselves. Beginning with the Viking II in 1977, there have been several visitations to Mars using Earth-based probes, the most recent of which, the Phoenix probe, landed on Mars on May 25, 2008. There is more data on this planet than on any other in the solar system, so it follows that it is the most likely next stop in human space exploration. The combination of scientific knowledge and public interest in Mars all but guarantees that more funding will be channeled into the project with the ultimate goal of sending a human expedition to its sandy surface. Still, there are detractors to the project who think that the expense outweighs the benefits, so the campaign to gather public endorsement continues.

Another tactic used to garner support for the Mars mission is more historically based, designed to tap into a collective consciousness that views Americans as great explorers, the tamers of wild frontiers. Kalfus describes a NASA-sanctioned promotional show that was exhibited in Baltimore in the 1990s. The setting was a future Martian settlement, one that

> [e]nvisions that all life-support supplies will someday be produced by the settlers themselves. It declares that "only luxuries" will be imported from earth, modeling a Martian future along the same lines as America's past. According to this popular, space-age version of Manifest Destiny, the human race will explore and eventually colonize the solar system and the galaxy. This treacherous analogy, virtually a commonplace in our expectations of the future, minimizes the differences between seafaring and spacefaring, and between the fertile plains of the American continent and the airless, sterile, radiation- and crater-scarred surfaces of other planets. (145)

So just as the distinction between life, multicellular life, and intelligent life is underplayed by the "space-propaganda" machine, so too is the distinction between exploring the earth's oceans and continents by sailing ships and exploring space's star systems and galaxies with spaceships. Few people truly understand the hazards of space travel, much less the serious limitations given our present state of technology. Much of this misinformation can be attributed to science fiction scenarios that portray space explorers as confined within infallible technological biospheres, complete with artificial gravity and all the comforts of a luxury liner. Space travel, when presented this way, seems not only to lack the hazards that really exist but also to be an inviting and comfortable activity. Space, in reality, is a cold, uninviting, extremely dangerous place for human beings, and its unimaginable vastness means that the practical time necessary to travel even to destinations as cosmically nearby as Mars must be counted in years. As Sagan has carefully pointed out, traveling to the nearest star, under our present technological state, would be impractical and temporally prohibitive. One must also understand that, if Einstein was in fact correct, that the speed of light is the "cosmic speed limit" beyond which physical spacecraft (or anything else) cannot go, then the likelihood of traveling to more distant places like stars becomes practically impossible. Some stars are hundreds of light-years away, others thousands. Other galaxies become, then, completely out of the question as some of them are hundreds of *millions* of light-years distant. Such restrictions make selling space-travel to the general public even more difficult, but the obvious irony is that, without the support and funding needed to either confirm the impossibility of interstellar travel or the research and development to solve the problem, we will never really know if Einstein's predictive speed limit is, in fact, insurmountable.

The "final frontier" that is space, then, becomes a rhetorical project that must work to identify the motivators that Americans respond to best. Space station MIR, for example, was sold as an exercise in international relations as much as a scientific laboratory, and it can therefore be validated as an important step toward lasting peace and trust between nations. Other motivators include economics; as we quickly deplete the finite resources of our own world, it is important to consider the advantages of mining other planets for their raw materials. This argument, if it holds, guarantees that the real space pioneers of the future will not be governmental, but entrepreneurial and corporate. Those corporations with the capital and the vision will stake out their claims in much the same way that our ancestors "tamed" the American West. As Kalfus warns, however, the challenges of space travel, exploration, and colonization—much less the engineering and logistic difficulties of constructing a sustainable and cost-effective infrastructure in space and/or on remote planetary outposts—means that the risks involved are as enormous as the potential payoffs. Americans, historically at least, have always risen to the challenge implicit in a skeptical claim that something is impossible. The "Manifest Destiny" analogy is therefore an apt one, for it contains within it the associative incentive to prove the skeptics wrong, to show the world that American innovation and adventurousness is still a part of our national identity. David E. Nye, in his book, *America As Second Creation: Technology and Narratives of New Beginnings,* traces the history of our technological manifest destiny by exploring the culmination of this attitude in the 19th century narratives that emerged as the nation spread westward in search of new opportunities, land, and resources. It would be a simple symbolic transference to use the same ideas prevalent in the 19th century to those of current and future space exploration. As Nye explains:

> The versions of the technological foundation story that emerged and circulated in the nineteenth century were literally about creating society by applying new technologies to the physical world; the mill, the center of new communities; canals and railroads, used to open western lands to settlement; irrigation, which convert-

> ed "worthless" desert into lush farmland. Each narrative described a process of community creation. (4)

To draw a 21st century equivalent to the 19th century technology, the mill has become the self-sustaining planetary colony, the canals and railroads have become space ports, space stations, spacecraft, and flight paths to these new colonies, and the irrigation of deserts becomes the "terraforming" that Kalfus mentions, a way of making lifeless planets capable of supporting life. It takes no great leaps of the imagination to tap into these already well-grounded cultural images nor does it require great effort to coax the ideological interest in exploring and "taming" new frontiers. These are part of the American heritage, symbolic of the pride we take in ourselves as risk-takers and adventurers. While the technology changes, becomes more advanced and sophisticated, the underlying drive is exactly the same. The inner force that propels American ambition makes space, in the last analysis, a very attractive and exciting frontier indeed.

A decidedly more modest proposal for embarking on this frontier is the colonization of the moon. Thomas F. Rogers, former head of the Air Force and MIT Research and Development Laboratories, catalogs an attractive list of benefits for establishing colonies on the moon. The broad categories are safety, economics, peace, and diplomacy. Under each of these he explains the relative contribution achieved by moon colonization toward the larger objective. For example, he sees moon colonization as an attractive option for dispersing large populations on earth (56). He sees the moon as an important source of economic supplementation because of its many mineral resources (56–7). He even goes so far as to suggest that moon colonization has implications for lasting peace on Earth: "we could use human development of the Moon to learn how to change our culture. To achieve peace is one of civilization's most important goals and activities, one of the greatest in history and pre-history" (57). Specifically how moon colonization will contribute to world peace, however, is left largely to the reader's imagination. Rogers does offer this argument:

> In human terms the Moon is essentially a *tabula rasa*—a blank space. There are no lingering hatreds, or even recollections of war, slavery, genocide, or religious persecution. There are no political or economic alliances or political boundaries, no land ownership,

> no weapons of any kind, especially weapons of mass destruction. (57)

As noble as the ideals in this argument may be, the level of naiveté is most surprising. The moon may in fact be a *tabula rasa* (except for the American flags that have been left there as a nationalistic symbol of our technological domination), but people clearly are not. The expectation that the unspoiled condition of the *place* will somehow prevail over the flaws in human character is clearly unrealistic. A better argument might involve the premise that lasting peace can be achieved through a common goal, one in which nationalistic and ideological allegiances are realigned, renamed, and reciprocated, not erased altogether. Clearly, Rogers is trying to forward an argument almost exclusively for persuasive effect, but his logic falls short of the rhetorical mark. While people may accept the notion of international cooperation for a common goal, it is unlikely that any marginally-informed member of the public will believe that the moon will magically transform human nature *because* it doesn't currently have a discernable human footprint on it (metaphorically speaking—literally, of course, it has a *lot* of human footprints on it!). If history is any teacher, we see that virgin lands are the *first* to fall to human greed and self-interest.

So it must be understood that the limitations of what can actually be accomplished in space may force us to re-evaluate whether it is so attractive after all. If we look at the logistics of mining resource from other worlds, for example, it becomes clear very quickly that the cost-benefit ratio is decidedly prohibitive of such entrepreneurial endeavors. While some celestial sources like asteroids and the moon may prove to be cost-effective options for space mining, from a research standpoint, they are not nearly as interesting as Mars or other outlying planets, meaning there is one less rhetorical rationale for exploring that particular planet. It has been predicted that the types of usable raw materials available on asteroids or on the moon is significant, mostly metals like iron, nickel, magnesium, aluminum, and cobalt ("Asteroid Data Sheet"). But the cost of locating, extracting, processing, and transporting these materials from their celestial cradles to the Earth is less than encouraging. The Space Shuttle, as a point of reference, costs about $12,000 per kilogram to operate in a normal orbital mission. This figure makes the traditional use of solid rocket fuel for vehicles launched from Earth an unlikely option for space mining, since the cost of operating such spacecraft for mining purposes would not be recoverable through sales of the material on Earth at market value. More promising is the use of alternative propellants like (believe it or not) steam for use in vehicles built and launched in space ("Transport of Earth, Lunar, and Asteroidal Materials"). This would require permanent space stations or colonies on the moon as transfer points for returning the materials to Earth. These are, however, the very sorts of challenges that Americans pride themselves in taking on and solving, another holdover from our earlier pioneering days that reflects a tradition of owning the wilderness and taming nature.

Another interesting facet of this same cultural inheritance is its "democratic" and "community-oriented" foundation. As part of the original conception of Manifest Destiny, Americans sought to expand not only their land holdings, wealth, and progeny, but also their politics and their religion as well. The West became a great equalization opportunity for many Americans who could not or would not fit into the prescribed social orders of the Old World or its east-coast equivalent in the United States, social orders that were often from old money and a pseudo-aristocracy and often, ironically, very rigid in their faith-based demands on their citizenry. Nye notes that

> Americans constructed stories that emphasized how particular tools and machines enabled them to inhabit new places and to create new communities. Importantly, they expressed a belief that these technologies would enable them to preserve their egalitarian difference from the Old World. (5)

The West, as a symbol that made "manifest" the "destiny" of Americans to create their own unique identity, was "democratic" in the sense that it represented absolute freedom. But as we all know, absolute freedom is a dangerous condition indeed, for it can carry with it the menace of self-interest at the expense of one's fellow humans. Such a situation means that certain institutional controls are needed to establish and maintain order, institutions like civic bodies, religion, and law enforcement. We can assume, human nature being what it is, that the same safeguards must be applied to the frontiers of space lest an unregulated exploitation of the worlds we "conquer" leads to a kind of State of Nature. Nye describes how the 19th century narratives once again relied on the intrinsic powers of technology and old institutions to maintain law and order in the western frontier:

> These narratives sought to explain how Americans could use new machines to transform the land while establishing communities similar to those they had known before. In these progressive narratives, machines were dominant yet democratic, transformative yet conserving. The narratives naturalized the technological transformation of the United States so that it seemed an inevitable and harmonious process leading to a second creation that was implicit in the structure of the world. (6)

Such diverse acts of "creation" carry with them both the utilitarian construction of new machines and technological networks and the sense that such frontiers have the potential, like Roger's moon, for a cultural and spiritual *tabula rasa.* Given the present state of US international relations, one that insists upon an ideological compliance to the apparent political superiorities of a capitalistic democracy, we can only assume, contrary to Roger's contention, that a transference of these ideals will naturally be extended to any space colonization that is bankrolled by the US government and/or American corporations and their allies. In this sense, democratic ideals will simply assume a more pristine form, a perfect democracy made possible by the many new "promised lands" of the solar system and beyond. Widespread colonization and development is, in many respects, considered the American birthright; it is what gives us both our national identity and our sense of purpose.

Alexis de Tocqueville remarked:

> The Anglo-American race fells the forests and drains the marshes; lakes as large as seas and huge rivers resist its triumphant march in vain. The wilds become villages, and the villages towns. The American, the daily witness of such wonders, does not see anything astonishing in all this. This incredible destruction, this even more surprising growth, seems to him the usual progress of things in this world. He gets accustomed to it as to the unalterable order of nature. (qtd. in Nye 9)

The American propensity toward the domination of nature, so in contrast to the Native American notion of natural unity—a conjoining of forces within the natural world—is, in part, what made the clash of ideologies so destructive for the Native American in the 19th century. A frequently cited example of these contrasting world views is the mythological story of the "purchasing" of Manhattan for some beads and deer skins. The story is often used as a way to underscore the "savagery" and "ignorance" of the native population, and ignored is how the Native American would have responded to such a transaction. They must have thought the new visitors to be naïve indeed. How does one "own" the land? At most, one occupies it because the land itself is willing to allow our presence, but it also resists us when we don't pay proper homage to its power. The concept was so foreign to them that they must have thought the Americans the ones who fared the worse for the encounter.

Tragically, of course, this attitude became the undoing of the Native American populations in this country. When confronted with a people to whom the land is merely a resource to be conquered, occupied, and exploited, ripped asunder and altered, surveyed and transformed, the spiritual guidelines of the Native American were not binding on this new invader. de Tocqueville's observation is so on the mark that we can see in these few short lines how the new Americans could as easily transform the landscape as they could brush aside the people who inhabited it. We were invaders, but more than simply occupying the land that we took, we had to change it into something that symbolized our dominance. Space, so vast and so rich in potential, must seem like a developer's dream, if only we could work out the details. What is the loss of one rock in space if it means the improvement of living conditions for millions of people on Earth? Who is to stop us from transforming worlds so that they submit to our needs? These godlike self-impressions guarantee that our destiny in space is one of rapid manipulation and restructuring. We have transformed whole continents; a whole planet should be the next logical and achievable step. And these are the standards that we, as a people, unapologetically live by. To resist "progress" is to be labeled a liberal "tree-hugger" who doesn't recognize our innate right to alter that which we encounter into our own image. The term "progress" itself functions as a powerful rhetorical argument for things like space travel, exploration, and colonization; it is what has given us our unique signature on the American landscape (and the global one) in the past, and to be the first to lay claim to prizes in space is part of this heritage.

The purpose, then, of this discussion has been to establish that there exists a mindset in the American people that is highly receptive to certain kinds of rhetorical approaches, approaches that are well-understood by the scientific pundits who practice them. The orientation of Americans is one that identifies with often conservative aims: with the maintenance of traditional cultural myths, with claims of progress, with the notion of manifest destiny, with the idea that technological advancement is almost always a positive thing, with the idea that it is our inherent right to spread democracy and capitalism all over the world and beyond, often introducing science and technology as a means to this end, and with the fiction of rugged individualism, a myth that has been extended, interestingly, well beyond the individual to become a patent feature of individual companies or corporate interests (corporations are, interestingly, considered "individuals" under American law through the corporate use of the 14th Constitutional Amendment, an amendment designed to expand the notion of "citizen" to newly freed slaves and women). All of these associative ideals have strong rhetorical currency for things like space travel and exploration, and when coupled with a serious lack of formal training in critical discourse and the standards of argumentation, much can be accomplished using ersatz scientific language, data, and arguments for many of our nation's most important projects. I have also tried to show that the objections leveled against the American public's inability to distinguish between "real" science and that which merely mimics it is a symptom of our faith in all that sounds scientific and in an educational shortcoming that prevents us from using a standard of scrutiny necessary for sifting legitimate claims from the merely specious. Some further discussion of this last issue is necessary in order to see where we are as a nation and to consider educational remedies to the cultural trend toward resisting anything that seems academic or intellectual. First, however, we must further establish the patterns I have been describing by examining other examples of influential popular sources for the various scientific projects that emerge as our society grapples with problems, investigates solutions, and establishes dominance through the sanction and power of science.

7 More Popular Sources for the Scientific Project

Another important component of popular science is the impact of influential scientific periodicals, especially those aimed at a general audience whose continued understanding of and support for scientific endeavors is informed by magazines aimed at educating the public about the state of scientific research. Two of the most popular are important to examine if only for the distinction in their respective approaches. The first, which provides a much more general treatment of scientific developments, is *Popular Science.* The second is in some ways far more reputable as a mouthpiece for pioneering, sometimes even radical, scientific developments, a magazine that has been around since the mid-19th century, *Scientific American.* Its circulation worldwide was 688,441 subscriptions in 2005 ("SciAm's Circulation is Red Hot"), suggesting that it has relatively world-wide appeal if not an enormous number of regular readers compared to other popular periodicals. The main distinction between these two publications is emphasis: Whereas *Popular Science*'s main focus is on showcasing the latest technological gadgetry and the championing of the unusual side of technological research, *Scientific American* is far more concerned with scientific problems and the latest developments in addressing them. "Problems" here takes on two meanings, in the sense that the magazine tends to deal with theoretical problems that face scientists from a wide array of scientific sub-specialties while also addressing the environmental, political, technological, and economic issues that have an impact on all of us. From this perspective, *Scientific American* is a much more important representation of the scientific concerns that touch scientists and laity alike, and its educational qualities are far more relevant to a general reader serious about understanding science on a deeper level.

Popular Science, while a fine publication in its own right, tends to appeal to those possessing a "what-will-they-think-of-next" curiosity about scientific applications. Despite its attention to technological oddities (and perhaps because of it), the magazine contributes to the overall awe and wonder that informs the American imagination regarding scientific progress, and it is therefore worthy of some discussion. As a reflection of our public infatuation with all things new and novel, it carefully orchestrates the intersection between technological advancement and consumer marketability. In many ways, it reads like a catalog of what's new on the technological marketplace, and it could easily be mistaken for one large, in-depth infomercial on the latest technological toys. But perhaps it would be more just to describe it as a brochure of applied science, a billboard celebrating our innovations and supporting a view of science and technology as a juggernaut of solubility (in the colloquial sense), embracing every problem (whether it exists or not) with the same energy that helped Americans "tame" a continent. Just as the transcontinental railroad opened new vistas in product and people transportation, blazing the trail that opened up the West, so too do the new technological gadgets open new pathways to a never-ending stream of gadget and counter-gadget. Once the consumer public becomes enamored with the potential of the computer age, the possibilities are endless—a kind of cyberspace frontier where untapped opportunities are popping up everywhere. There is certainly this sense of the pioneering spirit in *Popular Science;* one feels the power of technology and sees these new innovations as part of our nationalistic identity. Our might as a nation, we sense, is grounded in our ability to solve technological problems. But these are only the surface impressions one gets from glancing through *Popular Science.* It's real interest as a source of popular information can be seen in *how* it constructs a high-tech rallying point for our collective feelings of accomplishment and pride.

Consider, for example, this passage from an article posted on the *Popular Science* website on December 31, 2004 entitled, "The Daring Visionaires [sic] of Crackpot Aviation":

> In the bucolic hinterland North of San Diego, on a hillside shaded by eucalyptus and pine, Attila Melkuti pulls open the doors of a barn. Inside, light gleams off a strange and marvelous contraption he's been piecing together for the past seven years. He pushes his nearly

> finished creation into the sun: a curvaceous, cherry-red flying machine that looks like nothing so much as the unholy union of a UFO and a Corvette. "The idea came to me in a childhood dream," he says. "I believe it could transform aviation." (Wise)

The language is reminiscent of the early days of aviation and those daring young men in their flying machines, defying both physical obstacles and the ridicule of detractors to forge a new era in travel and transportation. One can anticipate the direction that this innovation and this article hopes to take: the development of an affordable, easily operable, and accessible personal aircraft, a dream as old as aviation itself. When research and development of helicopters was in its infancy, for instance, many designers hoped to see not a Buick in every garage, but a "whirlybird." This dream never came to fruition, obviously, but the reality of the helicopter did, and its versatility makes it standard equipment for the military, hospitals, police departments, news stations, and transport companies. Just as interesting, however, is the reinforcement of the mythos of innovation, the rhetorical buttressing of legendary inventions that echo the sentiment, "They laughed at Edison, too." *Popular Science* gives the idea further credence by pointing out:

> For the most part, it's not crackpots, or even eccentrics, who populate the fringes of aviation design. These are dedicated engineers and entrepreneurs who see a gap and wish to fill it, using technology that they have tweaked and honed, improved designs that may finally be optimal and strategies that big aero firms may simply have never thought of. Also in the mix: some obsessed amateurs and, yes, a few smooth-talking rogues. All are in hot pursuit of great ideas that just might—or might not—take off. (Wise)

This too is a uniquely American form of self-expression, the principle that an innovative American will take a good invention and make it better. A similar attitude was pervasive during the period of unprecedented affluence that was enjoyed by the middle class during the early post-war years. The "hot rod" was one such form of inventive expression, the idea being that a technically savvy amateur could take a stock automobile and transform it into a personal manifestation of beauty and performance. H.F. Moorhouse argues in "The 'Work' Ethic and 'Leisure' Activity: The Hot Rod in Post-War America," that activities such as the construction of hot rods reflect the work ethic *and* the leisure ethic inherent in American culture. The hot rod (or the hot plane, in this case) represents a hobby that is not trivial, that has intrinsic labor value, and is a working class form of artistic expression. It aligns the notion that idleness is the devil's workshop with the idea that all labor must have a productive purpose; hot rodding contained many of the values Americans had about work: it was constructive in the practical sense, it required technical know-how and the ability to modify existing designs (innovation), it was a superior form of consumerism, and it generated community among the "grease monkeys" who practiced it. So successful was the hot rod phenomenon that whole magazines, clubs, shows, and contests were (and are) dedicated to advancing the hobby. According to Moorhouse, the cultural message underpinning hot rodding "was taken up with the ideologies of activity, involvement, enthusiasm, craftsmanship, learning by doing, experimental development, display, and creativity, all of these circulating around the motor car in unpaid time" (285). In much the same way, the garage aviators showcased in *Popular Science* have continued this cultural tradition. It is clear that the activity of amateur aviation, while certainly not as widespread as hot rodding because of cost and the technical knowledge necessary to build aircraft, is an extension of the earlier hot rod culture. Amateur aviation serves to remind the public that productive hobbies, working with one's hands, learning new techniques and skills, and improving on existing ideas are still part of the American labor-value system.

In the above passage, we see a decidedly capitalistic articulation of the ultimate motive behind amateur aviation design, however, one that didn't exist in the same degree with the hot rod culture (in fact, one of the appealing things about hot rodding, oddly enough, was its emphasis on non-paid labor). The text's stress on "engineers and entrepreneurs" suggests that these two populations are on equal footing—that the designer and the marketer are important allies in the success of any new technology. In fact, it is implied, pragmatism dictates that one cannot exist without the other. The underlying motivation, then, is sales; it is important to have a "vision" that, while it may indeed change the way Americans travel, commute, and tour, will also have the happy side-effect of making the designer and his or her associates very, very rich. Imagine the profits generated by patenting the first aircraft that is as easy to operate as a car and that requires little more than the equivalent of a driver's license to legally fly. While countless copies would indeed follow, opening up a whole new market for personal flying vehicles and allowing the free-market economy to run its social-Darwinian course, the inventor of the first personal aircraft would be in the enviable position of enjoying both fame *and* fortune. We see again the overlapping of the mythos of American know-how with the dream of American success and its contribution to American freedom. Like the mass-production that made possible mass-ownership of the automobile that ultimately made Henry Ford both rich and famous, the personal airplane is the next logical step in the vision of freedom, wealth, and celebrity that has given America its unique signature. Here, in this short article, we see the features that make America distinctively American: the innovation, the recognition of opportunity, the persistence, the vision, the sense of liberty, and the entrepreneurial spirit, all wrapped up in one comprehensive rhetorical and ideological package. And, clearly, a key component of this larger hegemonic lens is the narrative of scientific advancement through which this vision of the American ideological landscape is transmitted.

But what is the effect of the message? How does the reader decode the significance of the ideological nostalgia that this article invokes? In most cases, textual patterns revealed in documents such as this are subtle, relying on the collective orientation that the audience possesses and the symbolic overlap that such an orientation implies. In this instance, there are a number of vaguely familiar images that would act as reminders of common principles implicit to the reader. The photograph of the aircraft itself (as well as the description, an "unholy union of a UFO and a Corvette") is reminiscent of the bizarre contraptions often showcased through history documentaries (we have all seen, for example, the grainy, turn-of-the-century film of the hopping parachute-car, or the "airplane" with twelve layers of wings that folds like an accordion as it attempts its take-off run). These images are quaint parodies of the "real" technology that has become so familiar to us. But consider just how strange and ungainly the space shuttle would look if we had no knowledge of aircraft or of flight dynamics. That it can fly at all—if visual aesthetics is any indicator—is a marvel in itself. But more to the point is that the image of the personal aircraft *is* evocative: it does suggest something, remote as it may be, in our shared cultural consciousness. We have seen machines like this before. Some of those machines worked, others didn't, but the real suggestiveness is symbolic—and what the machine symbolizes is the dream, the sense that Americans (from an American's point of view) have always taken the risks, set aside the ridicule, and forged raw ideas into the refined machinery that ultimately married the concepts of form and function. We consider ourselves the great manipulators of our environment, and not in an uncomplimentary sense; our survivability as a nation (so the cultural myth goes) has always been manifest in our ability to adapt, to find new resources, to use that around us in its most efficient and productive way.

However distorted this myth about ourselves may in fact be (we rarely, for example, relive our myriad technological failures except for the purposes of satire), from the standpoint of ideological underpinning, the article taps into this feeling of national self-worth. The use of the word "unholy" is interesting, though I am hesitant to put much stock in the precision of the author's choice of words. Is the author, Jeff Wise, implying that the aircraft is some abomination against "nature"? This seems unlikely, since nature has little to do with technological development, except as it resists our attempts to harness it. Is he suggesting that the machine's reliance on divine sanction is irrelevant, that faith is a superfluous concern? Or is it that the machine challenges our sense of proportion and aesthetics, an inherent sense of what an aircraft "should" look like? In any case, the word does describe a sharp departure from the norm, and in order for it to have meaning, the reader must have a sense of what the "norm" is. As well-acquainted as we all are with the classic airframe, the juxtaposition of such a norm to our usual sense of a functional airplane reveals our very familiarity with the old technology that is being advanced in this new design. In other words, our understanding of technological form itself has symbolic grounding in our expectations about what a machine must look like in order to perform its specified function. It looks "weird," but, of course, an F-15 jet fighter would look extremely weird to the Wright Brothers. So inundated are we with the technology that dominates our lives that we share a common experience about not only what technology is and does, but what it *looks* like if it is to work. Just as free verse as a form must have seemed bizarre and chaotic to more conservative poetic constitutions that preferred sonnets and blank verse, so do new technological forms assault our sense of what is proper, standard, conventional, and serviceable.

Yet, the pace of our technology in the past fifty years has at the same time required us to overcome our initial sense of skepticism and doubt. While our preliminary reaction to the personal aircraft might be one of dubiousness and wonder, our response to other, more likely technological influences—like the reaction we might have to future automobiles—is one of acquaintance. We have seen radical design changes to the automobile in the last fifty years, and these changes reflect a number of issues confronting automobile engineers, from economic and political considerations like fuel efficiency to comfort and ergonomic concerns that have made the automobile much more of an extension of our bodies than in earlier times. If one looks at the design of a "Tin Lizzy," the Ford model T, compared to, say, the new hybrid cars that are becoming more visible on today's market, we find it hard to imagine that they both have basically the same function—to transport one or more passengers and cargo from one point to the next via four wheels. The superficial design features of the Model T and the Honda Insight, for example, are really very much the same; that is, they both look like cars: Both machines have four wheels, headlights, windshields, roofs, radiator grids, steering wheels, etc., yet they appear to be entirely different species. What's changed to make these cars looks so different from one another? In the 98 years that separate the two designs, our understanding of aerodynamics, fuel economy, stability, ergonomics, space efficiency, and engine reliability has increased considerably. But practical design considerations are not the only factors in engineering new automobiles. The Honda Insight is a hybrid car, which, as most of us know, means it runs on both petroleum fuel and electricity. It gets an estimated 66 miles per gallon on the highway because of its small, three-cylinder engine augmented with electrical power ("The Honda Insight"). (This design was discontinued in 2006.) Fuel efficiency is the result of more than just an attempt to save the consumer money; as fossil fuels are "non-renewable" resources and electricity, for all practical purposes, is infinitely renewable, from a pragmatic point of view hybrids are an excellent move in the direction away from reliance on oil as a source of transportation energy. In this way, the design reflects our increasing social and cultural attention to the environment. We can also speculate that Honda, and other automobile manufacturers, are using the hybrid as a test-bed for the eventuality of cars that are exclusively powered by electricity. It is, therefore, a prudent, even responsible, business move. It has the added benefit of reducing toxic emissions, a move that helps keep environmental agencies (and the world as a whole, we would hope) happy. The widespread use of such vehicles, if it ever comes to fruition, would also mean decreased dependence on so-called "foreign oil" reserves, making the hybrid a highly politicized manifestation of certain international relations policies.

Even in terms of the internal design of the vehicles, the basic theory of internal combustion and electricity drives—conceptually and literally—both of these machines. The technology, on its most basic level, hasn't changed; the principles of internal combustion still apply—both engines require ignition, fuel delivery, cooling, lubrication, exhaust, a drive train, and timing, but the level of efficiency in the hybrid car—coupled with the complication of synchronizing an alternate electric drive system, a process that really isn't practically possible without the aid of a computer—make it a much more advanced, complex mechanism. This level of complexity, one where the component parts become more abundant without creating a corresponding level of redundancy, but instead serve to coordinate the increasing complication of individual but interrelated mechanisms, reflects the evolution of an idea and the application of it for a practical outcome. As obvious as all this may seem, what is not so apparent is the degree to which the machine is a manifestation of the culture—and the evolution of that culture—that both conceived of and assembled it. The machine, in other words, is a cultural artifact, a *text* that provides clues about its makers and its users. It mirrors our goals, our desires, our expectations, and our priorities. By observing these two autos, we even get a sense of our own history. Machines like these are, therefore, texts that not only give us as sense of who we were and who we are, but also offer clues about the trajectory of our future objectives. *Popular Science* functions in this way as a conduit through which we come to terms with our technological identity. It provides a source of industrial distinctiveness, and it uses the machines that we create as a way to remind us of our uniquely technological selves.

Technique, not in the Ellulian sense, but as the skilful interface between our bodies and the technology we use—*how* we interact with the machines we create—has a cultural signature as well. Edward Tenner, who is masterful at noticing the minute details of technology and how these relate to our cultural idiosyncrasies, spends an entire chapter in *Our Own Devices: How Technology Remakes Humanity* describing the cultural significance of such simple technology as footwear and how it impacts the everyday activity of walking:

> [B]efore their country was opened up to the West, Japanese children learned to "walk from the knees," moving the hips as little as possible, whereas European and North American children are trained to "walk from the hips," maintaining their culture's prized erect posture by keeping the legs straight. . . . [This] is the result of a developmental program that includes an entire society's technologies and techniques. For example, traditional Japanese walking gave a good footing on the often uneven local landscape; it was linked with the Japanese method of transporting heavy things by tying them to shoulder-borne poles. (11)

The ways we interact with our technology have unexpected outcomes and a reciprocal impact on both technology and technique, subtleties that impress themselves on the civilization in which they were developed. Clothing is a particularly obvious example, and it is interesting to note how the simple technology of clothing becomes an expression of technique that gives each culture its own unique mark. And the more sophisticated the technology becomes, the more important our attention to technique and its influence on our interaction with the objects of the world become as we change to adapt to the new technologies.

To illustrate this point further, we might look at another period in history roughly halfway between the Ford Model T and the Honda Insight, a period that saw the present-day manufacturers of these respective vehicles on the opposite sides of the same conflict: World War II. To anyone familiar with the hardware used during this war, it is clear the design signature between a Japanese tank and an American one was easily distinguishable. Because Japan was a country that was forced to import the vast majority of resources that it required both to survive and to wage war (a major reason for the conflict to begin with, in fact), along with the terrain of the theaters of combat for which they were designed (usually thick jungle in tropical and sub-tropical climates), Japanese tanks look flimsy and vulnerable when set aside their American counterparts. The Japanese Type 97 medium tank, for example, arguably one of Japan's better pieces of armor, was still no match for the American Sherman or Russian T-34 and KV-1 tanks, partly because the Allied tanks were simpler designs and were therefore less prone to breakdown in combat situations. The latter were also more heavily armored and powerful, whereas Japanese tanks, by contrast, reflected a conservatism necessary given the limited production resources available to an empire already stretched to capacity. Perhaps more tellingly, the ease with which Japan blazed through China in the years prior to Pearl Harbor created an ultimately devastating sense of superiority and complacency about armor design (among other things). Because China was largely rural and possessed virtually no counter-measures against modern tanks, the Japanese enjoyed great success against sparse Chinese forces with inadequate equipment. They mistakenly believed they would encounter the same results against any other land forces they were to confront. This example illustrates the manner in which a culture's technology functions as a document of its mindset. For Japan, the engineering challenge was to create a machine from limited resources in massive numbers, a machine that could make the most of the finite quantities of iron and oil at Japan's disposal. Likewise, human concerns were viewed as incidental to the Japanese military authorities; to waste valuable materials in order to protect expendable soldiers would be a travesty of practicality and not in the best interests of the Empire, for which each individual soldier was expected to give his life. The result was a functional, though not especially effective, compromise in armor design, a text that supplies evidence about Japanese thinking in this historical and ideological context.

Likewise, if we compare German armor to that of the Japanese, we see an entirely different set of beliefs articulated in this country's design choices during the same period. Legendary German engineering produced radically experimental designs, not all of which were successful, but all of which *were* reflections of a willingness to break through the membrane of conventional thinking. This is perhaps even more obvious with German aircraft, but the point is this: If one places an image of a German machine (say a Tiger II, arguably the most formidable tank designed by the Nazis during WWII) against that of an American or Japanese one, the design signature for the respective nations is clear to the extent that a deeper understanding of the people can be gleaned beneath the veneer of the machine. Careful scrutiny reveals, at the very least, certain trademark features that suggest the thinking and logistics that governed the engineering. The machine is not, therefore, just a device of convenience designed to meet an immediate need or to perform a particular duty. The machine becomes a mirror image of a mindset operating within the people who designed it, and by association, the people for whom the machine was designed to perform. It is an extension of a country's cultural and national ideology. The Tiger II tank, for instance, reflects an almost aesthetic marriage of form and function; despite the reality that the tank was designed to kill and to destroy, there is a certain beauty to it, almost as if the pride of a nation that had for centuries produced some of the most beautiful music, poetry, literature, art, and philosophy in the world was extended to this device of war. It looks formidable (which it was), but it also possesses a certain elegance that is discernable in its at once traditional and progressive shape. The "lines" speak Germany; we can almost see its "Germanness." It has, of course, a practical function, but this did not mean that its designers would sacrifice aesthetics or even a nationalistic insignia in the process of creating a functional piece of armor. The Sherman, likewise, has a certain "Americana" appearance about it; its lines are rounder, less harsh, than the German panzer, and its design is conventionally practical. There is no indication of radical engineering here—the tank was developed with a certain degree of haste. Americans, thrust suddenly into the global conflict that had been afflicting the world for some years, needed a tried and trusted design that could serve the immediate purpose. In times of war, unlike times of peace, Americans historically have a habit of doing what they need to do in only the most practical terms. There is no room for experimentation until the immediate requirements are fulfilled. This principle is also reflected in the Sherman's design; it was not an particularly impressive example of armor, especially when pitted against machines like the Tiger II (which tore the Sherman to pieces on the battlefield), but it was reliable and easy to mass-produce, and it served its primary function admirably in both the European and Pacific theaters of operation. The same might be said of the Japanese Type 97, with even more evident conservatism. But its design also reveals the Japanese preference for smaller machines, perhaps the result of attitudes inherent in a people who live on a densely-populated island nation where space is at a premium.[1]

In all three examples, the idea of the need for mobile armor was a direct lesson from history, which taught that the trench warfare stalemate situation that made the First World War drag on for five bloody years must be avoided. Tanks, introduced during this conflict, were a direct reaction to the need for infiltrating well-entrenched troops without the massive casualties that traditional infantry attacks had created. But, again, these mechanistic forms are far more than the sum of their designers' practical requirements. The machine, any machine, is a symbolic extension of its maker's imagination, belief system, thought processes, and cultural, historical, political, and social makeup. Machines become an expansion of who we are—they are, symbolically speaking, created in our "image" (this is perhaps where Biblical fundamentalists get into a literalist bind, for to say that God create humanity in "His image" is not necessarily to say that "He" made us to "look like" Him. Rather, it could mean that God made humans as a product of a divine *imagination* that had particular design parameters in mind for the world in which such a deity was planning to place us. It is apparent, at least from the examples that I have provided, that just because something is created "from" us and reflects our being does not mean that it must resemble us in literal appearance). Those things designed to be used by human hands and bodies, interestingly, often possess a negative impression of the body segment for which it is intended, much like a mold. The seats of a car, for example, are contoured to allow our torsos to fit within it; and ergonomically structured command consoles are bent and fitted to make them not only within easy reach of the operator, but with comfort and ease in mind as well. A well-designed vehicle, then, is formed to fit around our bodies—we "wear" the automobile that we drive so that in a sense we are part of the machine and it is part of us. In the respects that I have just outlined, machines of all sorts become impressions of our minds and bodies in a way that make them both literal and symbolic extensions of ourselves.[2]

The whole of *Popular Science* is dedicated in one way or another to a kind of vicarious technological self-aggrandizement that helps buttress the American fascination with gadgetry. *Popular Science* is intended for an educated, though lay, audience interested both in the trajectory of scientific advancement and in the technological manifestations of scientific progress. It, like its companion publication, *Popular Mechanics,* taps into the American cultural preference for things tangible, practical, and commodifiable. The magazine offers a mirror image of ourselves as we like to imagine the ideal practical American—as innovative, inventive, inquisitive, and entrepreneurial. It is celebratory in its representation of science and technology, and it shows little restraint in the way it glorifies American know-how. With its flashy graphics, brief articles, and unapologetic immodesty about new technological products, it is a uniquely American publication. Beyond simply finding a "niche" within a certain segment of the population, *Popular Science* carries on the historical mythos of the progressive American scientist, engineer, and industrialist for the entire population. It furthers the fable of technological prowess that helps define this country, rarely commenting on the extent to which our failures outstrip out successes. Failure, the publication implies, is part of the technological game, and it is therefore unimportant to dwell upon it. False starts, conceptual dead-ends, problems with design or materials, or simple ill-conceived ideas are eclipsed by the results of those courageous enough to undertake the creation of something new and important. *Popular Science* plays a key role in the maintenance of scientific and technological dominance in American culture.

A more sophisticated publication in terms of its treatment of scientific topics is *Scientific American,* a periodical that has been the mainstay for science enthusiasts (and even professionals) since the middle of the 19th century. What distinguishes *Scientific American* from the market-oriented glorification of technological advancement characteristic of *Popular Science* is its use of professional scientists; whereas *Popular Science* tends to rely on "science writers"(the scientific equivalent of sports journalists), *Scientific American* lends itself credibility by featuring professional scientists who usually do their own writing. While on the one hand this practice provides readers with a more complete and accurate picture of scientific developments, ideas, and theories, it also means that the articles are often dense, despite the widespread use of graphics, theoretical summaries, and carefully compartmentalized article subheadings. The challenge for the professional scientist writing to the general public is daunting; she must present information in an accessible, fair, and balanced way without being overly technical while, at the same time, trying not to sacrifice either clarity or precision. For highly theoretical topics like Loop Quantum Gravity (the focus of *Scientific American*'s January 2004 issue), this balance of accuracy and accessibility can be problematic.

Kenneth Burke continually reminds us that seeing is also a way of limiting; when one focuses his or her gaze upon an object (in the literal sense) or an idea (in the more abstract sense), one is necessarily excluding, literally and symbolically, those objects and ideas that do not fall within the field of the focused gaze. Visual rhetoric—which can be described as the production of an image or series of images for their ability to reflect (and sometimes to forge) a symbolic value that both the producer and the recipient of those images share—is a sub-field of rhetorical studies that has unlimited potential. Therefore, nearly every symbolic act that uses imagery is at some level a rhetorical act, but, as such, it is also a limiting act. The popularization of science is no exception. Graphs, graphics, tables, charts, videos, digitization, and other visual effects are used for everything from clarifying an esoteric scientific concept to selling a product using the ethos and authority of science. Some contextualization of the more relevant theoretical and scholarly precedents for questions about tropes and figures in scientific expression is necessary, as is an examination of two selected visuals from a single edition of the *Scientific American*. I will focus on a single theory, Loop Quantum Gravity, to suggest interpretations of the images this magazine uses to construct identification with its audience, the ostensible goal of which is to popularize bona fide scientific concepts. The reading also implies two key questions that may be directive of future inquiry: what are the shared motivations driving these images that make them successful (if not entirely accurate) representations of scientific ideas; and how does, in Burke's words, the "focus on object A involve the neglect of object B" (*Permanence & Change* 49), such that the viewer comes away with a fixed and restricted grasp of the concept that these images are designed to capture, thus supplying the viewer with a finite understanding of the conceptual meaning that can be extrapolated from the theory. This is not an exhaustive primer on the rhetoric of science or of the thinking written about the tropes and figures to be used here, but rather, a specific reading of two specific images that reflects characteristics that may be useful in other interpretive situations.

Theoretical Context

To situate the scholarly treatment of popular science, some background may be useful. Perhaps one of the most important early examples of how popular science is packaged for a non-expert audience has been

Ludwig Fleck's *Genesis and Development of a Scientific Fact,* translated into English in 1979. What makes Fleck important is his observation that the manner in which scientific information in visually presented is as cognitively influential as the information such visuals are meant to contain. Thomas Kuhn himself, in his forward to Fleck's book, notes that his reaction to the now-famous duck/rabbit image did not acknowledge that the duck/rabbit was neither a "real" duck nor a "real" rabbit, but rather, a linear representation of the animals taken *in absentia* of the creatures themselves (ix). This realization is a radical if obvious observation, permitting other thinkers to see and comment upon the symbolic nature of orchestrated scientific information. Fleck also echoes Burke's ideas of orientation, trained incapacity, and terministic screens when he suggests the presence of "thought collectives," what Fleck describes as "the result of a social activity, since the existing stock of knowledge exceeds the range of any one individual" (38). This impression is in contrast to the accepted scientific (and/or sociological, perhaps even positivistic) idea that cognition is not the product of an individual process or a "particular consciousness," but a shared knowledge that comes together from many different intellectual, historical, and personal quarters.

Fleck reaffirms this sentiment when he challenges what Burke might have termed a "naïve verbal realism" in the process of scientific discovery. The usual scenario is what Fleck characterizes as a moment of "conquest" wherein "a person wants to know something, so he makes his observation or experiment and then he knows," the assumption that the discovery was obtained independently of any extraneous influences, hatched and pursued only in the researcher's mind (84). Instead, Fleck insists, such a situation does not come to fruition

> [. . .] until tradition, education, and familiarity have produced *a readiness for stylized (that is, directed and restricted) perception and action;* until an answer becomes largely preformed in the question, and a decision is confined merely to "yes" or "no," or perhaps to a numerical determination; until methods and apparatus automatically carry out the greatest part of our mental work for us. (84)

The impact on the development of scientific discoveries according to Fleck can be summarized as a collective process wherein "even the simplest observation is conditioned by thought style and is thus tied to a community of thought." Thinking in general, and scientific thinking in particular, becomes "a supremely social activity which cannot by any means be completely localized within the confines of the individual" (98). The relevance of this to visual rhetoric is that images, as symbolic substitutes for words, also elicit directed interpretations and, more importantly, define a concept in preordained ways, a process that underscores the suggestive nature of images and creates obstacles to alternative interpretations and descriptions.

One perception has been that science threatens to condense our most difficult and significant philosophical questions about the nature and origin of humanity to a series of dry technical queries that merely produce data but overlook the "soul" of the human organism. Both of the most recent major influences in psychology—evolutionary psychology and cognitive science—have gained a substantial following in the scientific community and, to a lesser extent, among the general population. Whereas evolutionary psychology argues for a Darwinian explanation of the origin of the modern human mind, cognitive science posits that the mind, like any other anatomical system in the human body, is a machine and can thus be "mapped," studied, and, one must conclude, *repaired* like a machine. According to Kenan Malik, there are three "principles" or underlying doctrines to the cognitive science of the mind: the brain functions like a computer; the brain is composed of separate "modules" like an array of inter-working processors; and these modules are innate in *Homo sapiens* (271). There are many other sources that boil down cognitive science, but Malik's is one of the more accessible and retains the primary theoretical descriptors.[3] What is interesting is that all three of these principles see the mind in mechanistic terms—even if they afford us the dignity of having complex machines for brains in the model of the computer. The problem is that the metaphor of the computer parallels the state-of-the-art of human technology; it is probable that we are simply superimposing the metaphorical template onto the brain and seeing patterns that fit this template, and if this is so, it ironically belies a much more rigorous interpretive mechanism for describing the complex functions of the brain.

There are many instances of how metaphors can dictate and therefore limit the way we understand certain scientific ideas, and this is another example. But it isn't my purpose at the moment to challenge the theory underpinning cognitive science so much as it is to point out this trend in scientific (and humanistic, for that matter) thinking and to show how cognitive science fits in the historical narrative of scientific development. As soon as machines dominated the human landscape and material philosophy dominated human thinking, our search for answers to burning questions tended to take form in these technological terms. It is reasonable to assume that the more machines are a part of our everyday existence, the more likely it is that they will influence our search for connections that reflect our metaphorical preconceptions. But the brain's assumed commonality with computer operations (another example of the inverted comparison—wouldn't the development of computers more closely resemble the workings of the rational brain?)—is simply a convenient way to reflect our most prevailing mode of thinking in a specific period in history. The shift from a human-centered society to a technology-centered one is a disturbing prospect because it is an arrangement that favors the thing over the living. This might be easily compounded by a scientific school that, in effect, posits that humans *are* machines. If we consider "human" or "humanity" in the strictly material sense, then this is hard to deny; our anatomical bodies are electro-mechanical in nature, and this is more than a convenient metaphorical description. Bones, muscles, tendons, ligaments, circulatory systems, nervous systems, digestive tracts, respiratory systems, and the like are all mechanical—they use the laws of physics to keep the organism alive. The heart is *literally* a pump; the plumbing of veins and arteries are literally pipes for moving fluid; the liver and kidneys are literally filters for removing toxins from the body. In fact, many of our most common technological systems are in some way based on systems that already existed in the body. It should then be no surprise to conclude that the brain is *literally* a "computer."

As interesting as this path of inquiry may be, the conclusion we can draw is relatively straight-forward. A human being, even beyond our assumptions about the superiority of humanity or our self-assigned post at the top of the evolutionary summit, is more than the sum of his or her parts. The human body, as wondrous as it is, is nothing more than a mass of unanimated matter without the brain, the most complicated, intricate, and poorly understood system in the human body. And bringing something to "life" in the human sense is far more than making the machine walk and talk and perform robotic tasks; there is something about the brain that breathes not only existence in a human being, but substance as well. Part of that substance, ironically, is a need for certainty, a drive so strong in some of us that we are willing to reduce the human animal to a set of "parts" that can be disassembled and analyzed like any mechanistic system. This is a failure of science. Malik, as a psychologist, recognizes that cognitivism, while at least going beyond behaviorism in its favorable depiction of human activity (for a behaviorist, as most of us know, any animal, humans included, merely responds passively to stimuli, not through any will of its own), still views humans as a remarkably predictable lot, requiring only careful dissection of the integral systems to understand the workings of the mind, a perspective Malik sees as limited and limiting. More importantly from the standpoint of a rhetorical examination, one might question the favoring of a scientistic school of thought in the psychological community and its impact on the population at large. How is the language appropriated and by whom, and what are the ripple-effects of this way of viewing humans?

In other words, we in rhetorical studies are interested in how a field of study like cognitive science evolves in its particular historical framework and how such an idea symbolically influences the mind-set of the culture in which it flourishes. Another, companion school of thought in the science of the mind is evolutionary psychology. Like the cognitive sciences from which it springs, evolutionary psychology claims, in the words of David Buller, "that human psychological adaptations take the form of modules, special-purpose 'minicomputers,' each of which is dedicated to solving problems related to a particular aspect of survival or reproduction in the human environment of evolutionary adaptedness" (127). If we consider the rhetorical impact of this school of thought, the implications are alarming because the language used to describe the modules of the brain are thoroughly materialistic, a situation not necessarily problematic in itself unless the metaphor becomes too completely literalized. Consider this explanation of modules from Buller's book *Adapting Minds: Evolutionary Psychology and the Persistent Quest for Human Nature:*

> Modules [. . .] are characterized by the following properties. First, they are domain specific, functionally dedicated to solving a restricted range of very closely related adaptive problems. They are like highly trained specialists who are incapable of performing effectively outside their area of specialization. Second, they de-

> velop in the absence of explicit instruction in the problem domains in which they specialize. This is because, third, modules "embody 'innate knowledge' about the problem-relevant parts of the world." That is, they are "equipped with 'crib sheets': They come to a problem already knowing a lot about it." Fourth, a module is, to some degree, informationally isolated from cognitive processing occurring in other parts of the mind, "operating primarily or solely with its own specialized 'lexicon'—a set of procedures, formats, and representational primitives closely tailored to the demands of its targeted family of problems." In other words, modules tend not to access information employed by other modules or in nonmodular cognitive processing. Finally, informational isolation enables modules to be comparatively fast at solving problems in their special domains. (128)

If Burke's assertion that a way of seeing is also a way of not seeing (i.e., to focus one's gaze upon one thing means necessarily that other things are excluded within that literal and figurative field of view), then we can see how the metaphorical language of evolutionary psychology can dictate an overly-narrow field of view, if not for the scientists themselves, then certainly for those on the receiving end of the more colloquial description. An important feature of Buller's description (which he is merely recounting, not endorsing) is the apparent "will" that individual modules appear to possess. That modules do not need "explicit instruction," that they "embody relevant information," that they "come to a problem," that they are "highly trained specialists," all indicate a misleading manifestation of what these metaphorical modules do when "solving" a particular problem. The notion that they have their own "lexicon" (i.e., specialized genetic "instructions") or are equipped with "crib sheets," and have their own territorial "domains" all gives one the impression of a little office space in the mind, or, perhaps more appropriately, a network of tiny classrooms coalescing into one microcosmic university in each brain.

It's not that this is or isn't "true" in the strict sense of the word; it's that such a description leads both the evolutionary psychologist and the non-specialist to take on the metaphor in incongruously literal terms. And this strange little world in our heads seems like a perfectly acceptable representation of the workings of the mind. If we hold that such a process in the working brain is reality, we are likely to operate under the constraints of this depiction of the mind when applying measures of a psychological, sociological, or educational nature. In other words, if we think of the mind as a network of mini-processors, as essentially a "wetware" (a term sometimes used to describe the anatomy of the brain) configuration of computer programming, we are likely to treat it as we would a computer. This has implications for everything from "fixing" a damaged brain to "training" it to use new programming and may ignore the inventive aspects of the mind that make it uniquely human. One may notice that I have been using the words "brain" and "mind" interchangeably here as well, another result of treating the computer metaphor too literally. Whereas the "brain" has been traditionally seen as the organ responsible for housing the thinking and the autonomic functions of the body, the computer metaphor makes no distinction between the physical organ that is the brain and the more imaginative processes that make up the mind. John Searle has pointed out that our conception of the brain has almost always paralleled the most recent development in technology, thus changing the perspective we have on the brain as different technology becomes available:

> Because we do not understand the brain very well we are constantly tempted to use the latest technology as a model for trying to understand it. In my childhood we were always assured the brain was a telephone switchboard. ('What else could it be?') I was amused to see that Sherrington, the great British neuroscientist, thought that the brain worked like a telegraph system. Freud often compared the brain to hydraulic and electro-magnetic systems. Leibniz compared it to a mill, and I am told that some of the ancient Greeks thought the brain functions like a catapult. (418)

Is a computer any more apt a comparison than a switchboard? On some levels, perhaps, but on the more basic rhetorical level, any metaphor comparing the brain to technology is going to provide misleading information about the complexity of the mind. While it is true that computers have capabilities that outstrip those of the human brain—reliable memory and speed in processing, for example—this does not in itself make it appropriate for calling forth similarities. It may, in fact, demonstrate how radically different computers and brains really are. Yet, we frequently operate under the assumption that the computer/brain analogy is appropriate, and we act on the comparison—through educational policy, psychological treatment, and rehabilitation—as though it were literally real. This should demonstrate the power of a single rhetorical trope for guiding and misguiding our ongoing attempt to make sense of the most complex product of evolution.

Language by its nature denotes an appropriation of meaning in its ability to abstract the objects of description, and this leads to an imprecision that is the bane of scientistic thinking. Because of this, Western culture has a tradition of searching for language that addresses the inherent "problem" that language can only function as an approximation of the signified concept, event, theory, or thing. Burke favors a dramatistic approach to language on the grounds that it is a fundamental misconception underscoring the nature of language to assume that there can be a direct correlation between the word and the thing that the word symbolizes; language, for Burke, is rather a process of approximations, and it is in those approximations that meaning shifts depending on the vantage point of the user and the recipient. The distinction for Burke between the dramatistic and scientistic approach to language is that the latter "culminates in the kinds of speculation we associate with symbolic logic," whereas the dramatistic approach "culminates in the kinds of speculation that find their handiest material in stories, plays, poems, the rhetoric of oratory and advertising, mythologies, theologies, and philosophies after the classic model" (*Language As Symbolic Action* 45).[4] Even some psychoanalysts recognize the depth of language and the human need to exploit it and the folly of pursuing a "pure" language unencumbered by the natural traits many (scientists in particular) see as undesirable. Ernest G. Schachtel, for instance, has observed that anyone serious about using language must contend with what he terms the "temptation of language" that seduces us into settling on the familiar and convenient words (Carlston 146). He adds that the quest for a language that offers a direct correlation between the thing described and the symbols used to describe it is an illusion. Once we put something into words, the words symbolically manifest the meaning and significance of the object or subject it replaces:

> The danger of the schemata of language, and especially of the worn currency of conventional language in vogue at the moment when the attempt is made to

> understand and describe an experience, is that the person making the attempt will overlook the discrepancy between experience and language cliché or that he will not be persistent enough in his attempt to eliminate this discrepancy. Once the conventional schema has replaced the experience in his mind, the significant quality of the experience is condemned to oblivion. (qtd. in Carlston 147)

An admission of this sort—one where it is understood in so-called "postmodern" terms that language literalizes an experience or idea and touches off a symbolic transformation that substitutes the words for the signified object of those words—should tell us that no language is bereft of this phenomenon. Science, too, is subject to this transformation, perhaps more so than other forms of inquiry since it deals so often in things that are unfamiliar, with phenomena that have no precedent in experience, such that the very act of naming corrals a scientific discovery and envelopes it in a linguistic holding pen.

Some Notes on Scientific Writing in General

Some rhetoricians of science, most notably Alan G. Gross, have viewed the enterprise of scientific writing in polemic terms. Gross defends science writing against charges that hold most of it to be impenetrable, esoteric, and jargony. He sees science writing as an archetype that "has evolved as a master narrative that makes visuals central to its arguments, is organized as a master finding system, and is written in a style appropriate to its subject matter and audience" (Gross and Harmon). On the other hand, many less sophisticated readers than Gross see the seemingly impenetrable science article as a question of "honesty" with the public, and he rightfully defends against claims that "the scientific article seriously misrepresents the way science happens, authors of scientific articles are also mendacious, predators of a fraud" (Gross and Harmon). These charges, according to Gross and Harmon, trickle in as commentary even from reputable publications such as *Nature, The New York Times Magazine,* and *The American Scholar.* Moreover, he says facetiously, "we have it on good authority that scientific prose is execrable, scientific communication dishonest" (Gross and Harmon). As the authors certainly realize, the concerns raised by the readers of these periodicals do not reflect a problem of professional integrity per se, but whether it is more important to complicate the truisms we have inherited about the intrinsic reliability of language to see how it is, exactly, that it conveys meaning and directs interpretive responses. The issue is how to package scientific ideas in a manner that is both accessible and precise, and for highly technical and innovative theories, this is no easy task. As it turns out, like in any profession, some people write clearly and succinctly and others do not. To engage in the debate in terms of honesty and integrity suggests that some view the issue as one where scientists are having their character attacked in a manner that must be reflexively defended. But this is to perhaps state the process of scientific communication in only its starkest terms. It's not that scientists "lie" about or deliberately obscure their findings so much as it about how severely they must sacrifice precision in order to explain scientific activity to the non-specialist. More importantly, it's about the form that scientific ideas take in order to allow the public a glimpse into the sequestered arena of scientific action.[5]

In the October 13th, 1997, issue of *The Scientist,* for example, an entire article in the "Professional" section is dedicated to the effective presentation of scientific data. The goal of this article was to combat bland, even boring, presentations aimed at both scientific peers and the public:

> In the worst cases, researchers simply read from transparencies, piling monotone fact upon fact. Another frequent gaffe is information overload. Cramming too much material into a talk, scientists either saunter past the time limit or speak too quickly. Jargon, too, foils many a presentation. And cluttered visuals—such as posters or transparencies—bury main messages in a sea of prose. (Brown)

Much of the subsequent advice—like speaking conversationally and using "short, punchy language"—contributes to audience identification and makes talks of a scientific nature far more engaging. Of course, visuals are the mainstay for many scientific presentations. In the same article, the advice is to "keep graphics minimal and offer the audience copies of the paper, which relates the data in detail," and some scientists, like Geraldine Richmond, a chemist at the University of Oregon in Eugene, say that they "try to give colorful presentations" and "use pictures whenever [they] can" (Brown).

Though this overview describes professional advice about a process in the community of scientists that is actually conducting science, the more interesting examples of the delivery of scientific ideas occur in that intersection between "expert" and "popular" science that Gross identifies, wherein the former can be characterized through Fleck's distinction between "journal science" (science that is published in specialized scientific journals) and "vademecum science" (established scientific fact that might appear in manuals or textbooks), and the latter is defined as "science for non-experts, that is, for the large circle of adult, generally educated amateurs" (112). Fleck's description of the presentation of material for a popular audience is, in many ways, the foundation for further analysis of popular science, demonstrating as it does the intersection between specialized science, popular science, and the rhetorical methods used to convey ideas and information. According to Fleck:

> Characteristic of the popular presentation [of scientific material] is the omission both of detail and especially of controversial opinions; this produces an artificial simplification. Here is an artistically attractive, lively, and readable exposition with last, but not least, the apodictic valuation simply to accept or reject a certain point of view. Simplified, lucid, and apodictic science—these are the most important characteristics of exoteric knowledge. *In place of the specific constraint of thought by any proof, which can be found only with great effort, a vivid picture is created through simplification and valuation.* (112–3)[6]

The effect of presenting material in this manner is two-fold: it suggests that the minutiae of scientific ideas are neither accessible nor important for most general applications, and it creates an inaccurate view of science as a static body of knowledge that is only added to, not reformed or rethought in any significant way. Fleck further suggests that the metaphors that scientists rely on are more than reductionistic visual conveniences; through the use of metaphors like the "organism of the economy," they are drawing upon "their fund of popular knowledge. They build up their specialized sciences around these concepts" (112).

Lee Smolin, author of the article, "Atoms of Space and Time," for example, is a theoretical physicist and a founding member and research physicist at the Perimeter Institute in Waterloo, Canada. He is the author of *The Life of The Cosmos* and *Three Roads to Quantum Gravity* ("Lee Smolin"). His contribution to *Scientific American* regarding the theory of Loop Quantum Gravity helps to explain this highly esoteric vision of the physical world with remarkable conciseness, yet the nature of the subject itself is such that close treatment of the material is elusive in a mere six-page article. According to Smolin, one persistent question in theoretical physics has been whether or not space is continuous, smooth and unbroken, or whether it is separated into discrete pieces, "like a piece of cloth, woven out of individual fibers" (Smolin 68). This is a question about the "granularity of space and time," and it has vexed Smolin and his colleagues for years. Why is it important? Because competing (or, perhaps, compatible) theories such as quantum mechanics and string theory are scientifically untestable through the use of physical experimentation—there is, in other words, no hard evidence to support these theories, a drawback that Loop Quantum Gravity is able to overcome by combining Einstein's theory of general relativity (which deals with the effects of gravity on massive scales) and quantum mechanics, which deals with physical processes at the sub-atomic level. In another example of the reliance on analogy that informs this theory, Smolin offers this experiment:

> An important test is whether one can derive classical general relativity as an approximation to loop quantum gravity. In other words, if the spin networks are like the threads woven into a piece of cloth, this is analogous to asking whether we can compute the right elastic properties for a sheet of the material by averaging over thousands of threads. (72)

In the past, finding a method that combines the diametrically opposite poles of the physical universe (i.e., the very large and the very small) to create a consistent picture of the cosmos has been impossible—until, that is, the introduction of Loop Quantum Gravity (68). Physicists have long been preoccupied with finding "a theory of everything"—the so-called "unification theory"; this was, in fact, the driving motivation behind Einstein's work. Because most scientists are drawn to equations that are elegant, beautiful, and even simple, there has long been a suspicion among physicists that the universe is in fact gracefully uncomplicated, that the right single equation could unlock the mysteries that have baffled us for centuries. The fabric metaphor satisfies both this quest for elegance and a material means for testing an otherwise untestable hypothesis.

Without going into the minute details behind this theory (which are fascinating, but, in many ways, beside my point), my purpose is to examine the mode of delivery propelling this information for public utilization. The format of Smolin's article is typical in many ways for *Scientific American,* especially for its feature stories. The first observation, which is one that runs throughout this study, is the ubiquitous use of the metaphor. As one of the "Four Master Tropes," Burke sees metaphor as a "perspective," something that "brings out the thisness of a that or the thatness of a this" (*Grammar of Motives* 503). He says that metaphor "tells us something about one character as considered from the point of view of another character" (503–4). A metaphor has the capacity, then, to appropriate the meaning of the thing it is designed to explain, giving it primary terministic authority, implying a symbolic point of reference and access to the idea (whatever it may be) through the lens of the words (and in this case, images) used to compare the metaphorical terms to the "actual" state of the thing being compared. It is a device for constructing meaning that, coming from a more familiar point of view, must acquire the properties and vantage point of the thing used for comparison. Burke further notes that, by using both empirical data and literary tropes, "it is by the approach through a variety of perspectives that we establish a character's reality" and that "characters possess *degrees of being* in proportion to the variety of perspectives from which they can with justice be perceived" (504). As Jeanne Fahnestock has pointed out in her book, *Rhetorical Figures in Science,* metaphor is so central a trope in science studies that scientific revolutions can actually be described *as* metaphorical revolutions (5). For Fahnestock, however, attention to metaphor has been advanced at the expense of other, equally productive rhetorical figures in science. She explains that

> the fixation on metaphor as a stimulus in scientific creativity, as a device in scientific argument or pedagogy (Boyd 1993, 485–486), and, finally, as an explanatory resource for textual scholars overemphasizes the role of analogy in human reasoning and begs the

> question of whether all or even most scientific cases can be "explained" by a core metaphor . . .. The tight focus on metaphor in science studies, like the fixation on metaphor and allied tropes in textual studies, has taken attention away from other possible conceptual and heuristic resources that are also identifiable formal features in texts and that also come from the same tradition that produced metaphor, the rhetorical tradition of figures of speech. (5–6)

Fahnestock is in fact closing a conceptual gap in attending to the other figures that inform our understanding of scientific concepts, but for science I would argue that the metaphor remains the most ubiquitous trope. It is difficult, I think, to overemphasize the role of metaphorical analogy in human reasoning because it is this symbolistic capacity that allows language to function on its higher levels. Without the analogical form of metaphor, it would be nearly impossible to abstract ideas for either communication or further intellectual inquiry; we would have to make due with literal identifiers that merely name the things with which we have immediate contact, where even the apparently simple act of naming objects carries with it symbolic distortions that prevent any absolute correlative between word and object. If George Lakoff and Mark Johnson are correct in their assessment that metaphor is the trope that we use to cognitively wrestle with unfamiliar concepts, and if the metaphor then becomes not only a useful device for understanding but actually forges the limits of our understanding, it is very important to see how this trope functions for highly abstract ideas for which we have no other meaningful point of entry (see *Metaphors We Live By.* Chicago: U of Chicago P, 1980). Burke assigned primacy to metaphor in his Four Master Tropes, and for good reason; it is metaphor that gives humans, who rely so heavily upon visualization, the imagistic stimulus necessary to understand phenomena that are otherwise beyond our sensory perception.

For other scholars, metaphor is especially central for science, relying as it does on sensory—especially visual—observation and verification. Donald Carlston has pointed out that the centrality of metaphor in scientific theory has been long acknowledged by philosophers of science, even those of a more positivistic orientation like Lachman and Nagel. He notes in particular psychological studies that have been designed to show a predisposition for one metaphorical interpretation over another depending on the experiential, educational, and cultural background of the observer. For Carlston, such studies indicate an obvious affinity for certain metaphorical descriptions, but they also have farther-reaching rhetorical implications, both positive and negative:

> Scientific metaphors are useful because they can aid in the interpretation, recall, and generalization of complex facts. But they are also hazardous because they can obscure alternative interpretations and encourage selective memory, usually for facts that support the accepted metaphor. Such a metaphor may also lead scientists to assume, without evidence, that certain things are true because the metaphor implies they should be true. (153)

That metaphors can actually dictate scientific thinking to the degree that they govern how the results of data can be interpreted is an important component to understanding how visual images direct our attention to one manner of thinking while concurrently discouraging more substantive thought on the theory. If the visual, in other words, literalizes the concept, then it allows the metaphorical description to manifest the meaning and the interpretation of the concept it hopes to make clear. Burke realized how central this symbolic activity was to directing the gaze toward predetermined orientations and how this, in turn, could both reify and obscure the essential qualities of the theory under examination. One can articulate a concept from any number of possible orientations, all equally "correct" in one sense and "incorrect" in many others:

> Any performance is discussible either from the standpoint of what it *attains* or what it *misses*. Comprehensiveness can be discussed as superficiality, intensiveness as stricture, tolerance as uncertainty—and the poor *pedestrian* abilities of a fish are clearly explainable in terms of his excellence as a *swimmer*. (*Permanence & Change* 49)

What this means, for the purpose of this reading, is that the "reality" of something as esoteric as Loop Quantum Gravity can only be made accessible to an uninitiated reader by drawing attention to certain visual characteristics that in truth are physically *in*visible. Scientists also rely heavily on the meaning generated by the metaphors they use, and because the metaphor in question can also be easily visualized, it can be assumed that they tend to think about these ideas in visual terms.[7] All visual representations are, by necessity, metaphorical representations—they are facsimiles of the thing described, a means of capturing for the viewer the similarities between things known well and things not known at all. For Burke, this means that metaphor is "a device for seeing something *in terms of* something else" (*Grammar of Motives* 503). In the case of the visual metaphor for Loop Quantum Gravity, the "seeing" is literal.

To further extrapolate on this, in the introduction to his Loop Quantum Gravity theory, Smolin employs the "fabric" metaphor, a common trope used when describing the continuity of space and time. In one graphic, there is a frame representation of a star field (designed to represent space); toward the bottom of this "square" of space, the picture begins to "unravel" into a single winding thread connected underneath to what looks like a cosmic ball of yarn. The caption reads, "Space is woven out of distinct threads" (68). This image, while intended to clarify a highly theoretical claim in physics that would normally be described in complicated mathematical equations, gives the reader pause. For a reader accustomed to thinking about space as a "vacuum" (i.e., vast stretches of nothingness—no air, no matter, only pinpoints of light from distant stars) this graphic might seem a bit counterintuitive. In fact, without the theoretical foundation necessary to understand the physical context that drives this representation, it just seems weird. One would need to know, for example, that Einstein's understanding of gravity is not as a force that large bodies in space exert themselves, but the effects of these large bodies on the space around it—the "warping" of the "fabric" that creates indentations in space, much like a bowling ball resting on a trampoline (the reader will forgive my own use of metaphor in this description). Only with an understanding of Einstein's explanation of gravity (an idea probably not widely known) would this unraveling tapestry image make any sense. Even then, of course, it is limited. At what level does the metaphor break down? How does the use of metaphor both enhance and restrict our understanding of this theory?

The relationship between metaphor and perspective that Burke establishes in *Permanence and Change* and later embellishes in *A Grammar of Motives* is as an "'incongruity,' because the seeing of something in terms of something else involves the 'carrying-over' of a term from one realm into another, a process that necessarily involves varying degrees of incongruity in that the two realms are never identical" (*Grammar of Motives* 504). It is easy to see how Smolin's visual/metaphorical explanation of Loop Quantum Gravity can become a literalized "reality" for those unfamiliar with the idea. The visual metaphor becomes, as a literal perception of the idea, a way of limiting as much as a way of understanding. The viewer/reader can only process the similarity sufficiently if he understands it in metaphorically literal terms, a cognitive processing of the image that ultimately results in a restrictive depth of understanding. To use a perhaps better-known example, the same cognitive inadequacies occur in evolutionary theory where the process of evolution is represented as a "ladder" rather than the more accurate "branching," leading to questions like, "If we evolved from apes, why do apes still exist?" The ladder metaphor, in this instance, does not reflect the concept of gorillas and humans sharing the same evolutionary ancestor as the branching metaphor does—it implies a progression, even a transformation, from one species *into* another. Branching, as an alternative metaphor, however, has its limitations, too. It implies, for example, that the trajectory of evolution produced by natural selection is one that "grows" and fills out like the branches of a bush, suggesting a progression from simpler to more sophisticated species. As mentioned in earlier chapters, while this is often the case, it is just as likely that a species will experience a genetic anomaly that is not beneficial to an organism, and may in fact inhibit its likelihood of survival, thus making the branching metaphor inaccurate for representing this biological reality. Likewise, people may easily think of space literally as a "fabric" that shares all of the characteristic of cloth, not as a convenient means of describing the infinitely more sophisticated set of physical conditions necessary for having a full grasp of Loop Quantum Gravity.

The fabric metaphor does, however, enhance our understanding by giving us access to the idea in the first place. As an otherwise intellectually off-limits theory for most of us, Smolin helps to give us a glimpse of this otherwise remote, impenetrable idea partially through his ability to reduce the theory to a manageable visual package that the non-specialist can grasp and partially through his use of the very metaphorical illustrations that aid in over-simplifying the idea for lay accessibility. Herein lies the difficulty for the scientist who hopes to popularize science for an uninitiated public: in the process of packaging an idea so that a wide range of interested agents can comprehend it, the scientist must sacrifice one of the underlying doctrines of the scientific process—precision. From this point of view, it is understandable that most scientists choose not to capitulate to the lowered common understanding that must occur in order to make science available to everyone. This may be one of the reasons they sometimes have a reputation for arrogance, over-refinement, and self-importance; they seem above the practical concerns that motivate and preoccupy the rest of us. Discovering the origins of the universe or the one formula that will explain everything does have a certain loftiness to it, perceptions notwithstanding. These are the questions that, if answered, get us closer to God, whatever or whomever that might be. It is not surprising that the public sometimes views scientists with a combination of awe and suspicion, and yet, the scientific explanation is compelling to viewers when accessed through the terministic screen of the fabric metaphor because it gives them an insight into this unifying theory that would otherwise be unattainable. In fact, Burke notes a fundamental connection between evangelism and science: In both, he says, "it is held that certain important aspects of foretelling require a new orientation, a revised system of meanings, an altered conception as to how the world is put together" (*Permanence & Change* 80–1). It should not be surprising, therefore, that so many people have the same reaction to unusual scientific theories as they do to the counter-intuitive outcomes of magic and the inspirational enticement of religion.

But rhetorically, while all the symbolic motivators, qualifications, and warnings that I've just described lurk beneath the surface, we are as fascinated by the questions as the scientist is, and we respond to that most pervasive of symbolic tropes, the metaphor. In the example cited above, the trope works on one level because it creates a point of intersection where that which we know (fabric, thread, weaving, looms) and that which we don't (the composition of space-time) intersect. The traversal between the familiar and the foreign, the common and the exotic, however, has many pitfalls. The typical American today is not preoccupied with a sophisticated knowledge of how language works, rhetorically or otherwise. For most Americans, language is a transparent, sign and signifier relationship; language *can* be made clear and precise and only cryptic academics make it more complicated than it really is, mostly for professional reasons (like publishing and tenure). Language is literal, and "plain" language is best. This intellectually literalist outlook, one that assumes a word is just a convenient moniker for a thing or an action, overlooks how easy it is to make a metaphor not only *represent* an idea but to actually *become* it. Burke refers to this unsophisticated perception of language as a "naïve verbal realism," and if we are given a means for thinking about space-time in terms of a massive piece of cloth, that metaphor "literalizes"—time-space is not *like* a massive piece of cloth, it *is* a massive piece of cloth, just as evolution is not "like" a ladder—it is a progressive, upward-moving hierarchy with humans at the top. Here is where the usefulness of the metaphor breaks down, for Smolin is not suggesting that the features of cloth and the features of space-time are identical; he is invoking a framing device that gives the reader a metaphorical point of entry into the theory which is best described through mathematics (which also has its symbolic limitations, but is far more precise—if less colorful—than its literary equivalent). Without the rhetorical boarding pass of the convenient metaphor, any access to a theory of such complexity and cosmic proportions would be theoretically prohibitive for the average person. Metaphor, particularly one that has the visual properties needed to literalize it, in this respect, is the great equalizer, the most democratic of all tropes. While the calculus (in the colloquial sense) that scientists use to represent their ideas is beyond most of us, the metaphor is not only understandable, it is a fundamental device for everyone who uses language. But it has severe limits, and a discerning reader who has at least an intuitive sense of these metaphorical limitations (and, even rarer, an understanding of how the metaphor creates a cognitive reality that is illusory) will see that the explanation is rudimentary, designed as a springboard for further, more complete inquiry. If nothing else, the careful viewer will realize that a metaphor is only a glimpse, a facsimile, a reflection, or a model that is an abstract re-enactment. For the rest, the metaphor is enough to complete their understanding, and they will process this concept on the literal basis of the metaphor alone.

A couple of pages later in this same issue of *Scientific American,* another artistic/metaphorical rendition is offered, this time in the form of an hourglass. In the upper compartment, we again see a field of stars meant to represent space; through the small opening separating the upper from the lower compartment we see hundreds of tiny clock-faces trickling through. The caption this time reads, "Time advances by the discrete ticks of innumerable clocks" (72). Like the fabric metaphor, the hourglass metaphor/image attempts to compress a sophisticated concept into one comprehensive image that marries our common understanding of time (as the progression of a mechanical time-keeper like a clock or an hourglass) to the counter-intuitive notion that space and time are fluid, not separable in the experiential sense that humans perceive it at all. This is reflected by the framing device of the hourglass—suggesting that space and time are the same. In this way, the literalization takes on another "dimension" in that it creates a parallel between two totally antithetical notions of time. One is our daily, somewhat clunky measurement of the passage of time through the use of instruments that regularly mark off intervals related to our perception of temporal movement, and the other is the theoretical abstraction of the nature of time as the universal fourth dimension. "Time," in the latter sense, relates to Einstein's notion of relativity, the idea that time, space, and matter are interdependent (DePree and Axelrod 633).

Through this model, "time and space become merged in a four-dimensional continuum in which neither possesses more absoluteness" and "time- and space-separations between events . . . depend on the choice of motion of the observer" (633). So in these two rather simple metaphorical images, we have the theory of loop quantum gravity in one comprehensive package—space can be reduced to measurable chunks just as time advances in separate "ticks." The "Overview" of Quantum Spacetime on page 68 boils down the theory for fast, bulleted reference:

- To understand the structure of space on the very smallest size scale, we must turn to a quantum theory of gravity. Gravity is involved because Einstein's general theory of relativity reveals that gravity is caused by the warping of space and time.
- By carefully combining the fundamental principles of quantum mechanics and general relativity, physicists are led to the theory of "loop quantum gravity." In this theory, the allowed quantum states of space turn out to be related to diagrams of lines and nodes called spin networks. Quantum spacetime corresponds to similar diagrams called spin foams.
- Loop quantum gravity predicts that space comes in discrete lumps, the smallest of which is about a cubic Planck length, or $10–^{99}$ cubic centimeter. Time proceeds in discrete ticks of about a Planck time, or $10–^{43}$ second. The effects of this dis-

> crete structure might be seen in experiments in the near future. (68)

Such overviews are typical devices in *Scientific American* articles, and they are useful because they provide an access point for the "physically challenged" (in the sense that the principles of physics are mentally elusive) who possess formal training in neither mathematics nor physics. However, it is difficult for the reader to see the real connection between this description and the metaphors previously offered. Articles like Smolin's, despite being cast for a lay audience, can still become rather dense rather quickly, so the rhetorical combination of metaphor, image, and overview provides a convenient set of learning tools while at the same time creating an incongruity between the theory and the simplicity of the metaphor used to describe it. This incongruity may be irreconcilable for the common reader, who is more apt to rest on his or her understanding of the metaphorical representation. Of course, the practical significance of understanding theories such as these is also elusive for most of us; theoretical physics lays down a wonderful foundation for the imagination, but what do we "do" with this knowledge? This is a question that may discourage people from moving beyond the limitations of their metaphorical understanding. *Scientific American,* in many ways, is predicated on the philosophy that knowledge, regardless of its practical value for the general population, is never wasted, and if it aids in educating the layperson about scientific matters on any level, it has fulfilled its role as a popularizing medium.

Be that as it may, what Burke provides us with in terms of how best to understand this rhetorical process is a vocabulary for showing the relationship between the symbolic trope (in this case, the metaphor) and the phenomenon or concept being filtered through the terministic screen of this trope. Both the fabric and the hourglass metaphor act as terministic screens that allow a very general level of detail to enter while sifting out information and conceptual frameworks regarding these theories that are fundamental to truly grasping its significance. Furthermore, the utilization of selective images and metaphors may have the effect of "directing the intention" toward certain elements of the theory that allow the viewer to take part in the discovery process without having the necessary internalization of the theoretical details to refute its claims (*Language as Symbolic Action* 45). Burke defines terminisitc screens, interestingly, in visual terms, making this notion even more useful for the decoding of scientific visuals:

> When I speak of "terministic screens," I have particularly in mind some photographs I once saw. They were *different* photographs of the *same* objects, the difference being that they were made with different color filters. Here, something so "factual" as a photograph revealed notable distinctions in texture, and even in form, depending on which color was used for the documentary description of the event being recorded. (45)

Burke's point, of course, is that, had other filters been used (or any other photographic techniques), other effects would have resulted, changing the perception of the objects and the meaning they elicit. What we "see" is not always what we "get" in interpretive terms; "reality" is altered, even constructed, on the basis of how we choose to symbolically enclose it. Here is where a literalist interpretation of any metaphor (or any use of language) breaks down; the same "reality" is represented in different ways by manipulating the symbolic point that resides in-between the viewer and the viewed. The result is a changed perspective and, in Burke's words, a "deflection" provided by "any nomenclature [that] necessarily directs the attention into some channels rather than others" (45). Interestingly, even in the meta-linguistic arena of language theory, Burke necessarily relies on metaphor to underscore the importance of this critical observation. Words, and in the case of the Loop Quantum Gravity example, pictures, construct a unique point of reference that would be different if different words or pictures were used to describe the phenomenon. The mistake both the viewer, and, frequently, the scientists as well, make is to substitute the representation for the thing itself. Imagine if a different metaphor had been used, say, one that described Loop Quantum Gravity not in terms of a fabric, but in terms of an ocean. How would this change both our understanding of the theory and the methods we used to examine it? According to Burke, as a new set of terminologies, we would necessarily treat the theory in those new terms. We would have little choice, since such metaphors require that we grapple with these new ideas in the language that we have supplied ourselves with.

The relationship between the phenomenon, idea, or theory and the metaphors (both visual and linguistic) that define them, in the case of the Loop Quantum Gravity example (and in all cases that rely on symbolic figures) has both illuminating and obscuring properties. Burke notes that, even in its metaphorical imaging, that a metonymic reduction occurs, and a reduction is still a *representation* (*Grammar of Motives* 507). Hence, any attempt to condense an idea as comprehensive as Loop Quantum Gravity to a series of metaphorical images or comparisons is to shrink it to a manageable size (or, conversely, enlarge it to a visible size). The fabric and hourglass metaphors become, in this sense, mere appendages of the larger theory. Just as a map of North America is only a reduced relief of an entire continent, so too is the notion that space/time is a cloth enhanced by the discrete ticks of innumerable clocks simply a sort of cartographic image of a theoretical concept that deals with both the minutely small and the infinitesimally large at once. As such, these metaphors focus the attention on the aspects of the theory that can be enhanced by the similarities between known objects and processes and unknown—to the lay person, at least—details of the phenomenon.

Burke's observations about the function of metaphor are useful as a way to interpret explanatory images in science writing. While I have focused on only one such example, the methods of illuminating complex theoretical concepts in science through the use of visualized metaphors is ubiquitous in the writing of *Scientific American.* They are, in fact, used extensively in almost all descriptions of scientific ideas that are intended for a general audience, and even scientists themselves rely on the models provided by imagistic metaphors for their own point of reference. While mathematics is of course the language of science, as beings who rely on both their visual aptitudes and their natural tendency to respond to symbolic representations, humans need visual metaphors to forward their perceptions of complicated natural phenomena that cannot be seen directly. This same responsive benefit is, however, also a liability in that we have a habit of confusing the model for the thing itself, leaving us with a literal interpretation of what happens when atoms collide or quantum strings unravel. Language, by its nature, is an imprecise apparatus, and no one was more attune to the limitations of language and its tropes than Kenneth Burke. His guidance helps us realize that the symbolic methods we use to make sense of the world are also important contributors to our own trained incapacity and to the selectivity of our choice of terministic screens.

Yet, despite the sacrifices in clarity or accuracy that *Scientific American* is forced to accept by using these images, there does seem to be a practical agenda to the science-writing they produce, namely, to inform the public of advances in scientific thinking. This enterprise alone might aid us in making more informed decisions when we are confronted with the opportunity to do so, and educating the public on matters of scientific interest has been a project that dates back at least to the mid-nineteenth century when publications like *Scientific American* first appeared on the newsstand. But "education" through popular scientific sources is at best limited, and at worst vulnerable to misinterpretation and binary thinking. Consider, for example, the implications of a psychological theory that challenges the notion that encouraging "self-esteem" in children can have many positive outcomes. In the January, 2005 issue of *Scientific American,* Roy F. Baumeister et al., wrote an article entitled "Exploding the Self-Esteem Myth," in which the authors posit that, "[b]oosting people's sense of self-worth has become a national preoccupation. Yet surprisingly, research shows that such efforts are of little value in fostering academic progress or preventing undesirable behavior" (84). Those of us who are professional educators have suspected for some time that the emphasis on self-esteem—especially in primary and secondary public schools—has been the bane of a realistic outlook for many students when they finally reach college because, in short, students have often been told they are much better at a given skill than they really are or that they have been more thoroughly prepared for the rigors of college study than they actually have been. Perhaps the most common rebuttal to low grades from students in my composition classes has been, "But I was in honors English in high school" or "I always got 'A's' in English on my writing before." This opinion of one's own abilities can only come from grade inflation predicated on the theory that low grades translate into low self-esteem.

However, we also know, intuitively if not scientifically, that continuous blows to one's self-esteem likewise translate into discouragement that is antithetical to improvement. In an attempt to make this sociological issue a scientifically verifiable phenomenon, the overview for the above-mentioned article reads:

- Self-esteem is viewed as a communal problem for Americans, who worry that inadequate self-esteem leads to various undesirable behaviors.
- Bullies, contrary to popular perception, do not typically suffer from low self-esteem. Neither do those who become sexually active at an early age, nor do those prone to abusing alcohol or illicit drugs.
- Raising self-esteem is not likely to boost performance in school or on the job.
- People with high self-esteem tend to show more initiative and appear to be significantly happier than others. (92)

This is a useful example for illustrating the sociological and psychological effort to scientifically validate what most of us already intuitively know to be true. It is interesting to note that the second bullet, the one "challenging" the bullies-don't-feel-good-about-themselves myth is attributed to "popular perception," not to the earlier psychological/sociological theories from which the myth actually originated. This assumption about bully self-esteem is not, however, currently in vogue, at least if *Psychology Today* (another source of soft-science popularization) is any indication of consensus among psychologists. According to Hara Estroff Marano:

> Bullies are also untroubled by anxiety, an emotion disabling in its extreme form but in milder form the root of human restraint. What may be most surprising is that bullies see themselves quite positively—which may be because they are so little aware of what others truly think of them. Indeed, a blindness to the feelings of others permeates their behavioral style and outlook.

It is interesting to note that Marano considers the finding that bullies think highly of themselves is "surprising," implying as it does that we have long held the belief that bullies pick on those weaker than themselves in order to compensate for some inadequacy they perceive in their own character. However, one must wonder how well this idea has ever corresponded with actual experience. Bullies, at least outwardly, certainly project an air of confidence and superiority—that's part of what makes them bullies. However we may view the feelings of self-worth that bullies experience, the real issue here is how the public perceives self-esteem as a condition for social equilibrium and financial and academic success, and it is more than a little compelling to note that the "popular perception" is attributed to the public, not the psychological community that dispenses advice to it.[8]

Other interesting language in the "Overview" contributes to the sense that public perception about self-esteem formed in a vacuum, that psychologists, sociologists, and educationists (all of whom use a quasi-scientific methodology, even if the results of that method are often mistaken, misapplied, or otherwise unscientifically utilized) contributed not at all to the notion that self-esteem aids in the development of the individual. Referencing the bulleted list above, for example, the claim that "self-esteem is viewed as a communal problem for Americans" suggests that this is a concept that Americans arrived at independently of any outside influence and have obsessed over to the extent that they "worry that inadequate self-esteem leads to various undesirable behaviors." But where has this apparently erroneous idea originated, and how has it been validated? It should come as no surprise that one source is—so far as the public is aware—the psychological community itself (or those who pass themselves off as members of it). Dr. Barbara Becker Holstein, for example, is a "positive" psychologist (apparently) who espouses the benefits of high self-esteem, but in a rather lengthy search I conducted attempting to pin down her credentials, I could find nothing about how she was schooled or where she received her PhD. In her "bio" on a website calling itself "Psychjourney" is the following information about "Dr. Barbara":

> Meet Dr. Barbara Becker Holstein. Positive Psychologist & Happiness Coach Dr. Barbara Becker Holstein is the originator of *The Enchanted Self.* She has been a positive psychologist in private practice and licensed in the states of New Jersey and Massachusetts since 1981. Her book *The Enchanted Self, A Positive Therapy* was published in 1997 by Harwood Academic Publishers and is now in its second printing through Brunner-Routledge. This book is Dr. Holstein's pathfinder book where she outlines the best in Positive Psychology treatment techniques, two years before the term was coined! Her second book *Recipes for Enchantment, The Secret Ingredient is YOU!* has received rave reviews

> as a wonderful inspirational story book with space to journal.

The good doctor has a license (so we're told), but beyond this, her only other articulated credentials are her books. The public, usually not well-schooled in critical discourse, may have difficulty here discerning valid credentials from mere claims of expertise (the "Dr." moniker is often enough substantiation for many people) and is left with the daunting task or sorting out the genuine article from the merely declarative. Psychology, perhaps more than any other discipline that uses science to enhance itself professionally, is ripe for exploitation and professional abuse because it is a "science" that is applicable to the layperson in the most intimate ways. The general public may indeed be embracing a myth about many psychological "truisms," but they are myths that are handed over by what appear to be valid authorities. Even if Dr. Barbara is for real, it is the psychological community itself (or a professed member of it) that has given the American public the idea that self-esteem is key to a successful life—not wives' tales or urban legends as Baumeister et al. imply in the *Scientific American* article dispelling this myth.

While it is true that such "knowledge" about popular psychology and the experts who peddle it can be distorted or even misrepresented, the sheer attractiveness of psychology as a panacea for all public and private ills is very often enough to legitimate even the most absurd path to "happiness." Here the overlap between consumerism and science (or, perhaps a more accurate term for the psychological venue is "The Clinical") is at its most palpable. People seeking change and meaning in their lives are in an extremely vulnerable position for the consumption of emotional snake oil. As any cult leader will testify, people experiencing low self-worth are easy targets for quick fixes, magic therapies, and social conditioning. The practicing psychologist is in an extremely powerful position, one that requires the utmost discretion and professionalism. A good psychologist takes this responsibility very seriously and recognizes the importance of being realistic with her patients. Achieving life satisfaction is difficult, life-long work. One does not simply wake up one morning after a long spell of gloom to an attitude of personal fulfillment. "Happiness," whatever that is exactly, means different things to different people, and to sell formulas like "positive thinking" (an approach that goes back at least as far as Norman Vincent Peale in the 1950s) as a mystical way of reaching this elusive goal is the worst kind of charlatanism. In the book (cited above), *Recipes for Enchantment: The Secret Ingredient is YOU!,* Dr. Barbara has hit upon her own recipe for happiness in the form of book sales, but whether this state of being can be extended to the reader is questionable. The blurb on the text tells us that *Recipes for Enchantment*

> is a culmination of over ten years of work and research in the field of personal enchantment by Dr. Barbara. She teaches us how to recall positive memories, regardless of our history, recognize our potential, and bring into our lives more of what gives us pleasure and joy. This is all combined with daily positive actions. A joy to read, you will learn how to "cook up" delicious days for yourself and create a "Recipe for Enchantment." (http://www.enchantedself.com/books-by-dr-barbara-holstein.html)

Again, the language is telling. "Personal Enchantment," so far as I'm aware, is not a "field" recognized by any reputable psychological organization. "Enchantment" itself is hardly a clinical phrase, and conjures notions of a mystical, miraculous transference that begins with feelings of despair and leads to feelings of bliss. We are told we can "cook up" "delicious days" like a witch's brew for warm thoughts, which is in keeping with the "enchantment" *and* the "recipe" metaphor, emphasizing as they do both the mystical process and the idea of a magical potion. Nor is this choice of metaphors some underhanded means of hiding the professed intention of the book. The metaphors are clearly designed to give one the impression that happiness is a supernatural waving of a psychological wand, as we see in the next paragraph:

> This is a breakthrough book combining psychology/self-help with inspiration. Psychology/self-help activities are beautifully interwoven with inspirational stories that teach the most basic recipe: incorporate within our daily life positive actions, combined with our unique positive thoughts and feelings! The rest will take care of itself *as if by magic!* (http://www.enchantedself.com/books-by-dr-barbara-holstein.html, emphasis added)

Here, we see an unfettered combination of psychology, "self-help," and "inspiration." In other words, Dr. Barbara has ostensibly merged the best of several knowledge systems into one comprehensive cookbook; where science ends, inspiration begins, and self-help and "recovery" inevitably follows. She has concocted a recipe for normalcy.[9] For a person made prone to such promises through misery and desperation, her guarantees must seem attractive indeed. In all fairness, I must admit, it may be my own "negativity" that keeps me from seeing the splendor of this arrangement, but I do recognize what a sweet rhetorical ploy it is on Holstein's part: she has conveniently included in her program a built-in mechanism for dismissing her detractors. All she needs to claim is that those who don't embrace the wisdom of her project are suffering from chronic negativity; they cannot see the beauty of her system because they are mired down in their own feelings of desolation and their own cynical outlook on life. This is, in fact, a common conditioning tactic in many psychological and/or therapeutic programs. To resist the help that others wish to give you in a therapeutic session is the product of your own "denial," a brilliant strategy considering that any suggestion that you might not be experiencing the problems with which you are diagnosed can be classified as a symptom of the very problem you are denying. Through such discursive appropriation, psychological talk-therapy replaces negative or damaging discourse with positive, "healthy" language. While people certainly do "really" engage in denial, the charge is easily abused. For example, a therapist convinced of his diagnosis that Patient A suffers from an addiction to soap operas can charge that patient with denial if the patient insists that, while she may in fact enjoy watching them, they are not hampering her life in any deleterious way. Both the definition of "hamper" and the assignment of denial to the patient are rhetorical mechanisms under the control of the therapist. If Patient A puts off doing her schoolwork for an hour so she can watch "Days of Our Lives," under a strict definition of addiction, the procrastination has had a negative effect on her life—it has interfered with "work." If she feels that such procrastination is minimal, and she gets her work done regardless, her denial may only exacerbate the problem. Moreover, the therapist controls the discourse, and whenever it is the case that one agent has privilege over the language that is acceptable, that agent also possesses all of the power. We see here how easily a therapist can manipulate responses to fit loose hypotheses that may or may not be accurate. Discursively, the therapist is in complete control of the interpretive situation by having both the power to establish definitions and to analyze the language given by the patient.

"Dr. Phil," of Oprah fame, is far more of a household name than Holstein. He, too, casts himself as a "positive psychologist," but unlike Holstein, he has not hidden his credentials. On his very own web page, Dr.Phil.com, his bio mentions his five books first, then, at the end, where he studied: "Dr. Phil has a B.S, M.A. and Ph.D. in clinical psychology from North Texas State University with a dual area of emphasis in clinical and behavioral medicine." Dr. Phil is cashing in every bit as much as Holstein, but in a different and, in some respects, more insidious way. His advice is based upon training, we can assume, but is so general that it has a "one-size-fits-all" effect; rather than treat clients (a more appropriate term than "patients") individually, he offers sage advice to everyone as if their lives are cookie-cutter patterns that require only the most perfunctory of solutions. How far can one get in solving life's individual problems with the slogan, "Get real," advice that Dr. Phil might well follow himself (Dr. Phil is now, incidentally, getting involved in the matchmaking game after having witnessed the spectacular pseudo-scientific success of eHarmony.com)? Of course, I am oversimplifying the program; one must also "Get smart" and "Get going," professional counsel that can be placed on a bumper sticker. Such sloganizing has been popular in therapy groups for some time. Twelve-step programs are particularly fond of slogans, recommending that we "Keep it simple, stupid," take things "One day at a time," or "Change the things that we can." Catchphrases like these have the effect of commodifying therapy, like a jingle for a product, a branding mechanism for healthy discourse.[10] We attribute the idea that someone needs a "reality check" to Dr. Phil in the same way we attribute "My boloney has a first name" to Oscar Meyer wieners. It is a packaging device, and what makes it most objectionable is that it treats genuine, adult, *human* problems as if they are something that can be treated with the right product. Do you have lime on your bathroom stall? Use Limeaway; do you have trouble communicating with your "significant other"? Keep it real. Such slogans have gained cultural currency because one successful media personality has endorsed the man who popularized them, creating a media union that has a distastefully nepotistic feel to it. What's worse is that this media phenomenon does not stop at the peddling of psychological tidbits; actual merchandise (not just books and videos) is sold as well. From Dr. Phil's website, we can shop for t-shirts, sweatshirts, scrubs, photo frames, baseballs caps, and even latte mugs (see http://drphilstore.com/ and http://drphilstore.com/merchan2.html). As most self-respecting psychology professionals would likely agree, Dr. Phil has done the most to level blows against the credibility of psychology as a serious, scientific field of study since John Gray, author of *Men Are from Mars, Women Are from Venus.*

Obviously, I consider Dr. Phil just as dubious as Dr. Barbara,[11] degrees notwithstanding, but taken together they do raise an interesting question: When does the public, especially those most vulnerable in the population—those searching for satisfaction, stability, or just plain answers in their own lives, often undereducated and underemployed—know when to trust the seemingly endless stream of "experts" who are happy to sell them such answers for $24.95 (in paperback)? And what makes the *Scientific American* writers more credible than their mainstream counterparts? Naturally, *Scientific American* doesn't deal strictly with psychological issues in the same way that, say, *Psychology Today* does, but it does enjoy credence with the scientific community at large. Perhaps this is because *Scientific American* seems to possess an honest agenda, one that is not above profit, certainly, but one that will not sacrifice the integrity of its project for celebrity and the bottom line. A regular feature of *Scientific American* is a section known as "Skeptic," written by Michael Shermer. Shermer is the publisher of a magazine of his own by the same name, and his curmudgeonly critiques of anything New Wave, Alternative, or otherwise Pseudoscientific create both an ethos of credibility and candor and a source of amusement.

One interesting example of Shermer's dauntless positivism occurs in the February, 2004 issue where he reviews a "romantic" and "revisionist" account of the famous mutiny on the *Bounty.* The author of this revisionist account, Caroline Alexander, concluded that the real cause for the mutiny was "a night of drinking and a proud man's pride, a low moment on one gray dawn, a momentary and fatal slip in a gentleman's code of discipline" (qtd. in Shermer 33). Shermer's skeptical, "deeper" account relies exclusively on the "hard" historical evidence and scientific fact he can bring to bear on the event. He says that his explanation is more "intellectually satisfying" because "it is extrapolated from scientific evidence and reasoning." He further asserts that there were two levels of causality that contributed to the mutiny: "proximate (immediate historical events) and ultimate (deeper evolutionary motives)" (33). According to Shermer, the first cause, the one that challenges the notion that Bligh was overly harsh on and cruel to his men, can be explained thusly:

> A count of every lash British sailors received from 1765 through 1793 while serving on 15 naval vessels in the Pacific shows that Bligh was not overly abusive compared with contemporaries who did not suffer mutiny. Greg Dening's *Mr. Bligh's Bad Language* computed the average percentage of sailors flogged from information in ships' logs at 21.5. Bligh's was 19 percent, lower than James Cook's 20, 26, and 37 percent, respectively, on his three voyages, and less than half that of George Vancouver's 45 percent. Vancouver averaged 21 lashes per man, compared with the overall mean of five and Bligh's 1.5. (33)

The numerical effort to redefine the relative declaration of "harshness" that has been leveled at the corporal punishment used by Bligh is an interesting attempt to quantify a cultural norm for lashings during the period that Bligh commanded. His point, of course, is that Bligh was no harsher—and in some cases was more lenient—than his peers, dispelling the commonly-held myth that Bligh was an oppressive commander with a streak of sadism. However, such quantification does not include the reasons for the lashings, only the number administered. This is an important oversight, because certainly feelings of abuse might be exacerbated by feelings of injustice that are absent from the historical records or subject to a variety of interpretations (Mr. Christian's journal would be more credible as a piece of evidence for the mutiny than Shermer's quantification of lashings, but it of course could not be reduced to fixed number of countable offenses). But as Shermer attempts to torpedo the theory that abuse was the cause of the mutiny with statistics, one is left wondering what could have been the actual cause? According to Shermer, it was simple biology: The men were horny for Tahitian women, and since nearly all of them were of an age that we would today consider the "sexual prime," the choice between living in a tropical paradise with scantily-clad women or returning to the stuffiness of 18th century English society was, as they say, a no-brainer. Shermer describes the neuroscientific requisite for a sexually-triggered mutiny by noting that chemical motivators like oxytocin, which is secreted into the blood by the pituitary gland during sex, could have put the crew into a state of mutinous frenzy. He adds:

> Ten months at sea weakened home attachments of the *Bounty's* crew. New and powerful bonds made through sexual liaisons in Tahiti (that in some cases led to cohabitation and pregnancy) culminated in mutiny 22 days after departure, as the men grew restless to renew those fresh attachments; [Fletcher] Christian, in fact, had been plotting for days to escape the *Bounty* on a raft. (33)

Shermer's account is designed to give us a rationalistic explanation for a crime of passion. But does it work? Shermer concludes that "[p] roximate causes of mutiny may have been alcohol and anger, but the ultimate reason was evolutionarily adaptive emotions expressed nonadaptively, with irreversible consequences" (33). Human motivation, under this interpretation, can be reduced to "ultimate" causes, and the details of the "proximate" causes are incidental, minor contributors. It is difficult to deny that our biology has more of an impact on our behavior than we usually assume—certainly sexual drives contribute to our search for mates and our interactions with the opposite sex—but Shermer seems to be suggesting that the mutiny on the *Bounty* was an *evolutionary* event, that the crew was simply "adapting" to a changing environment, even if that adaptation proved (in the short term, at least) to run counter to the best interests of this representative group of the species *Homo sapiens.* Such a reading is simplistic at best; surely the proximate causes had far more of an impact on the specific behaviors in this context, and certainly the men were moved at least as much by nurture as by nature. Furthermore, as a scientist, it is irresponsible for Shermer to make any biological claims about the men involved in the mutiny unless he can examine them himself. He does not know the level of oxytocin secreted into the men's bloodstreams and should therefore draw no conclusions about it.

One of the fundamental aspects of disparate knowledge systems that seem contrary on their face is the idea that different ways of knowing are necessarily incompatible. Shermer, representing science, thinks that his interpretation of the events is more "intellectually satisfying" than Alexander's. But it seems likely that Alexander would surely disagree, and she would perhaps even concede to the idea that both the proximate and ultimate causes are not mutually exclusive. However, when studying human motivation, one might question how satisfying (or even accurate) a strictly biological motivation is. We could, for example, describe greed and its manifestations as simply an extension of our territorial instinct to carve out a piece of space that we can call our own. The more money and power we possess, the more space (in both a literal and figurative sense) we can occupy. But this explanation, in my view, at least, is hardly "satisfying" at all. It is plausible to suggest, for example, that the baseball player who is offered a $92,000,000 dollar contract but is holding out for $97,000,000 is not especially interested in the money, since $5,000,000 more makes little difference when one is pulling down this kind of scratch. What motivates this person? Is it the "territorial instinct," a need to accumulate as much as possible in order to secure one's survival in the world? On a rudimentary level, perhaps. But it is more likely that the status afforded being a top-paid player is a stronger impulse, and this is certainly the result of *social* more than biological influences.

Regardless of how much fun Shermer must be on a date, the level of credibility he lends to the publication and the furtherance of the positive scientific ethos are undeniable. He is the archetypal rationalist, the person who believes (or, rather, reasons) that rational, scientific causes are the only ones that matter. However rational an outside interpreter may be, imposing that rationalistic template onto irrational humans is bound to skew the results, and Shermer should know this. Still, *Scientific American* maintains its positive ethos in just this way—by showcasing well-established rational thinkers, those who the general public views with trust (and perhaps a twinge of envy). Shielded from hoodwinkers, charlatans, humbuggers, quacks, swindlers, counterfeiters, imposters, and intellectual grifters (including "revisionists" who are, these days, generally referred to in the pejorative) with the impenetrable armor of reason, figures like Shermer stand as the last bastion between us and the crooks who would sell us everything from happiness to everlasting life.

All facetiousness aside, people like Shermer do serve a valuable public function by showing us the wizards who are really just men behind curtains. In the February, 2005 issue of *Scientific American,* Shermer goes after those claiming to have been abducted by aliens, people who fall into a different category altogether: the self-delusional. In the April, 2003 issue, he allays people's irrational fears about cloning; in the February, 2003 issue, he explains why scientists do not believe that extrasensory perception or psychic phenomena exist. In the November, 2003 issue, he dispels myths about the nature of dualism, the so-called mind/body problem in philosophy, and explains how such a separation is probably wrong. In the March, 2004 issue, he challenges the reliability of eyewitness testimony with the results of perceptual blindness experimentation. Clearly, there is no shortage of pseudo-scientific theories or popular misconceptions to debunk—his feature on skepticism runs every single month. Shermer is a rationalist, and while he may challenge (and often, successfully disprove) many of the ideas we truly *want* to believe in, such a service is important in the battle for democratic argumentation. Without certain standards of evidence on which to base argumentative discourse, whether they be scientific, legal, or rhetorical, people tend to make poor choices. In a democratic society that relies on a general public to vote on important issues, policies, and political candidates, those who are ill-equipped to distinguish between sound arguments and nonsense are dangerous indeed. One way that Shermer and *Scientific American* aid in this ongoing battle is by relying on a methodology that supports the inclusion of sound evidence and excludes claims that cannot meet this threshold. However, as the mutiny on the *Bounty* example illustrates, we must question whether strictly scientific standards of proof are always appropriate or accurate, especially when we are dealing with often irrational, nebulous motivations that drive human behavior.

Popular sources for scientific information revolve around several key approaches. They tend to access an already existing sense in the public that science and technology are a positive part of our cultural make-up. Likewise, they maintain a pervasive sense of our own innovative, rationalistic heritage, the ability of Americans to face challenges with pragmatism and efficiency, key components of the scientistic ideology. Such sources also capitalize on the inevitable connection between science, technology, and consumerism, making the motivational impulse to stay on top of technological developments an important part of owning and mastering the latest electronics. In the areas of personal fulfillment, therapy, and counseling, popular sources become self-legitimizing and self-perpetuating aspects of the entire mental health industry, though it is interesting to note that, despite the widespread popularity of therapy, there seems to be little professional consensus on the causes of certain maladies while there is a concurrent tendency to adopt a boiler-plate methodology for addressing issues of emotional well-being. Popular sources, then, have the doubly-vexed effect of generating continued support in a scientistic worldview while maintaining an influx of funding for all sorts of professional endeavors. While sources like *Scientific American* pursue the laudable objective of informing the public on the most recent scientific developments, it is imperative to a periodical such as this to continually enlist support for the sciences in general. Moreover, all popular sources are in some way driven by the same economic factors that beset other professional areas. But what of other motivators for science or its alternatives? Where does the pursuit of professional security end and the need for ideological certitude begin? The next chapters will provide some possible answers to these questions.

8 Intelligent Design, Creationism, Evolution, and Darwinian Descents

In the last forty years, one major court issue continues to resurface over and over again—an issue that one might have expected to be put to rest generations ago but which has, in fact, been the most persistent major educational question of the last half of the twentieth century. The issue, contentious as it is, and as the following rulings will attest, has in fact been repeatedly settled legally, if not socially:

Epperson v. Arkansas (1968)

The Supreme Court found that Arkansas' law prohibiting the teaching of evolution was unconstitutional because the motivation was based on a literal reading of Genesis, not science.

McClean v. Arkansas (1981)

A federal judge found that Arkanas' "blanced treatment" law mandating equal treatment of creation science with evolution was unconstitutional.

Segraves v. California (1981)

A California judge ruled that teaching evolution in public school science classes does not infringe upon the rights of any students or parents to the free exercise of their religion, even if they sincerely believe that evolution is contrary to their religious beliefs.

Edwards v. Aguillard (1987)

In a 7–2 decision, the Supreme Court invalidated Louisiana's "Creationism Act" because it violated the Establishment Clause.

Webster v. New Lenox (1990)

Seventh Circuit Court of Appeals ruled that school boards have the right to prohibit teaching creationism because such lessons would constitute religious advocacy and, hence, such restrictions do not constitute an infringement on a teacher's free speech rights.

Peloza v. Capistrano (1994)

Ninth Circuit Court of Appeals decision that a teacher does not have a right to teach creationism in a biology class, that "evolutionism" is not a religion or world view, and that the government can restrict the speech of employees while they are on the job.

Freiler v. Tangipahoa (1999)

Fifth Circuit Court of Appeals found that a disclaimer to be read before teaching about evolution ultimately had the effect of furthering religious interests and was therefore unconstitutional.

LeVake v. Independent School District (2001)

A federal district court finds that a school may remove a teacher from teaching a biology class when that teacher, a creationist, cannot adequately teach evolution.

Kitzmiller v. Dover Area School District (2005)

The suit was brought in the U.S. District Court for the Middle District of Pennsylvania seeking injunctive relief. Since it sought an equitable remedy there was no right to a jury trial; the Seventh Amendment did not apply. The eight Dover school board members who voted for the intelligent design requirement were all defeated in a November 8, 2005 election by challengers who opposed the teaching of intelligent design

> in a science class, and the current school board president stated that the board does not intend to appeal the ruling. (Cline)

In the last few weeks leading up to the presidential election of 2004 the current Commander-in-Chief, George W. Bush, kept deadly pace in the polls with his rival, Mr. John Kerry. They were daily jockeying for position, and the outcome of the election, at that precise moment in time, was very uncertain. The current magistrate was a "born again" Christian, and it is telling that he had a constituency that in large measure bought his theological swindle. They believed he was (and is) a man of his word, a revived sinner, and a man of action. Of these qualities, the last is undeniable. He does act, often impulsively, like a little kid who wants everything his way and wants it *now,* and uses the machinery of other human beings at his disposal. Are people incensed by this? Are they outraged that a man could be so cavalier with other peoples' lives (in many cases, sons, daughters, brothers, sisters, and friends)? Some are, but if you talk to people on the street, as it were, you realize that many are not only not indignant about this, but they applaud him as "steadfast"—not a "flip-flopper" like Kerry, who has the annoying habit of thinking about things before making a decision and even changing his mind if new information becomes available, who might actually have a crisis of conscience over the prospect of sending other people's families into the intense dangers of a war zone (as a Vietnam veteran, Kerry has first-hand experience of this; Bush, by contrast, used his privileged status to avoid combat service, and therefore never saw the horrors of war with his own eyes).

One of my students said I was a "liberal" because I pointed out that I saw a vast contradiction in the notion that Bush likes to send people to war on the very principles of freedom that he concomitantly likes to take away from his own people. I was also a "liberal" because I saw an inherent contradiction between the life-affirming convictions of anti-abortion and the indifferent blood lust of sending young men and women to die in a far-off desert. The point here is that, in a rhetorical climate that breeds an anti-intellectual, action-oriented, under-educated population, zealousness can easily abound, especially if a conservative agenda keeps people in their SUV's and the government out of their pockets. It is no surprise, therefore, that the religion that people do seek is one of convenience and simplicity. Spirituality, when people do in fact seek it, must be in a form easily identified and easily digestible. It must not unduly muddy the issues with questions of doctrine, dogma, or tradition. It must not try to reconcile contradictions in Scripture because to do so is blasphemous and, in the end, futile. We must face evil head-on because God is on our side. In the absence of any "mystic" religious conviction, however, popular science provides at once an acceptable surrogate and a plausible alternative. It, like fundamentalism, is no-nonsense. What you see is what you get.

If we were to be honest with ourselves, we would have to admit that the issue of whether a Creationist or an Evolutionary explanation of human origin is more likely was settled over a century and a half ago, and to give it more attention than I do here is to lend the debate more credence that it realistically deserves. It does, however, illustrate the power of belief systems and the rhetorical nature of the language that contributes to their stasis. At the center of the religion "vs." science issue is the question of our origins, so naturally people have a lot at stake in the answer to this question. It should be clear to any educated, critical person that evolution is far more that just a "theory," that it enjoys scientific consensus unlike almost any other scientific idea in history and is demonstrable on both a micro and a macro level, yet the detractors of evolution (preposterously referring to themselves as creation "scientists") know that tapping into the ethos of science will help further their credibility. At the same time, creationists deny the very science that they use to give their own movement (at least in name) credence. What this tells us is that people are easily confused about what is within the province of science, what is in the province of some other knowledge system, what is neither in the province of science *nor* another legitimate way of knowing, and what, given our current state of intellectual aptitude, is and is not knowable at all. That creationists can confuse the issue by speciously turning science against itself (which it *never* successfully does) only speaks to the general public's inability to distinguish between sound scientific argumentation and that which only sounds intuitively reasonable.

At the same time I would be loathe to give the reader the impression that I think the battle for intellectual primacy between science and religion (and all the other domains of knowledge that scientists often ignore) should continue—the Gaza Strip of the religion/science struggle. We could be willing to consider Gould's resolution of non-overlapping magisteria if the ramifications were not so extensive. The question, however, is whether this is at all possible. And an even better question, insofar as it relates to the subject of this book, is how this argument might be received by the public. The creationism/evolutionism debate has resurfaced at several points in this book, because it is an important and timely example of the intellectual battlefield (sometimes referred to as the "culture wars") that not only defines America as a nation, but has far-reaching consequences if we don't educate ourselves about the state of knowledge regarding this "issue." We must pretty much accept the idea that "intelligent design" has nothing to do with science, yet from a public relations perspective, it is science that has been put on the defensive. This is strange, since acceptance of the doctrines of faith-based alternatives to evolution means the insistence that no evidence is required, whereas the arguments against evolution are all about the evidence (or supposed lack thereof). In other words, creationists hold science to a standard that they will not bring to bear on their own belief system. Verlyn Klinkenborg puts it well in a *New York Times* editorial entitled "Grasping the Depth of Time is a First Step in Understanding Evolution" when he states that

> [i]ntelligent design is not a theory as all, as scientists [or any other rational being] understand the word, but a well-financed political and religious campaign to muddy science. Its basic proposition—the intervention of a designer, a.k.a. God—cannot be tested. It has no evidence to offer, and its assumptions that humans are divinely created are the same as its conclusions . . .. This isn't a triumph of faith. It's a failure of education.

Indeed. And one of the biggest obstacles to a full understanding of evolution, as I've mentioned elsewhere, is a failure of the human imagination to grapple with the staggering amount of time necessary for evolution to produce multi-celled organisms, much less human beings. Klinkenborg echoes the same sentiment when he states:

> One of the most powerful limits to the human imagination is our inability to grasp, in a truly intuitive way, the depths of terrestrial and cosmological time. That inability is hardly surprising because our own lives are so very short in comparison. It's hard enough to come to terms with the brief scale of human history. But the difficulty of comprehending what time is on an

> evolutionary scale, I think, is a major impediment to understanding evolution.

The central problem, Klinkenborg suggests, has as much to do with conceptual shortcomings as it does with any of the facts in evidence. As corporeal beings (beings that are prone to the limitations of evolutionary endowment), we are simply not equipped with the sensory or contextual apparatus to experience geological time. We can attempt to imagine such vast seas of time, but we are never able to truly understand it. What makes this deficiency interesting is that evolutionary time represents what, for all practical purposes, a creationist might call an "eternity." The tens or hundreds of millions of years necessary for evolution to create complex organisms simply escapes the capacity to intellectually envision it, a failure that is especially profound in a literalist society.

The other significant motive driving the creationist agenda, of course, is human primacy. So accustomed are we to being the dominant species on this planet that we are more than little hesitant to admit that perhaps this was an accident of biological environmental adaptation—as evolution so convincingly shows us. It is far more comfortable to assume that our dominance is part of a grander plan, that we are in the driver's seat of planetary destiny because that was God's intent all along. Science is an often uncomfortable practice because it so frequently humbles us, reminds us that we are as susceptible to the fickle whims of an indifferent nature as every other creature on this planet. But it is far more comforting in another respect to realize that natural "disasters" like hurricanes, earthquakes, floods, volcanic eruptions, tsunamis, lightening strikes, epidemics, tornadoes, draughts, blizzards, mudslides, avalanches, or infestations do not occur because God is "punishing" us, but because this is the way a planet—even one so benign in comparison to the rest of the solar system as ours—naturally operates. In our rush to validate our own importance and guard our own luxuries, we tend to be highly selective in what we will accept as true and what we will dismiss as uncomfortable. Education is as much a burden as it is a liberator, a curse as well as a gift, but the alternative is abject ignorance, denial, and all of the unsavory activity that grows out of these conditions.

Let us look at this issue for a moment, if only to illustrate how erroneous arguments against evolution can command such cultural currency in the enlightened twenty-first century. In a recent *New York*

Times editorial called "The Crafty Attacks on Evolution," the editors cite the current attempt to discredit evolution by religious advocates in Cobb County, Georgia, who refer to evolution as little more than a "theory" that is just as plausible as any other theory. The first blunder here, obviously, is that a "theory" in scientific parlance is not some mere musing that is hatched in-between the first and second half of a football game, but a "carefully constructed framework for understanding a vast array of facts." Vast array, indeed. Scientists have cataloged hundreds of thousands of fossils, identified thousands of extinct species, traced evolutionary patterns from one organism to the next, and meticulously mapped species mutation over eons. Secondly, that evolution has not been officially upgraded from a theory to a natural law only speaks to the rigorous standard of evidence a scientific idea must undergo in order to achieve this elusive this status. Creationists use the word "theory" as a way to imply disagreement within the scientific community, as if there are many alternative theories vying for recognition about the origin of life. While the details of evolution are not universally agreed upon by scientists (Stephen Jay Gould and Edward O. Wilson may have some differences in opinion about minutiae within the *process* of evolution), no respectable biologist, paleontologist, botanist, or zoologist denies the essential principles of natural selection.

But because the board of education in Cobb County cannot reconcile the notion that a scientific theory and a religious legend are two entirely different modes of discourse, they have, in their wisdom, decided to place on all textbooks the following "parental advisory": "This textbook contains material on evolution. Evolution is a theory, not a fact, regarding the origin of living things. This material should be approached with an open mind, studied carefully, and critically considered" ("The Crafty Attacks on Evolution"). This might be funny if the consequences weren't so stark. As the editors rightfully ask, why is evolution the only idea so precarious that it needs such a warning? Shouldn't all knowledge and every idea be "approached with an open mind, studied carefully, and critically considered," *especially* ones like "creation science" that have absolutely no scientific or logical merit? Part of the problem is that critical consideration isn't generally encouraged unless there is deemed a compelling reason to apply it, such as Christian fidelity, and even then it is only applied to the *opponent's* position. This story is disturbing because we view this kind of zealous tactic as something from our checkered past, from a time when public education was the exception, not the rule. More unsettling, for Constitutional purists, is the threat to the separation of church and state, a tradition many of us view as foundational to any true democracy. For our purposes, the debate underscores an important and unfortunate departure from the utopia of an informed society, and it constitutes a woefully sinister attempt to tailor public education to particular beliefs and to deliberately misinform a segment of our society that is not well-endowed to deal with the material "critically." In other words, students in a high school biology class, let alone their parents or board of education members, clearly do not have either the scientific or humanistic education necessary (or perhaps the common sense) to see how the curriculum is being abused. Nor, apparently, are they being given the apparatus needed to approach *any* system with an open mind, careful study, or critical consideration. If they were, this would simply not be an issue. On a larger scale, the debate is indicative of how creationists feel threatened by the encroachment of scientific explanations of our origin; if evolution weren't so convincing, it is likely they wouldn't feel compelled to wage a holy war against it.

Belief, like theory, is a word that is often misused in these contexts. For instance, in a web-based opinion site called "Wondering About Evolution," the anonymous author asks in the title piece, "Do You Believe that Evolution Is True?" The title alone is misleading, because it implies that science is a belief system in the same way that Hinduism is a belief system, as if consensus in belief is enough to determine validity. When science is operating in its proper venue answering the questions it is equipped to deal with methodologically and epistemologically, "belief" is not an issue. It matters little whether I "believe" in gravity or photosynthesis or the speed of light; these things are "real" regardless of my belief systems and they will operate with the persistent indifference to my feelings that is the defining force of nature. The same holds true for evolution and the doctrine of natural selection. My "belief" about its processes has no relevance to its ancient algorithmic pattern—species adapt, change, and either survive or perish. This process will continue long after humanity as a species has shuffled off its mortal coil and the debate surrounding the reality of evolution has been silenced. But creationists are fond of finding inconsistencies with natural selection where none exist. Take, for example, the "quiz" offered from the above-cited web-based opinion site, designed to challenge your "faith" in evolution:

Do You Believe that Evolution is True?

If so, then provide an answer to the following questions. "Evolution" in this context is the idea that natural, undirected processes are sufficient to account for the existence of all natural things.

Something from nothing?

The "Big Bang", the most widely accepted theory of the beginning of the universe, states that everything developed from a small dense cloud of subatomic particles and radiation which exploded, forming hydrogen (and some helium) gas. Where did this energy/matter come from? How reasonable is it to assume it came into being from nothing? And even if it did come into being, what would cause it to explode?

We know from common experience that explosions are *destructive* and lead to *disorder.* How reasonable is it to assume that a "big bang" explosion produced the opposite effect-increasing "information", order and the formation of useful structures, such as stars and planets, and eventually people?

Physical laws an accident?

We know the universe is governed by several fundamental physical laws, such as electromagnetic forces, gravity, conservation of mass and energy, etc. The activities of our universe depend upon these principles like a computer program depends upon the existence of computer hardware with an instruction set. How reasonable is it to say that these great controlling principles developed by accident?

Order from disorder?

The Second Law of Thermodynamics may be the most verified law of science. It states that systems become more disordered over time, unless energy is supplied and directed to create order. Evolutionists says that

> the opposite has taken place-that order increased over time, without any directed energy. How can this be?
>
> ASIDE: Evolutionists commonly object that the Second Law applies to closed, or isolated systems, and that the Earth is certainly not a closed system (it gets lots of raw energy from the Sun, for example). However, *all* systems, whether open or closed, *tend* to deteriorate. For example, living organisms are open systems but they all decay and die. Also, the universe in total is a closed system. To say that the chaos of the big bang has transformed itself into the human brain with its 120 trillion connections is a clear violation of the Second Law.

It is necessary to perform a bit of systematic deconstruction on this line of argumentation to show its flaws, and my analysis is important to the focus of this chapter because arguments reflected in "critiques" (like the one above) of established scientific concepts muddy the waters needlessly, thereby creating an atmosphere of doubt in the general public about well-documented scientific doctrines, doubt that extends to other areas where science has shown its merit and accuracy. It is important to note that I am not arguing against the careful evaluation of scientific claims, especially when those claims originate with dubious sources and for questionable motives; however, attempts to discredit scientific ideas like evolution and natural selection serve only to create pockets of doubt in the general public that would be much better situated elsewhere. In short, such attacks against science create confusion about valid and invalid ideas in the uninitiated, further exacerbating whatever level of public ignorance about scientific matters already exists. My analysis is an argument for a proper negotiation between good science and science that is misused, abused, or improperly challenged. Without an understanding of this distinction, science becomes just another tool in the sales kit of the intellectual counterfeiter, another way to sell products that don't work and aren't needed, or worse, of furthering ideas that lead to misinformation, misplaced skepticism, and even disaster. The author of this opinion, however, is not furthering either critical thinking or an understanding the basic tenets of scientific inquiry; rather, the author is attempting to discredit that which has already been demonstrated—through decades of scientific research, countless fossil records, thousands of professional papers and conferences, courses, and educational programs—to be above question as a biological process.

The author's definition of evolution as "the idea that natural, undirected processes are sufficient to account for the existence of all natural things" has already stepped into the realm of plain and simple error, because this is neither an adequate nor an accepted definition of evolution, from either a logical or scientific point of view. It emphasizes elements in the definition that would not be there if not for the orientation of the person constructing the definition—i.e. it introduces charged words to redirect the attention of the reader, words like "sufficient" and "undirected." This is an effort to underscore the apparent absurdity of the evolutionary claim that given enough time, the natural selection process is capable of producing complex organisms by chance. The reader is asked to think about the idea in a particular, predetermined way. Reference to "undirected" processes suggests that the theory of evolution is mechanical, even chaotic, because it does not account for the teleological "necessity" of a prime mover, a god to design, supervise, and maintain biological activities. The definition contains within it implied sanction for the teleological argument, cleverly preparing the reader for the idea that apparent order must have a designer. The definition highlights features of the theory that the author will later attack, as if identifying a tumor that needs excising. Subtle as the insertion of this word may be, it still produces the desired effect and insists upon a peculiarly religious orientation: It makes the reader wonder in the very way that the author wants her to wonder. How *can* life progress "undirected"? Doesn't there have to be *something* controlling life's functions, advances, even fate? This leads to the second element of the definition, the implicit question about the theory's "sufficiency" to "account for all natural things." The egregious misrepresentation of the theory here lies in its allusive attempt to negate the idea of evolution in the definition itself. In this definition, the author questions how evolution is "sufficient" to "account" for "all natural things." The *Concise Encyclopedia of Science* provides two separate definitions, one practical and one theoretical, both of which are more neutrally presented. Evolution is, according to the editors of this reference:

1. The development of an organism toward a perfect or complete adaptation to environmental conditions to which it has been exposed with the passage of time;

2. The theory that life on earth developed gradually from one or several simple organisms (appropriate molecules) to more complex organisms. This is sometimes called *organic evolution.* The term *evolutionary biology* is also used. (DePree and Alexrod 285)

According to DePree and Axelrod, evolution is an explanation of the development of *organisms,* not "all natural things." Note the effort to superimpose elements of the author's belief system onto a definition for the theory of evolution; whereas creationism or intelligent design accounts of the "beginning of the world" include both living and non-living natural creations (that is, god created the heavens and the earth as well as all living creatures), evolution restricts itself to *life,* not "all natural things." This is laying the rhetorical groundwork for the creationist author's attempt to conflate cosmological concepts like the "Big Bang," which comes next in his "quiz," but it also allows the author to draw attention to "weaknesses" in the definition that he or she will later exploit, despite the fact that this is not even an agreed upon, scientific definition of evolution but a definition of the author's own. This tactic aids in confusing the two concepts, one of which is truly a theory with much less hard evidence than evolution to support it, while the other is a well-established, thoroughly documented biological phenomenon. In this rhetorical attempt to dismiss the validity of one theory by attacking the weaknesses of another, the Creationist author suggests—in just the definition—that a) evolutionary processes are chaotic and undirected; b) the idea that such processes are "sufficient" is ludicrous; and c) that evolutionary theory is an explanation not just for life, but for all natural things. In the definition alone, then, we see the occurrence of several serious logical fallacies, all of which are designed to get the reader thinking in terms that have been established by the author, not in terms that have been supplied by evolutionary evidence itself. It's an old rhetorical trick to distract an audience through symbolic sleight-of-hand, and the example in question also has the benefit of suggesting an alternative in its definition. If life must be directed, it must be directed by a god (specifically, a Judeo/ Christian God[1]). If evolution is the explanation for the existence of all things (which it isn't), it must account not only for life, but for the creation of planets, stars and star systems, galaxies, and the universe. Ergo, the only entity or force capable of doing all that is God.

Moving on to other fallacious features of the creationist's argument, we see an egregious example of the "false dichotomy," the assertion that there are only two possible explanations for any given circumstance. The implication in the above argument is that if evolution—as outlined by the author—is not sufficient to account for the variety of life or the existence of the cosmos, then there is only one other possible explanation. The argument, of course, isn't strong enough to discount evolution or natural selection, much less to provide affirmative evidence for the existence of a god, making the position even more untenable. The problem is further exacerbated because the Creationist author is willing to apply the strictures of logical verification and scientific evidence only to the idea of evolution (or, at least, what appear on the surface to be such strictures), not to the faith-based alternative of creationism. The author attempts to use the Second Law of thermodynamics against evolution by positing that "[e]volutionists say . . . that order increased over time, without any directed energy." Assuming this explanation of the Second Law of Thermodynamics is accurate (which it isn't)[2], we must make some further leaps of logic, experience, and philosophy to accept the idea that the universe within which we live is in fact "ordered" in the sense that the author means. Certainly, we can discern patterns and create accurate predictions about certain phenomena on the basis of those patterns, but to call the universe "ordered" in the sense that everything clearly meshes with everything else like a Swiss watch is overstating (and oversimplifying) our reality. The claim is an anthropocentric understanding of what "order" is, and the Second Law is designed to discuss this order in a closed system (i.e. one that is self-regulatory and self-maintaining). DePree and Axelrod define the Second Law this way: "In thermodynamics, a system so chosen that no transfer of mass can take place across its boundaries; as a system that can exchange matter but not energy" (157). The author attempts to counter this point by claiming that "living organisms are open systems but they all decay and die. Also, the universe in total is a closed system. To say that the chaos of the big bang has transformed itself into the human brain with its 120 trillion connections is a clear violation of the Second Law."

Living organisms themselves—and the environment in which they live—are not, in fact, "open systems" at all; the matter that makes up an organism and that is contained by the earth is finite and the energy is constant. Even if they were, the death of organisms hardly disrupts the principles of thermodynamics; death is simply a transformation of matter and energy from something animated into something inert. Closed does *not* necessarily mean "isolated," as the Creationist author claims. The contradiction is further revealed by the claim that the universe is a closed system, suggesting that we have an open system (organisms) within a closed system (the universe). This misuse of the relative terms "open" and "closed" is either deliberate—in which case the author is attempting to confuse the distinction between these two conditions—or misunderstood by the author. More interestingly, the author makes no attempt to apply the Laws of Thermodynamics to his creationist alternative, indicating that the attempt to turn science against itself is a one-way effort. If he had, it would become immediately clear that while evolution does not in fact violate the Second Law, the idea of Creationism certainly does. God, as an omnipotent entity, certainly meets the criteria of a "closed system"; given the very argument that the author seems to be forwarding, how is it more likely that a "god" can make matter and energy from nothingness when the Big Bang has just been dismissed as a thermodynamically impossible account of universal origins? In other words, the author conveniently accepts the logical/scientific dictates of thermodynamics for one system (evolution) and ignores it for the other (creationism). This double standard is further compounded by the question of whether something can come from nothing, the primary doctrine of the creationist ideology. God made the cosmos through a mere utterance, a much more questionable premise than the Big Bang, for which at least fragmentary evidence exists. And this, again, confuses the concept of the cosmic origins with an evolutionary premise that explains the existence, development, and variety of life (another matter that Creationism cannot explain). The Creationist author concludes:

> **Complex things require intelligent design folks!**
>
> People are intelligent. If a team of engineers were to one day design a robot which could cross all types of terrain, could dig large holes, could carry several times its weight, found its own energy sources, could make more robots like itself, and was only 1/8 of an inch tall, we would marvel at this achievement. All of our life's experiences lead us to know that such a robot could never come about by *accident,* or assemble itself by *chance,* even if all of the parts were available laying

> next to each other. And we are certain beyond doubt that a canister of hydrogen gas, no matter *how long* we left it there or what type of raw energy we might apply to it, would never result in such a robot being produced. But we already have such a "robot"-it is called an "ant", and we squash them because they are "nothing" compared to people. And God made them, and he made us. Can there be any other explanation?

This demonstrates yet another egregious example of a basic logical fallacy, this time in the form of the "false analogy," which dictates that analogous comparisons must share enough features with one another that the conclusion drawn about one example can be logically attributed to the other. When important shared characteristics are absent, the comparison is analogically false. The flaw here is that the author is comparing machines (robots) to life, wherein life has a number of demonstrable characteristics that are absent in machinery: it reproduces independently; it grows independently (so long as a food source is available); it is capable of finding and processing its own energy sources; and it dies. The author hastily concludes that because machinery has an obvious maker, so too must life. This is a simplistic rehashing of the age-old teleological argument, also known as the "argument from design." But machines and life are radically different, so much so that any comparison almost immediately breaks down. To suggest that throwing the parts of a robot on a table and expecting them to spontaneously evolve into a functional whole is another transparent attempt to get the reader to think of the evolutionary process in terms as reductionistic as the author's. However, if we throw certain bacterial cultures together with other key chemicals (antibiotics, for example), we see first-hand how life does in fact mutate into different forms. "Our life's experiences" *do* in fact show us evolution at work if we only look closely at such examples. Furthermore, "life" is a separate category of matter and energy than "minerals"—we don't expect to see rocks come together into some working machine any more than we expect the pieces of a robot to. But this is only part of the problem. The author is suggesting that direct observation tells us that a robot can't build itself, yet he is framing it on a timescale that is totally inadequate for the comparison. Given enough time, inert matter does in fact change, sometimes merging with other matter, sometimes by changing into something entirely different (carbon into diamonds, for example). But it never changes into "life," and evolutionary theory does not suggest this. Even when it does change, it takes millions of years, not the insignificant lifespan of a human being. The author has created another erroneous association, and then concluded that since science cannot fully account for the complexity of humans, there is only one possible alternative explanation: God. There are more than a few missing premises here, but the author is banking on the prospect that the reader won't notice that, either. But given the evidence, the logic, and the careful analysis that the concept of evolution has brought to bear on the creation and variety of life, the creationist argument fails all criteria of analysis by comparison. Conclusion: intelligent design is far from an intelligent argument, folks.

As I hope I have demonstrated, the creationist notion of a universe (actually, a "world") that was created in six days by a supreme being can barely hold up under the scrutiny of its own faith-based set of contingencies; it certainly cannot hold up under the judgment of science, assuming that a creationist would ever even attempt to put the argument through such rigors. The creationist author wants to be able to argue against evolution using the methodological exigencies of science, but fails; knowing that a theory of creationism would fare even worse under such scrutiny, the attempt to put it under the microscope of the scientific method is not even made. The hope is that whatever internal contradictions exist with evolution will reveal themselves through a pseudo-logical, pseudo-scientific argument, but as we have seen, the argument suffers so badly from its *own* contradictions, double-standards, fallacies, errata, and outright misrepresentations that it can only be swept aside as ridiculous. There remains one glaring problem, however: Most people are not adequately schooled in either logic or science to discern the seriousness of the problems with the Creationist author's argument to know that it is categorically false. Because the author *seems* to have some knowledge of science, and because the argument apes a logical sequence of points, many people will view it as not only plausible, but likely. Herein lies the realm of rhetoric, not in the pejorative sense, but in its ability to hermeneutically dismantle not only the logical points of the argument, but also the motivations that propel the creationist author, his readers, and the scientific community.

My assertion here is this: If a significant portion of the general public is insufficiently equipped to draw careful, skeptical, and logical conclusions from observations and data that do or do not meet the threshold of the scientific method, it is easy to see how pseudo-science, charlatanism, and grandiose claims of discovery, novelty, and progress can be used to fuel the consumer machine. While the evolutionist/creationism debate may not have a direct impact on "The Economy" (except where it is used to sell books or create religious converts), *it demonstrates how easily people can be persuaded to act using methods that only mimic the scientific process.* The results from an interesting 1991 Gallup Poll (redone, according to the authors, in 1997 with no significant change) on the Creationism/Evolution question, reprinted on "Religious Tolerance.org," gives us an indication of belief systems that Americans may in fact hold. The results of the poll show that the majority of people who hold onto the Creationist notion of a god-created world are both undereducated and relatively poor. 65 percent of those polled who possesses no high school diploma believe that God (and, again, we must assume a Judeo/Christian god here if only by the way the "Creationist View" category is identified) created the "universe" within the last 10,000 years. 59 percent of those under a $20,000 annual income level (in 1991 terms) believe the same thing (Robinson). The correlation between lack of education and income level is undoubtedly a close one, and this fact only underscores my assertion that people with little formal training in the art of rhetoric and the scientific method are most likely to be taken in by pseudo-scientific creationist arguments. Michael Shermer, mentioned earlier as the skeptical contributor to *Scientific American,* has dedicated his March 2005 "Skeptic" column to what he describes as the Creationist "fossil fallacy." He explains,

> When I debate creationists, they present not one fact in favor of creation and instead demand 'just one transitional fossil' that proves evolution. When I do offer evidence (for example, *Ambulocetus natans,* a transitional fossil between ancient land mammals and modern whales), they respond that there are now *two* gaps in the fossil record. (32)

Clearly, Creationists are more adept at muddying evolutionary evidence than they are at recognizing what constitutes hard data (and fossil evidence is pretty hard). And it works because those who are predisposed to believe in a god that cares about them and their fate on earth (which is, after all, the center of the universe) are both ignorant of and resistant to the body of evidence that might suggest otherwise. More than this, as Shermer notes, it is a clever, if fallacious, rhetorical device because

> it reveals a profound error that I call the Fossil Fallacy: the belief that a "single fossil"—one bit of data—constitutes proof of a multifarious process or historical sequence. In fact, proof is derived through a convergence of evidence from numerous lines of inquiry—multiple, independent inductions, all of which point to an unmistakable conclusion. (32)

Shermer also points out that the preponderance of evidence about evolution's reality is not the domain of a single, marginalized field of science, but is an interdisciplinary effort from "such diverse fields as geology, paleontology, biogeography, comparative anatomy and physiology, molecular biology, genetics, and many more" (32). From a standpoint of argumentation designed to convince the opposition, however, the evolutionist is fighting a losing battle. Even if a thousand other "transitional fossils" were found, the creationist camp would not suddenly concede victory to the evolutionist—it would simply find another point of contention just as fallacious and defend it just as vociferously. From this point of view, the debate is hardly worth having, at least insofar as the objective is to convince creationists of the error of their ways. Nothing, as they say, is more stubborn (and dangerous) than a zealot, and any attempt to convert the righteous is destined to fail, no matter how overwhelming the proof is.

However, the debate takes on greater significance for those whose verdict on the matter is not yet in, especially our children. I'm certain that the creationists are convinced that they are fighting the good fight to save their children's (and your) souls from eternal damnation and from the irretrievably decadent, atheistic scientists who would lead them to believe that they are insignificant atoms on a mote of dust in a vast cosmic vacuum. Viewed this way, it is easy to see why creationists are uncomfortable with (actually, devastated by) the notion that human existence not only was *not* predestined by a benevolent deity—but also was actually an incredibly serendipitous accident that took place in a remote corner of one galaxy in an even more remote star system in a time-frame that, from a cosmic perspective, was less than the blink of an eye. In other words, we are all just plain lucky to be alive. Compare this to the idea that humanity is the species chosen to carry out the will of God, and success in this task secures for us eternal bliss in the realm of heaven. We will not die like we would under the evolutionist model. If there is no afterlife, from the point of view of a creationist, then there is no "accountability." Armed with the notion that this life is all we have, they fear, no one will be anything but self-serving during the short time they exist in this world. However, if moral acts are required to get that seat in heaven, people will think twice before committing a multitude of sins. The gulf between the two arguments is so vast that there will never be agreement on it. Those who have not yet decided their position and, even more crucial, students not yet familiar with either argument, are the trophies in this public contest. That is why each side is so jealously inflexible—will future societies side with the reality of our existence, humbling (perhaps even nihilistic) as it may be, or will they choose an illusion that provides them with smug, if erroneous, comfort? The implications are enormous. Nothing less than our own self-awareness, our own *raison d'être,* is at stake, and on which side we fall will mean the difference between how we situate and understand our role on this world and in the cosmos.

Moreover, the trend toward a creationist view of world and life origins indicates to me that there is something significant missing in the worldview of many Americans, a void that science and capitalism isn't able to fill. It is a common truism that "Everyone needs to believe in something," a platitude that normally refers to a belief in a "higher power" (a concept borrowed from Alcoholics Anonymous, which should tell us something about how the therapeutic culture bleeds over into other areas of our life and, even more tellingly, how the conflation of religion and therapy has taken on a quasi-secular flavor). The idea that everyone must have a spiritual component to his or her life is forwarded as common sense, but the "spiritual" component is not fidelity to a God, but fidelity to an institution. It is reasonable, even obligatory, to question whether this claim is in fact true, and we should at least be asking how "spirituality" is defined and what form such spirituality should take. It is also commonly held that a person can only be "moral" (in the parochial sense of that word) if he or she has some sort of religious affiliation. This is important to question as well, since the assumption is that humanity is inherently bestial and unable to check its appetites, thereby necessitating the monitoring gaze of institutionalized religion to keeps us from wandering astray. The basic assumption we should challenge is not only whether we are really as wicked as the religious right sanctimoniously thinks we are, but also whether organized religion provides us with a safe harbor against ourselves in the event that we really are as bad as all that. If anything, it seems, we find the comfort of The Church a good reason to assume that we are forgiven no matter what we do, a thin incentive for good works if we are convinced of our own grace. We are even perhaps more likely to engage in thoroughly anti-Christian activities like war if we are convinced of our own posthumous place in heaven and our own moral rectitude on earth; it is much easier to "sacrifice" something as transitional as life under this belief system. Inconvenient scripture is freely reinterpreted (or ignored altogether) to match the doctrine in such cases, an interesting irony considering the literalist outlook many fundamentalist evangelicals hold regarding the Holy Scriptures.

But what really complicates this situation is that people revere, on the one hand, the benefits that science is able to afford them while challenging its authority on questions that it is truly capable of answering. Science, in this respect, becomes an epistemological convenience; we invoke its name for matters that it cannot realistically or adequately address alone (like educational standards) while ignoring its real power for matters that it can (like the origins of life). Why is this? Is it a basic matter of insufficient education, or is the sociological cause for this ambivalence more complex? Whatever is missing in our lives, we mistakenly run back to the institutions that we had successfully matured beyond in the past. We seem to be living in a time of intellectual regression, and what is most troubling about this is that science must assume much of the responsibility for it. This is not to say that science has "caused" our species to experience some form of social, cultural, and rational retardation, but it has made dependents of us, dependents who are reliant upon and subservient to its powerful influence. It frequently startles us with its bluntness and its apparent indifference. Many are now resisting what they see as the dehumanizing mechanism of scientific knowledge and method, but perhaps for the wrong reasons. Instead of questioning the "magisteria" that science can rightfully claim as its province, many people question science on issues—like evolution and the environment—that it is highly qualified (indeed, *uniquely* qualified) to address.

The real question that I think confronts us, then, is how do we negotiate the truth-value of science with the comfort and security of religion in a way that is at once honest about the physical world and our place in it and also honest about the nature of human beings? In other words, can we recognize the human limitations that are revealed via science and reconcile them with the notion given to us by religion that we are in fact special? Many of our most cherished belief systems have social currency in large part because they have a long history and rich cultural accoutrement, but this history complicates its interpretive purity. Kenneth Burke noted early on in his *The Rhetoric of Religion* that the study of formalized religion, from a secular standpoint, is as much about the nature of *words* as it is about the nature of God (1). This suggests that arguments pertaining to the origins of humanity fall prey to the verbal complications implicit in any attempt to express doctrines and principles through language. Perhaps the most important inversion of the true nature of religion from a rhetorical perspective is the academically backward principle that "man was created in the image of God." But, as Burke points out, the language used to discuss God, as a humanly constructed instrument of knowledge, means that we must necessarily discuss God in the terms consistent with the modes of knowledge *created by* man; hence, in terms of the verbal representation, description, and characteristics of God, He is made in man's image, not the other way around. This is important since it is therefore easy to extend this inversion to other doctrinal areas of the Judeo/Christian religion; if God is created in man's image by virtue of the human language needed to bring Him to life, as it were, then it is no stretch to then use these same imperfect language constructs to describe the primacy of human origins. Creationism is a belief system based solely in the authority of a human-generated text; evolution is also based in a text, but of a sort that is materially independent of human existence—the fossil records, biological and geological shifts, and fluctuations over the eons of countless species. Such a text has certain advantages for its interpreter: it is not, for example, deliberately (or inadvertently) distorted by human-centered motivations as it would be during the act of writing literature (which is not to say that the human/symbolic element does not distort the evidence because of personal motivations like egocentrism, reputation, results, etc., for the interpreter *viewing* and *analyzing* the fossil record). In the case of evolution, we are forced to admit that the data is simply too overwhelming, no matter how uncomfortable that realization. Nature has no incentive to write the text in a certain way or to purposely forward obscurities that might lead to flaws in interpretation. Such texts do still rely on the interpretive acumen of the researcher, but the natural texts themselves are not beset with the problems inherent in any human text, especially one so loaded with the cultural significance of a major belief system. Scripture, read as literature written by humans, is doubly problematic: it is a symbolically generated text that relies on human interpretation. The possibilities for misreading, misapplication, and just plain misunderstanding are therefore amplified exponentially. It is odd, given the textual disparities between written and natural texts, that any conversation about the origin of humanity can take place at all. It is not so odd that when it does take place there is little in way of any real dialogue that might advance a resolution. Hence we see surrender in a non-overlapping magisterial position like Gould's, who sees the debate as ultimately fruitless. Rhetorically, the texts and the language used to represent the respective sides of the debate may be just too radically different to reconcile.

One must wonder what might constitute adequate justification for imposing beliefs on the masses, to advocate a "faith-based" administration, to deny the reality of our beginnings, and to confuse the creationist's scientifically-worded objection to evolution with a merely specious one. Beliefs notwithstanding, we operate under a very unusual double-standard about beliefs in general: We give lip-service to tolerance of other beliefs as long as those beliefs are both recognized religious institutions and as long as they are cast in the affirmative. We do not, interestingly, consider agnosticism or atheism "beliefs" in the same way we consider Taoism or Hinduism beliefs because they are viewed as an absence of belief. The very first definition of "belief" in Webster's Dictionary (if this can be trusted as authoritative) is "the state of believing; conviction or acceptance that certain things are true and real" (132). Note that this definition doesn't insist that a belief hinge on a set of principles that millions of others share with you in an organized way. It is, in other words, a mistake to think of an atheist as someone who "believes in nothing." On the contrary, an atheistic belief is every bit as affirmative as theistic one; the atheist simply operates under the "conviction or acceptance" that God is a fiction, and that this idea is as "true and real" as that of the most zealous fundamentalist Christian or Muslim. Once again, those who control religious discourse have appropriated the language that can be used to express spiritual convictions, thereby appropriating the ways in which nebulous abstractions like "beliefs" can be discussed.

And the language of the religious quarter has great discursive power. The vast majority of the population, because of the limitations placed on how we can express religious ideas, views those without attachment to any given religious principle as somehow weak and impressionable. What a strange idea, considering that it is much easier to capitulate to the beliefs of others than it is to swim against the current of popular perception, to seek the truth in the face of so much denial and willful ignorance. This is not to say that I subscribe to any particular belief system in the way that is usually conceived; rather, as regards the debate over evolution, it seems to me intellectually irresponsible to deny the overwhelming evidence that, even if the minute details are still being ironed out, evolution is real. To say (in fact, to argue vehemently) that it is less tenable than the idea that some omnipotent being created the world in six days is to return to the days when we insisted that the world was the center of the universe, an idea that even the most ardent bible-thumper cannot hold (though some have tried). But I also want to be clear about the reason for this section: The significance of the evolution/creationism debate, from a rhetorical standpoint, is less about the origins of humanity and more about who controls the vocabulary of power. For example, in a book review of Michael Ruse's *The Evolution-Creation Struggle,* Judith Shulevitz of the *New York Times* summarizes Ruse's project by pointing out the (somewhat artificial) distinction Ruse makes between "evolution" and "evolutionism." The latter "ism" denotes a philosophical push to consider the moral implications of subscribing to the evolutionary theory. But what Ruse has apparently done, says Shulevitz, is to succumb to the language of creationism, to forward the, in Shulevitz's words, "assertion that evolutionism amounts to a latter-day postmillennialism" that "feels more like a clever metaphor than a genuine link between ideas" (3). According to Shulevitz, Ruse's position is an attempt at penetrating the creationist perimeter on their own discursive terms, such that "the notion that evolution equals progress still runs through many evolutionary theorists' works and public statements, giving them, at times, a curiously spiritual feel" (3). The trope that Shulevitz indentifies here has the effect of shifting the discourse of evolution to the language of creationism, allowing the Creationists control of the words that normally would be the discursive domain of the evolutionist. The implications of this rhetorical error provide further ammunition for religious influence in public schools. The stridency with which the Christian (mostly) contingent has attacked the public school system on this issue has to do with the attempt to wrest power from what they see as an increasingly detached, amoral scientific agenda.

One way to remedy an unexamined commitment to dubious and counter-scientific beliefs—those that affect not just the believer but the larger society—I submit, is both an increased scientific literacy and a more acute rhetorical acumen. The former is geared toward educating people on the content of science, and the latter is designed to supply a critical apparatus for judging the means by which that content is delivered. (It seems, for instance, that Gould's simple solution does have some merit, if applied judiciously; there is no necessary reason for religious advocates to mettle in a *biology* class, which is a course about *science,* for they can offer their view of the Origin of Man in their religion classes. If they want to change curricular policy, fight for representation in the form of courses.) The next chapter will provide one possible critical remedy for assessing scientific language, especially language that has gone through a series of translation processes before it reaches the public ear and eye. Essentially, this will be a vocabulary and a method of critical evaluation for sizing up scientific information that is prepackaged for public consumption. I call the method "residual field theory" because it attempts to decipher the cultural "static" that occurs when science is distorted through the various delivery systems of the modern information superhighway. With this metaphorical tool, I hope the general reader can work to make sense of the confusing and omnipresent white noise that constitutes modern media representation of science (and, I hope, a number of other germane fields of inquiry).

9 Residual Field Analysis

Randy Allen Harris, in his introduction to *Landmark Essays on Rhetoric of Science,* identifies scientists' resistance to admitting the use of rhetoric in their argumentative strategies as one that "[. . .] stems from the immensely opposite connotations of the two words [rhetoric and science] on either side of the proposition in that phrase [rhetoric of science], the tarring blame of *rhetoric* and the imperious praise of *science*" (xi). While the relative status of rhetoric and of science is undeniable, Harris's primary emphasis, as with most rhetoricians in the last three decades or so, is that the subspecialty we refer to as "rhetoric of science" is one that examines what scientists discuss amongst themselves when they forward a particular hypothesis, theory, or explanation. That is, the rhetoric of science, as a discrete focus in the larger study of rhetoric, has historically been concerned with the discourse scientific agents use to convince *one another* of the plausibility of a claim, idea, or enterprise, though this emphasis has begun to shift in recent years. Harris observes that one of the contributing factors to the shift toward the study of science and public discourse is that the word "rhetoric" is used pejoratively and the word "science" is used assertively, even audaciously, and the limitations he describes for the study of scientific rhetoric become even more clear when he posits that " [. . .] scientists argue, and their arguing is absolutely central to their success: science is rhetorical" (xi). Science *is* rhetorical, and the scholarship is increasingly targeting the relationship between the scientific and public communities. Harris adds later in the same introduction that the "[r]hetoric of science is simply, then, the study of how scientists persuade and dissuade each other and the rest of us about nature" (xii), acknowledging the shift in rhetoric of science scholarship that tends to examine the minutiae of scientific language as it manifests itself within the community of science itself or as it effects policy-making decisions on a legislative level is giving way to the adjustment in focus in broader, more inclusive terms and for a much larger audience. Because science is so well-regarded in its epistemological dominance, the rhetorical import of its discourse is not confined to the professional communities that practice it; scientific discourse is now also being seen by scholars as an important ideological apparatus, and the way scientific ideas are packaged for lay consumption is an element of the larger framework we call the "rhetoric of science." What is absent, however, is a methodology that is written for the people towards whom much scientific language is directed: the general population. The purpose of this chapter, therefore, is to re-align the project of what it is the non-specialist reader can hope to discover when encountering scientific rhetoric, to discover who the main players are in this rhetorical drama (that is, who the agents are and where the agency lies), to describe what methods scientists or scientific advocates use when "popularizing" their ideas, and to offer a counter-method that the non-specialist can employ to decode popularized scientific discourse.

The problem with studying scientific rhetoric, as with any system of interpretation, is one of discerning *meaning*. This rather nebulous problem is compounded in interdisciplinary studies, where one must construct an interpretive machinery sufficient to address the idiosyncrasies of the fields under examination while also attempting to apply a critical rubric that captures the symbolic intent underlying the text that is being scrutinized. In other words, while certain literary templates are very useful for unlocking some of the meaning embedded in a poem or a work of fiction, the same is not as easily said about a critical system that is meant to "reveal" the motives and purposes behind the furtherance of a specific scientific theory, such as evolution or quantum mechanics. Suppose, for example, that we read in the news that a new study has been released which lends credence to the idea that moderate alcohol consumption actually improves health, particularly for those who have high cholesterol or are at risk for heart disease. The study—somewhat isolated and certainly not yet universally accepted nor even reproduced independently—is actually given more media attention than it warrants because it is novel (challenging as it does accepted temperance wisdom), and it benefits certain parties to over-saturate the public with this finding more than is perhaps justifiable at this stage in the scientific process, a phenomenon that we have already touched upon. Responsible physicians are rightfully suspicious of such a premature claim, while less meticulous ones may quickly begin advising patients to have one or two glasses of wine daily for its perceived medicinal value. But doctors and patients are only one small part of the agency equation. Also involved are the researchers who released the study (and have an interest in its positive reception), the media (which always likes an unorthodox story), the alcohol industry (which can clearly benefit from this study both financially and in terms of its improved image in the public eye), members of the substance abuse industry (who may decry the evils of addiction but are, in fact, irrevocably tied to—and even reliant on—it), the clergy (which, like the substance abuse industry, maintains a large part of its livelihood on the indiscretions of others), and the general public (whose responses may range from indignant outrage to unequivocal support to plain indifference). And let us assume, for the sake of argument, that the scientists who conducted the study are doing so only for the sake of science itself and are hoping for the most altruistic outcomes. Once this study is released, it has essentially become public domain, and their rights to it, or even their caveats about it, while perhaps not forfeited, become, for all practical purposes, irrelevant. The study itself is no longer within their purview, no longer subject to their sphere of influence as a public document, because it has been commandeered by other agencies whose motives are quite different from their own and whose purpose for using the study are absolutely unscientific.

Enter the rhetorician, whose motive for studying this phenomenon is to understand better how the language of these distinct communities operates and how the utterances fit within a larger puzzle of communicative interaction and to what *end* the discourse is intended from its inception. A system of rhetorical analysis is needed, but the system must grapple with competing systems already firmly entrenched. As I.A. Richards puts it in "Towards a Technique for Comparative Studies," " . . . can we maintain two [or in this case three, or four, or five] systems of thinking in our minds without reciprocal infection and yet in some way mediate between them? And does not such a mediation require yet a third system of thought general enough and comprehensive enough to include them both?" (38). The problem for the rhetorician, then, becomes not one about the *truth* or *falsity* of the science that has been performed, but how science is being presented and received, to what ends different agencies may put scientific declarations, and what effect science is having on the knowledge claims and ideological underpinnings of the laity. Today's laity receives information from many different, often incongruent sources (radio, television, the Internet, and print material, to name a few), and the information is almost always fragmentary, making objective conclusions difficult to draw. In the "information age," the problem becomes one of assimilation and coherence, not a lack of "knowledge" in its broadest sense. Where the lay public has difficulty, therefore, is not in too little information, but in an inability to cohesively cobble together material from different sources into a comprehensive and meaningful whole and to judge the material with critical scrutiny.[1] It should come as no surprise that under these conditions, science reigns supreme as a source of information that provides the illusion of certainty, even if scientific findings frequently contradict each other or studies are released at some incomplete stage of their development.

The difficulty for the rhetorician in developing a usable system might best be illustrated through an example. Referring again to the book, *How the Mind Works,* Steven Pinker discusses the problem of intersecting ideologies that have been formed, in large part, by the received knowledge created and cultivated by science. The Standard Social Science Model (SSSM) dictates that there is a fundamental separation between biology and culture, and that because of this separation, in his words, "[c]ulture is an autonomous entity that carries out a desire to perpetuate itself by setting up expectations and assigning roles, which can vary arbitrarily from society to society" (45). Because the SSSM "has become an intellectual orthodoxy," it has also "acquired a moral authority" (45) that is at best strident in its doctrines, and at worst, anti-intellectual and even defamatory in its single-mindedness. Socio-biological challenges to SSSM have been met with contempt at scientific conferences that look more like political protests than the scientific interchange of competing ideas. Pinker explains that past challenges to the authority of this model led to

> [a]ngry manifestoes and book-length denunciations [that] were published by organizations with names like Science for the People and The Campaign Against Racism, IQ, and Class Society. In *Not in Our Genes,* Richard Lewontin, Steven Rose, and Leon Kamin dropped innuendoes about Donald Symons' sex life and doctored a defensible passage of Richard Dawkin's into an insane one. (Dawkins said of genes, "They cre-

> ated us, body and mind"; the authors have quoted it repeatedly as "They *control* us, body and mind.") (45)

Clearly, the political program in this example is directly linked to the science ("hard" scientists might say *soft* science, but the distinction is not an easy one for the general population to discern) that helped construct the Standard initially—to the extent that challenges to the SSSM are met with ideological, not scientific, resistance to form a pattern in thinking that Pinker calls "Psychological Correctness" (44). Pinker may be mistaken about the degree of understanding that exists about the human mind, but he understands political discourse in professional fields, no doubt for good reason. And whether or not one agrees with Pinker's particular theories of cognitive psychology is beside the point. From a rhetorical point of view, the "human" sciences are, under the SSSM model, used not as a method for uncovering truths about the nature of the human psyche in a climate of rigorous but collegial scholarly debate, but as a way to maintain a social and professional *status quo* that embraces a specific outlook on human beings as culturally, not biologically, determined ("determination" is, in fact, a question better left to philosophers than scientists). The history of the resistance should be relatively clear: "Social Darwinism," as I have already mentioned, has been used in the past for bigoted, unsavory, and even eugenic ends, and there is a real desire to avoid repeating (and one might say, a naïve desire to erase entirely) these devastating mistakes from our collective past. However, in the rush to intercept any ideas that might imply a biological determinism that is perceived by SSSM advocates as socially dangerous, the spirit of the scientific enterprise is compromised and even sacrificed. Scientists who challenge the dictates of SSSM have their ideas—and their character—systematically sabotaged. Daniel Dennett has recently released a statement about the dangers of oversimplifying the so-called "Nature vs. Nurture" debate in *The Chronicle of Higher Education* ("The Mythical Threat of Genetic Determinism"), and in his attempt to clarify the issue not as an either/or proposition but as a both/and one, he has furthered his role as a major player in the clarification of scientific ideas and debates for the non-specialist, a move that denotes the obvious rhetorical dimension of scientific debate.

The rhetorician should view this episode as an example of how science done in the field and in the lab is only a small part of the discursive equation. While the "debate" took place primarily between scientists, it was not exercised on the basis of scientific findings, and the implications that SSSM advocates feared were not so much within the scientific community, but with how those from outside might interpret and apply the results. In short, the resistance displayed by SSSM advocates wasn't "bad science" as many are tempted to claim; it wasn't science at all, but a concerted effort to maintain an ideological framework that was seen as a more humane way of looking at human determinism—the "nurture" camp run amok. As Dennett has asserted, the move to dichotomize the issue into a "nature *or* nurture" debate may be part of the problem, but the rhetorical reality is more complicated than that because there are a number of motives powering a number of differing beliefs. Nor is this merely an issue of scientific ethics, since the controversy is based on a strictly social agenda that deliberately interferes with scientific inquiry that does not hold up to the SSSM moral conditions. Moreover, the motives that drive this issue deal with science only insofar as new scientific ideas threaten an existing paradigm, and just as Darwin's theories were originally resisted on moral and theological grounds, so too is the biology/culture debate an issue that is driven more by belief systems than scientific data. The difference here, however, is that the political layers of the argument are furthered not by a competing area of knowledge, but by scientists themselves.

The next question, then, might be one of agency; that is, if we apply Kenneth Burke's pentad to this debate, we are able to recognize whose motives are driving which argument for which purpose, and we discover that the agents involved have motivations far beyond the scope of garden-variety scientific curiosity. Suppose that I am a sociologist who believes that studies in genetic determinism have dealt great blows to the progress made in many quarters of the social arena: feminism, affirmative action, criminal justice, and education, for example. Professionally, I am a scientist who is concerned with the acquisition and understanding of new knowledge, but morally, I am a conscientious opponent of ideas that have the *potential* to undermine decades of social work that have, I believe, made the world a more civilized, compassionate place to live. I understand that the nature or nurture debate is an oversimplified dichotomization of a much more complex argument, but I have little faith that non-specialists or the general public can make these subtle discriminations. Do I stick to my professional ethics regarding the scientific method and hope, when "all" the data falls into place, that the rest of the world will make the right choices? Or do I, in fact, practice my own form of scientific sabotage to uphold what I feel is a more imperative moral principle? If I choose the former, my professional conscious is intact, but unscrupulous (or merely ignorant) parties may run with a ticking time-bomb. If I choose the latter, I have compromised the goals of science, but I have helped avert what I feel is a socially disastrous trajectory, one that may irretrievably damage the political progress I value as an important part of a compassionate society. My choice will determine whether I guard my professional ethics or my social conscience with greater zeal. For purposes of this discussion, the choice is indicative of one of the most vital sectors of serious rhetorical study: the examination of scientific discourse as neither a vehicle of new knowledge nor a means for scholarly debate, but as an instrument for forwarding (or challenging, or deconstructing) specific ideological principles in realms outside of the scientific community.

And what happens when such misgivings actually become institutionalized, as in the case of SSSM? Here, we have agency on a bureaucratic scale that is far more powerful than a single individual or even a group of influential associates. What happens to the science in this quagmire of motive and counter-motive? As Michel Foucault more cryptically pointed out three decades ago, this question provides a clue about how to redefine scientific rhetoric as a power structure, as a discursive entity that people revere and respond to, and as a mode of discourse that contributes greatly to the modern mind-set, much as Catholicism contributed to the mind-set, principles, and cultural make-up of Europe for many centuries. This definition goes well beyond the sometimes cavalier assumptions we have about science's role in our lives; it suggests that science is not only an ordering presence, but can also be a dogmatic one. We routinely turn to science to answer questions about everything from how to best landscape our garden to which food is healthiest to consume. But we also use science to rationalize civic, social, and personal matters; when we debate the relative merits of a punitive versus a rehabilitative criminal system, we invoke science. When we assume that every minor psychological tick is treatable with therapy or drugs, we invoke science. When we obsessively follow every new finding on diet and weight loss, we are invoking science. I would like to offer a definition of the rhetorical use of scientific discourse (as opposed to the rhetorical study of science), then, as *a quasi-positivistic mode of discourse that touches all corners of modern life, civic, cultural, political, and moral, carrying with it the authority of science but not necessarily the same rigors, skepticism, and scrutiny that science* qua *science does.* It impresses itself upon the populous, and the populous is receptive to its message, but the general populous is often not taught to discriminate between "good" science and "bad" science, science that is conducted for the sake of inquiry and discovery, and science that is conducted (or merely used) to drive social, political, or institutional motives. And as one final condition of this definition, I should note that *the general populous does not know when scientific ideas are coming from the scientific community and when they are coming from other agents;* as we have already seen, there are many delegates for scientific "knowledge," and the vast majority of the ideas that trickle down to the public are not endorsed or "certified" by reputable scientists. In its worst (but, ironically, most benign) form, this leads to utter faith in pseudo-sciences like astrology or the paranormal. At its most common (and therefore most dangerous), it leads to a blind devotion to any idea that *appears* scientific.

A brief tangential discussion about what Robert McChesney calls the "Market of Civic Religion" will perhaps shed light on how the mythology of "market forces" is treated as a pious orientation by Western capitalistic economists (and their communities) and on how both the mass media and science are beholden to this orientation, being treated effectively as sub-denominations of the main Market-forces faith. The doctrine of the market-forces myth rests on some key but fictitious principles. McChesney restates these principles as follows:

> The argument goes that the self-regulating market is the most perfect, rational, and equitable mechanism for regulating social affairs known to humankind; efforts to interfere with it by governments or outside agencies will only reduce its sublime powers. [. . .] This pro-market argument remains infallible only to the extent that it is a religion based on faith, not a political theory subject to inquiry and examination. (44)

The argument, McChesney points out, is based on the fiction of "pure competition" where "an infinite number of small entrepreneurs battle to serve the public by lowering prices and improving quality in constant ferocious competition" (45). This presentation of the market-forces argument is "mythological" because, as we know, "small entrepreneurs" are hardly working on a level playing field with their corporate competition, which possess a seemingly bottomless pool of resources, political influence, and legal advantages that the small business simply cannot match. "Free Competition" is perhaps the most flagrant pretense in this doctrine, since large corporations can easily sweep aside small competitors through their almost unfettered access to power and mass marketing. In fact, McChesney notes, the faith-based market-forces model "is any corporation's worst nightmare, and successful capitalists invariably and quickly move to reduce their risk by increasing their size and reducing the threat of direct competition" (45). Yet the model is perpetuated because it implies equal opportunity, a lie that helps create the political environment that corporations need to thrive.

The religion of Market-Forces is central to McChesney's assertion that mass media, even more so than other sectors of private industry, are perhaps the most susceptible to this myth because, "In communication, this means that the emerging system is tailored to the needs of business and the affluent" (45). It is also thoroughly anti-democratic, since it guarantees that those with the most money and power will have the most influence on what is covered in the mass media, and consequently, what is adopted as policy—the effectiveness of political participation is therefore directly proportionate to the class and tax bracket that one occupies. McChesney uses this platform to dispel yet another myth surrounding the media of the Church of Market-Forces: that broadcasters, producers, and sponsors are simply "giving the people what they want"; rather, McChesney aptly points out, such media forces "give the people what they want within the range of what is most profitable to produce and/or in the political interests of the producers" (46). The approach is especially menacing considering it embraces a piety on the basis of erroneous doctrines: "free competition" does *not* exist because small, insignificant companies cannot compete with large corporations any more than a Pop Warner football team can compete with the New England Patriots; "market forces" do *not* determine social policy because only market forces that benefit the elite are seriously considered; the media do *not* give over to the people's will because the range within which that will can be expressed is severely limited.

Science has always been beholden to market concerns, but science as it is depicted in the media must suffer the doubly distorting effects of widespread market interests *and* those that are unique to the media industry. It is in this intersection between "pure" scientific interest and a "pure" market-forces allegiance that science must operate in the media sphere. It is easy to see, therefore, how scientific topics become even further diluted and how scientific motives become almost completely eclipsed by the motives of the corporate infrastructure and the media profit incentive. And the media, as a composite of private agencies that are themselves indebted to corporate advertising and marketing influences, must report on science in a manner that is both selective and reductionistic. One can see how quickly and thoroughly motives become buried, how information becomes fragmented and bent, and how individual interests instantly displace more democratic ones.

Since the rhetoric of science, as a field of study, is the systematic examination of the language structures that appear in the rhetorical use of scientific discourse, its main goal is to determine the whys, whens, wheres, and who-froms that comprise the events, claims, or issues that summon science as a primary argumentative, verbal, and rhetorical platform. I should emphasize again that the rhetorician studying the rhetoric of science is less interested in which claims are "true" and which are "false" scientifically (though this may well emerge in the examination) than he or she is in determining the motive and purpose behind the use of scientific discourse to achieve a particular goal. In simpler terms, when and why do people implement scientific language and what end(s) are they trying to reach by doing so? As a method for understanding the distinction between practical science and the use of scientific language for non-scientific objectives, a rhetorical analytical method is an important contributor to discourse studies because it supplies us with a method and a lexicon for distinguishing between the scientific process (during which new knowledge emerges slowly and spans a number of independent attempts to reproduce results) and what might be called scientific bandwagoning—the impulse to summon science when it suits a specific a-scientific need. There are seven main steps to the process I am offering, steps that the receiving agent (usually the reading and viewing public) may invoke to help determine the context of the scientific information that he or she is receiving. I put these in the imperative form so that it is clear that such analysis should be applied in a systematic order, though of course there are other equally productive approaches that one might adopt:

1. Define the issue, question, topic, or problem. Is it a political issue (funding for a new project; articulating a platform for election or re-election; solving a social problem like drugs or crime)? What parameters must be placed on the issue in order to manage it? Who is affected?
2. Determine whether the issue, question, topic, or problem falls—both realistically and practically—under the purview of science. Can science, in other words, solve the problem entirely? Is it the best method for addressing the topic? What other forms of analysis can be employed more usefully?
3. Note the agent: Who is invoking the scientific language? Is it a scientist? If so, who does the agent work for? If not, what credentials does the agent have for reporting scientific findings or using scientific data to support a given idea, argument, or agenda?
4. Speculate on possible motives, especially those that are not typically associated with science. This is perhaps the most difficult to pin down with any certainty, but some generalities can be drawn: Are the agent's goals scientific (that is, do they support a scientific mode of inquiry and understanding, or do the goals appear to be pecuniary, personal, or political)?
5. Distinguish between the scientific motives and the a-scientific motives. This is an important step since, while many motives are *basically* scientific, there may be other motives that push the agent to use scientific claims or data that are disproportionate to the intended goal or to exaggerate scientific claims for unannounced purposes. For example, a representative for the American Cancer Institute may be genuinely concerned with increasing medical knowledge in order to find a cure for cancer, but she may also be the family member of someone who has cancer. When raising funds for more research, she may exaggerate the success of different studies to suggest that medical researchers are closer to a cure than they actually are. Her motive to cure cancer is further driven by her motive to save the life of someone she loves and to further a cause that she strongly believes in; therefore, the data she presents may be more positive than it actually is, a pattern of wishful thinking that concurrently places an affirmative spin on the research for

those who might become potential contributors to the cause of stamping out cancer.

6. Pay close attention to the juxtaposition of scientific language with a-scientific language. Agents will frequently blend scientific language with common diction. Scientists themselves do this when reporting scientific information to the public, often relying on metaphor, analogy, synecdoche, and/or metonymy to explain "in layman's terms" the significance of research or the results of a study. Often, however, non-scientists will blend scientific language or data with the language of a belief system, particularly if there is moral issue at stake. Try to tease apart discourse the originates from different ideological sources to see how, or whether, they interrelate.
7. Analyze the language in connection with the other six steps. This, of course, is the most involved step of the seven, and requires a more detailed explanation of its content, as will be discussed next.

Analysis must involve the careful dissection of the discourse as it is articulated within its specific context. Of course, this can be said of any rhetorical interpretation of a particular discourse, but it is especially important when scrutinizing scientific rhetoric because the context and language summoned may not be as apparent as they at first seem. Since agents frequently enlist scientific discourse for nonscientific ends, the hermeneutic employed must assume a degree of distortion in the motive as well, as pointed out in step five above. Complete and reliable information is very important (but equally hard to come by). An understanding of the issue or problem is vital. Internalizing the goals of the agent is absolutely necessary. This interpretive method is a variation on Kenneth Burke's dramatistic tradition,[2] so that an understanding of the discursive network demands interpretation of its central players (agents), the site at which the discourse is operating (scene), what activities are transpiring (act), and the social/historical climate in which the discourse is taking place (context). Such a method also requires an inclusive knowledge of the issue and its scope (information) and for what end the various players in the network intend to use it (purpose) and why (motive). I term the issue, question, problem, or claim in question the "axis" upon which the debate turns, which may also involve the "agency"—the method of informational delivery.

What the observer—the readers, listeners, watchers—are left with is a distorted static that only gives us a partial picture of the situation, the "residual frame" which looks whole, but is in fact merely the leftovers of the discussion (the background noise, if you will, like the TV snow left over from the Big Bang) that took place at another place during another time after the event had already transpired. Even here, where the observer is located within the residual frame, only fragments of the discourse are available. If this network can be clearly discerned by the interpreter, it will look something like the diagram in Figure 2:

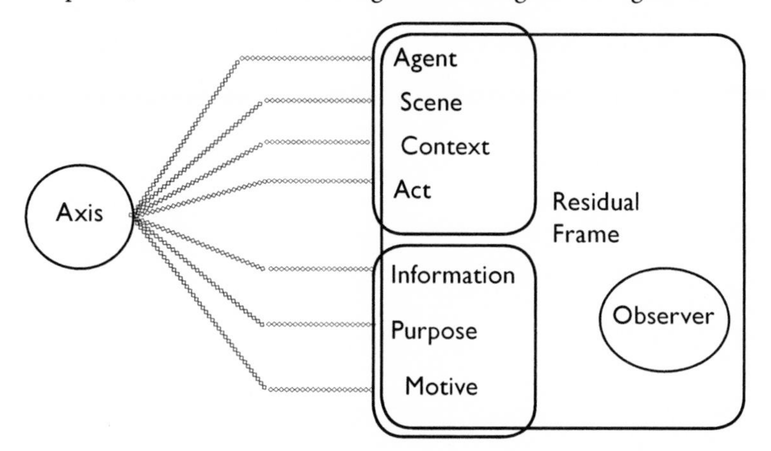

Figure 2. The Residual Frame

In order to more clearly demonstrate this method, let us use a working example. Several years ago, the space shuttle *Columbia* disintegrated upon reentry into the earth's atmosphere, killing everyone aboard and generating a torrent of questions about "what went wrong?" and "could it have been avoided?" These questions alone indicate a degree of faith in science and technology that should give us pause; space travel is a highly dangerous business, and that the shuttle program has had only two major accidents during its entire existence of some 25 years should indicate its successes, not its failures. However, far from wanting to belittle the loss the families of the crew and the nation as a whole suffered, I should point out that given the highly *unnatural* process involved in space travel, we should expect a level of risk associated with it that is unavoidable. Considering the number of things that *could* go wrong, it is amazing that this type of accident doesn't happen more often. But if we tune into the voice of the media, in their many and contrasting forms, we get an entirely different picture. In typical big news fashion, the news networks covered this story with its usual flair for the dramatic, calling upon an eclectic combination of experts from NASA and elsewhere and eyewitnesses from the lay public. For the sake of convenience, I shall focus on the printed news medium for my example.

Because shuttle launches have become routine, creating an illusion of security and, so far as the public is concerned, general indifference, news of this type tends to be highly charged and, to invoke a favorite word of news sources, "shocking." Put in perspective, shuttle launches should provoke neither apathy when they go well nor shock when they don't. If we consider that a shuttle launch involves the placing of human cargo atop what is essentially a directed explosion of atomic proportions, blasting a specialized cylinder into the sky with enough thrust and velocity to escape the earth's gravitational force, we begin to see how incredible this feat is. The stresses on materials and extreme physics necessary to achieve escape velocity push human and machine to the edges of endurance, leaving absolutely no margin for error. Even more amazing is that, once the shuttle has achieved its goal in space, it must reenter the earth's atmosphere, creating friction hot enough to burn most materials to cinders and fly, more or less, like a conventional aircraft—a process involving a controlled fall onto a small strip of concrete thousands of miles away. This is to say nothing of the hazards the shuttle is exposed to when orbiting the earth, not the least of which is that *any* collision is by its very nature going to be of extremely high velocity, and the ballistics of such a collision involving an object as small as a grain of sand can be enough to breach certain areas of the spacecraft's fuselage, causing a serious emergency. These are the basic technical and engineering problems of any shuttle flight; each individual flight has its own set of unique hazards (such as EVA's—Extra-vehicular Activities—to perform extensive repairs on an orbiting telescope).

The question, as per step one, then, is what of the thousands of things that could go wrong *did* go wrong, and can these and other hazards be minimized so that the likelihood of another shuttle explosion is considerably lessened? Because our nation's scientific and technological expertise has given us a false sense of security, it is necessary to re-imagine the true scope and nature of some of our more ambitious projects and note that accidents *will* happen, despite our best efforts to avoid them. The issue and/or problem must be framed by our own limitations—we should admit that the best we can hope for is a reduction in the number and severity of accidents, not the complete eradication of them. Note that this is also a political issue; many agencies and private citizens are beginning to question the "wisdom" and "economic viability" of the Space Program. This means that the allocation of funds is at stake, and that the language used by NASA and other supporters will reflect the insistence that the shuttle is safe and reliable, though these are, of course, themselves highly subjective, even relative, conditions (safe and reliable in comparison to what? having a Sunday picnic? playing catch with jars of nitroglycerin?). NASA's detractors will, naturally, argue that the nation is throwing good money after bad on a program which has questionable value, "value" being a word that generally translates into "return on an investment." What can we hope to gain by this investment (new scientific knowledge isn't enough; there must be a profitable application of that knowledge)? If there are none, or if the cost outweighs the return, many opponents will see the shuttle program as folly. Jobs and large sums of money are at stake, and these realities clearly influence the direction and stridency of the debate.

The next question is, does the shuttle debate fall strictly under the supervision of science (and its technical applications)? The answer is yes and no. Yes, the shuttle program is run by scientists, engineers, designers, and a whole smorgasbord of technicians, but it is also run by administrators whose first specialty may be, say, theoretical physics or even business. Like many professionals whose area of expertise ironically lands them in an administrative position, NASA administrators must take into account many issues that are only loosely relevant to science: how will public opinion influence political support? How will this affect future funding? Who will be blamed, and how will the accused be handled? What changes in procedure need to be made, and who will oversee them? How will the investigation be conducted, and who is the best front-person to report the findings? How should the press be dealt with? The press has an annoying habit of conducting independent investigations, and their findings (or rumors) may contradict official ones. Science is the philosophy that drives the shuttle program, but the "realities" of modern life interfere with the execution of scientific processes. The discourse that ensues must carefully weigh the benefits of the scientific idea with the reactions of people who are not directly involved with scientific enterprises. Such a conversation is, at its core, a rhetorical one, since it must take into account the goals of the program and the messiness of the current public and political climate to make a convincing case.

There are bound to be many agents impressing their views on this issue, but for brevity and demonstrative purposes, I will provide only one example here. On March 11th, 2003, the Associated Press released a story about a NASA engineer who was involved in "a furious internal debate about Columbia's safety" and had a "lingering uneasiness moments before the fiery disaster" (Bridis 5). These feelings were apparently exhibited by NASA engineer Robert Daugherty through a series of emails and phone calls that, taken together, "described with eerie precision what investigators believed probably happened aboard Columbia: searing temperatures penetrating the stricken shuttle's left wing and melting it from the inside" (5). The language choices made by the writer of this story, Ted Bridis, reveal his bias—that the uneasiness was "lingering," that the internal debate was "furious," and that the description was "eerily precise"—suggesting that the warning signs were glaring and, in typical bureaucratic shortsightedness, ignored. The language distorts the issue in some serious ways (aside from being thoroughly un-journalistic) in that it implies a coordinated effort by decision-makers at NASA to disregard the engineer's concerns. This makes identifying the agent difficult, because, although we do have some direct quotations from Daugherty himself, the journalist's language is superimposed on Daugherty's own, alluding to Daugherty's almost mystical acumen for discerning the explosion's cause (I can say for myself that, even as a non-scientist, I suspected that the explosion had something to do with a structural weakness that occurred at takeoff since the problem took place upon reentry when such damage could cause serious problems, suggesting that the concern was more a matter of common-sense than an act of prophetic wisdom or scientific insight), and the writer freely interprets Daugherty's feelings and analysis of the cause. Later in the story, Bridis characterizes Daugherty's reaction this way: "By the time radar sites failed to locate the inbound shuttle, Daugherty knew all was lost. He quickly recalled the debate inside NASA during the preceding days about how damage from extreme heat inside the shuttle's wheel compartment and left wing could prove catastrophic" (5). We have, then, at least two agents operating in this story, the writer and the engineer, but the writer has primacy because it is he who has control of how the language is used and the information is dispersed.

If we were to speculate on the motives for presenting the story this way, we might make some obvious observations about the press's tendency toward hyperbole and the impulse to find conspiracies whether they exist or not. This sells papers and gives certain publishing entities a competitive edge over others. The journalist's motives may be more individual, but the reasoning is essentially the same. He wants to report "breaking" news and uncover the "truth" about the accident. Delving a bit deeper, we see that the network of motives and the techniques for indulging them are more complicated than that. In this example, we have the journalistic motive of individual and institutional recognition (as well as the motive to do a job and report the news), the motive of the engineer who, as nearly as we can tell, is concerned about the cause of the accident and anxious about making sure that all technical resources are explored and acknowledged so that future accidents of this type can be anticipated and avoided. We have the public, whose reactions range from anger to sympathy, and is motivated by a need for reassurance. We have the political agents who wish to mollify the public but who may also have agendas in support of or opposition to the shuttle program. And we have the scientific community, which is naturally concerned with the integrity of the program, genuinely wishes to see it succeed, and wants to assure the public and the polity that they are capable of monitoring their own enterprises. This network, one where motives overlap and agree, but also contradict and challenge one another, makes this a complex relationship of emotive and rational impulses, fragmentary information, individual agendas, and contradictory language. Dismantling the scientific and the a-scientific is no easy task.

We can say, however, that the scientific explanation for the cause is reasonable enough. Weakened external materials on the shuttle caused heat great enough on reentry into the earth's atmosphere to compromise the structural integrity of the entire spacecraft. The shuttle did not explode, but merely flew apart from stress on weakened materials.[3] Reentry is one of the most hazardous moments of any shuttle mission, and it is at this time that any malfunction or compromised component is vulnerable to disaster. The collection of physical evidence bears out this scenario, as nearly as we can tell from the information provided, and in some ways, from the scientific end, at least, little more needs to be said. The next scientific step is to discern why it went wrong and to try to avoid it in the future. The incident may also dictate a more comprehensive structural re-examination of the shuttle, but this would be a cautionary measure more than a necessary one. As a result, scientific explanations are minimized in news reports, since even to the layperson the cause seems obvious. The reason for the failure, however, is less certain, and so technological rationalization must always have a human component; that is, someone must be at fault. There appears to be an intuitive belief that technology is only as reliable as the people who monitor its operation, and this is hardly an unreasonable expectation. On the other hand, accidents do happen, and when we consider the complexity of this particular technological feat, we must assume a degree of risk that is simply unavoidable. Such fatalism does not set well with the public, however, which is always preoccupied with "safety" and assigning blame, and the media feeds public panic by investigating presumed incompetence whenever it has the chance. Ideologically, this is an interesting rhetorical phenomenon, since the underlying belief is that humans are in control of everything they do, and we don't appear to suffer failure quietly, even if that failure is inevitable. The blending of faith in science, of human ingenuity, of a competitive spirit, of an economic mindset, and of an hysterical need to eradicate risk are at the heart of this incompatible attitude. Common media phrases and so-called rhetorical questions like "could this have been prevented?" or "is NASA cutting corners?" occupy the pages of newspapers and fill the airwaves, yet we want real answers to these questions. We need to be comforted by the idea that this accident was not random, that someone must have erred, and we ignore the physical reality that dictates the inherent perils involved in orbiting the earth in a tin can.[4]

In the final analysis of this rhetorical exercise, we realize that we are confronted with a resulting distortion of reality that is caused by the symbolic interplay of many different discourses—discourses driven by a variety of motives and a more or less common pool of assumptions. The residual frame within which we find ourselves is a surrounding, sometimes penetrating clatter that we mistake for authentic perception, when in fact it is only a fragmentary component of a much more complex discourse and a more complete reality, like listening to *The New World Symphony* but only hearing every third note from each instrument one at a time at arbitrary intervals and thinking that we are hearing the entire piece. The value of such an analysis, in my mind, is that it gives us insight into our frame of reference, which is the first step in removing ourselves from that frame so that we question, and ultimately, answer, our preconceived notions of who is speaking, the reason behind what is being spoken, and what the speakers hope to accomplish from speaking. In terms of the rhetoric of science, we have a hermeneutic with which to understand science's role in this ongoing loquaciousness. One hopes that this both validates science and shows its discursive limitations to a public that might otherwise be blinded by its indiscriminate invocation from non-scientific agents.

The residual field approach, as much as anything else, affords us an opportunity to see the limitations of the information we are receiving, though it can also supply enough answers to give us clues useful for direction in the next stage of our search. If, for example, we want to know more about what happened with *Columbia,* then the residual field template gives us two things: a healthy understanding of the incompleteness of the story that we have received and an intimation about how to proceed if we want to find out more. As for the former, we know that all the intricate details leading up to the crash are so remote from our vantage point that understanding how they all fit together is unlikely. It is unlikely, in fact, that those most closely involved will have anything like a comprehensive understanding of all the variables that contributed to the before, during and after events of this narrative. There are simply too many discussions going on in too many remote areas too far removed from one another. As for the latter, the interested layperson can begin digging deeper based on this knowledge and the indicators of promising leads offered by the analysis itself. Keeping in mind the motivations of all the agents participating in this narrative, the informed reader can read further and in a variety of sources, compare contradictory stories, and study the history of NASA's bureaucracy. Perhaps the most valuable lesson that the residual field approach supplies in this instance is the knowledge that what we're dealing with has almost nothing to do with science. Except for the explanation surrounding the technical problem that led to the crash, science is an incidental concern.

The reader might be asking what else this critical method can provide. The answer is that it depends upon the situation upon which it is being applied. Some events are complicated with so much rhetorical chatter that it is very difficult to discern anything like a comprehensive understanding of arguments, discussions, and recommendations from various agents. Other circumstances are so distant from the time and place from which we received the residual information that much of it has been distorted, translated, paraphrased, and fragmented to a degree that piecing it all together can provide only a partial picture at best. But in all cases, we at least get a better sense of who is involved in a particular aspect of the narrative for a particular possible purpose. I use the word "narrative" because what the audience receives when being provided with scientific information via the media is a story, a series of events and characters, and an interplay of actors and thing acted upon that involve a continuing dialogue that reveals motives and interpretations in the agents. In this "ongoing conversation," the audience is made up of tardy participants who have neither witnessed the event first-hand nor heard the initial discussions that coordinate the trajectory of the event. The "frame" or "field" of residual noise is that area in which the audience must tune their own receivers to sort out the static from the clear signals. Without some critical mechanism to make sense of what is being heard, the residual field sounds at best like an incomplete transmission and, at worst, like only so much indecipherable clatter. The field is a space of communication, but it almost always demands even further communication. Even the most thorough news coverage can only suggest possible interpretations, and we have all experienced coverage of a news item of special interest to us only to feel that there was much more to the story than was being revealed. A situation like this requires further engagement and activity on the part of the audience and some means by which to piece together scattered and partial information.

Though I have supplied only one example of how the residual field method might be used, I hope it is clear how this can be employed for any circumstance that involves separate and distinct discursive elements that must come together to generate meaning for the observer.[5] It is particularly useful for interpreting scientific discourse because it removes the highly technical aspects of science from the equation without sacrificing a more comprehensive understanding of the interchange in the process. Such a method could be easily applied to the creationism/evolution debate, media health reports, advertising, science education (or discussions of education that invoke science as an assessment instrument), public policy, environmental issues, economic analysis, or anywhere that the popularization of scientific information is central to interpreting a significant event. In the next chapter, I provide some further rationale for rhetorical studies in order to demonstrate how the residual field fits into this larger framework of academic inquiry. I have placed this discussion late in the book because it seemed prudent to frame the discussion with practical specifics before tracing the intellectual history that informs this rhetorical method of analysis. The next chapter is also intended to give the interested reader some further background and readings in the event that he or she wants to pursue a separate, detailed path of inquiry on specific questions about rhetoric and discourse analysis.

10 Postmodernism, Humanism, and the Science Wars

Some discussion of the theoretical basis for the residual field method might help illuminate the logic driving this critical approach. The idea of the residual field can be linked closely to developments in the scholarship of visual rhetoric and the so-called "New Media." These areas of inquiry, very popular in communication and rhetoric departments in the US at the moment, help shed light on the complex nature of mass communication, a process that is only getting more and more intricate as new forms, new interfaces, and new technologies become available. It is therefore important for anyone who is interested in how scientific information is disseminated to have some working knowledge of how mass media processes deliver information and have an impact on our collective attitudes. Marshall McLuhan has written much of the traditional scholarship and journalism on the technological delivery of information, while Kenneth Burke has given us a rhetorical analysis of how attitudes are formulated as well as a dramatistic method for decoding these attitudes, and Jacques Ellul has discussed in detail the "formation of men's minds" through an analysis of propaganda and the technological dissemination of ideological messages. However, in the last decade, scholarship in communication theory, especially as it has surfaced in visual rhetoric, has exploded. This is due in part, of course, to the exponential increase of computers by nearly everyone and the development of visual and text interfaces through computer programming and the Internet—areas of information that have been more or less traditionally separated are now freely interchanged for the fullest possible rhetorical impact. Rhetoric, as a discipline, has therefore changed drastically in recent years; whereas it was once considered, a la Aristotle, Cicero, and Quintilian, as "the art of persuasion" or under the more ethical description of "the good man speaking well," more attention has been paid in the last several decades to the function of rhetorical analysis as a study of "discursive" interchange. That is, while rhetoric once occupied the place in the curriculum that taught the craft of effective and persuasive speech (and, by extension, writing), it now has blossomed into a full-blown area of critical discourse. It has much in common now with philosophy, cultural studies, and literary criticism, traditions which it draws on and enhances with its own millennia-old corpus of intellectual material, and is used as a hermeneutic to "logologically" dissect all forms of human communication.

While this function of rhetoric has already been discussed implicitly (and more directly in the introduction), it is necessary to establish the disciplinary vantage point from which rhetorical analysis takes place. It is equally important to address those areas of scholarship and their representatives that hold a disparaging attitude toward any discursive analysis that challenges fixed meanings and complicates easy distinctions. Since rhetoric has taken on pejorative connotations in the American vernacular—i.e., uses by journalistic and political pundits such as "once we get beyond the rhetoric" or "Senator Jones's rhetoric aside"—it is necessary to dispel the myth that rhetoric is somehow separated from who we are and what we say and do as human beings. In other words, to say something like "once we get beyond the rhetoric" is to either willfully or through ignorance proffer a notion that there is a clear distinction between the words we use and the actions we engage in or to suggest that we can ever get "beyond" the language that drives our attitudes and ideologies. We never get "beyond" rhetoric as long as we are using language, and even our actions have decidedly symbolic significance, and it is certainly the purview of any system of thought that relies on language analysis in order to more fully understand the nature of human motivations to examine the degree of this significance. We are reminded that the two traditions of rhetoric that have been unwittingly conflated in the public mind—*rhetorica utens* and *rhetorica docens*—can be simply described as the difference between doing rhetoric (*utens*) and using rhetorical analysis (*docens*). Speaking generally, *utens* was favored in early treatises on rhetoric; such works as Cicero's *de Inventione* and *de Oratore* were concerned primarily with teaching students how to speak well and appropriately for specific civic purposes. *Rhetorica docens* was certainly part of this tradition—Aristotle made the original distinction—but rhetoric as a full-fledged hermeneutical system in its own right didn't gain real prevalence until well into the twentieth century. People interested in critical analysis of texts were "literary critics," philologists, or philosophers, not rhetoricians (or to use the more archaic phrase, "rhetors") in the sense that we in the field think of them today. However, today's public does not think of rhetoric as anything more than the flowery use of empty language; they certainly aren't acquainted with the long and once-respected humanistic tradition of rhetoric as a field of study. It is unfortunate that the word has been appropriated by an army of lawyers and politicians who, if they had any sort of substantive humanistic education, should certainly know better (a questionable assertion, to be sure). Rhetoric as a field is more than the coarsest application of *utens,* a sort of verbal Machiavellian obfuscation or manipulation of words. It is, rather, an intellectual apparatus for understanding the various ways that human language represents itself, how these different ways intersect, what they mean, how they are used, and why they are effective.

To reinforce what I've alluded to in earlier chapters, this means that rhetoric is a useful critical tool to apply within the sprawling domain of the media. I have saved this explanation for late in the development of the central premise of this book for several reasons: it didn't seem prudent to overwhelm the reader with a heavy dose of theory early on in the progression of the book; the "literature survey," while useful for those wishing to read further in the area of rhetoric of science and discourse theory, is more appropriate for specialists or students in the field who are already familiar with the material; the primary audience for this book is the educated general reader or student, not the professional rhetorician; and the material that follows serves as a theoretical buttress to a more practical application, namely, how to decode scientific language to increase general awareness of how scientific discourse is constructed for non-specialists. Finally, the body of scholarly material is so large that it was necessary to limit the survey to rhetoric of science as it relates to the specific details of this text and the overlap it shares with communication theory generally and media discourse in particular. The following examples provide an indication of the depth and breadth of material available on these topics. For example, while Jereon Jansz's essay on "The Emotional Appeal of Violent Video Games for Adolescent Males" may provide useful insight into how new and developing technology may provide an emotional, participatory outlet for those engaging in video games, thus further blurring our society's boundary between reality and fantasy, it cannot provide the sort of analysis that gives us revelation on the more passive activities of news-gathering. The point is that, given the sheer volume of material available, all of which is germane on some level and to some degree, I have chosen only selected work that is particularly relevant.

The valuable and fascinating work being done in the area of communication and media analysis frequently places rhetoric at the center of the critical field, and does so because of its facility in describing and interpreting media-generated phenomena. Perhaps the seminal work in the rhetoric of science, a book that revolutionized how we look at scientific discourse and processes, is Thomas Kuhn's *Structure of Scientific Revolutions.* This book opened the door for reading scientific discoveries in social and rhetorical terms—so much so that Randy Allen Harris describes its impact as having "brought rhetoric to the absolute heart of scientific studies" (xiii). *Structure,* released in 1962, introduced a whole new dimension in the philosophy of science, a field that already had a fairly lengthy tradition. It created a stir, therefore, among philosophers and scientists alike because it made a distinction between "normal" science, which dealt with the daily problem-solving of scientific problems within an existing "paradigm," and "revolutionary" science, which exposed processes and ideas that operated outside of the existing paradigm—a "shift" from one defining framework within science to another that required rhetoric in order to happen. Though Kuhn did not use the language of rhetoric per se to describe what happened during both the shifts in the dominant paradigms and in the maintenance of the existing ones, the process he illustrated was persuasive, dealing as it did with "a particular set of shared values" and how these values "interact with particular experiences shared by a community of specialists" (Harris xiv). Harris suggests that the most important impact this book had on all subsequent study in the rhetoric of science was to bring down the "wall of expertise" in a way that opened the doors for examining the "processes and practices" rather than just the products of science (xv). This is an essential component to understanding how scientific ideas are both formed and forwarded, and with this new access, the epistemological primacy science once enjoyed was as exposed to rhetorical critique as any other area of discursive consequence. However, Kuhn, and a number of other scholars who followed him—Feyerabend, Toulmin, Rogers, Bronowski—were themselves elite scholars whose ideas were reserved for the sequestered halls of the academy. It is only recently that scholars have made significant attempts to see how science and the language of science has been used in areas of discourse that fall outside the often self-contained domains of the scientific community.

Many scientists, as one might guess, are resistant to a rhetorical analysis of what they do, seeing any effort to describe scientific activity as anything less than pure and untainted by the linguistic messiness of other fields as an expression of professional jealousy. This is, of course, a defensive reaction to a condition of fallen grace; science, while having certainly exacted its influence on the orientation of most Americans, is both revered and suspected, a strange attitudinal paradox that lends science the same tenuous authority as almost every other institution in twenty-first century America. As much as many people—political pundits and academics alike—try to downplay the significance of the 1960s because it has taken on a legendary and Romantic status when viewed through the lens of popular culture, the decade did do one thing that is undeniable: It gave us permission to question the institutions and the authorities behind them in a way that hitherto had been both discouraged and repressed, sometimes violently. The authoritarian status of science had, during the early Cold War period, given science a degree of credibility that was in many ways deserved but in other ways quite dangerous. Such status gave scientists an unprecedented level of intellectual and policy-making authority, authority that would push some thinkers to question the content of that authority, to determine whether the practices and processes of science were, in fact, as untainted as many had been led to believe. The scientific whipping-post that would emerge in reaction to such impious questioning of scientific dominance would be the two-prong attack against "relativism" and "postmoderninsm."

This last word, which emerged as a result of literary and cultural theory, had, admittedly, an extremely fuzzy definition, if it had anything like a stable definition at all. Some scientists rhetorically seized upon this fuzziness as a way to discredit any of its intellectual contributions to understanding how social networks operate discursively, claiming that it was only so much relativistic nonsense because it didn't hold certainty in the same lofty position as scientific rationalism did and, worse, was used it as a way to describe scientific operations itself. One vocal (one might argue obnoxious) opponent to postmodernist critical methods, especially as it was applied to science, is a man I have already mentioned: NYU physicist Alan D. Sokal. Sokal wrote a "hoax" article and submitted it to the cultural criticism journal, *Social Text,* in 1996. The article, entitled "Transgressing the Boundaries: Towards a Transformative Hermeneutics of Quantum Gravity,"[1] was accepted by *Social Text* and subsequently reprinted in the Spring/Summer 1996 issue. Sokal later admitted to writing the article as an "experiment" to see whether a "leading North American journal in cultural studies" would publish a piece that was "liberally salted with nonsense" if "(a) it sounded good and (b) it flattered the editors' ideological preconceptions" (*Lingua Franca* 62). This smug "confession" created more than a stir; it validated the notion of "The Science Wars" to which the *Social Text* issue was actually dedicated. Pundits ranging in quality from Rush Limbaugh to Stanley Fish responded with passion to the hoax, and the variety of interpretations was wide and revealing of the profound miscommunication, distrust, and outright hostility that existed between the sciences and the humanities. Sokal's ostensible motive, he later admitted, was to validate suspicions that had "troubled" him "for some years," namely, "an apparent decline in the standards of intellectual rigor in certain precincts of the American academic humanities" (62).

Whether Sokal was truly "troubled" or he was simply trying to reassert the superiority of scientific disciplines over the humanities only he can answer for sure. The stunt was anti-collegial, to say the least, and anti-intellectual, to say the most, and the *Social Text* editors later tried rather pathetically to recover their lost face by arguing that they had accepted his article in good faith, and that, as essentially a physics paper, they were not qualified to judge the contents of the pseudo-theoretical argument. They were, they said, concerned with the apparent lack of mastery over the critical material, but because they were hoping to establish collaboration between the traditionally disparate disciplines of science and cultural studies, they had taken on faith that the physicist was trying to establish academic rapport even if he was a bit out of his league. Nobody outside of the editorial board was buying this explanation, however; instead, the incident was seen as an archetypal example of how much gibberish was passing for scholarship these days. For example, Limbaugh, quoting Stanley Fish (who immediately became involved in the debate, attempting to defend the misinformation circulating about the basic tenets of postmodernism), interpreted the following statement by Fish to mean that "the world does not exist": "What sociologists of science[2] say is that of course the world is real and independent of our observations but that accounts of the world are produced by observers and are therefore relative to their capacities, education, training, etc." ("Professor Sokal's Bad Joke"). Any careful reader, especially one trained in rhetoric or cultural studies, would view this statement as an obvious cornerstone of any claim made about the world. It does *not* say "the world does not exist"; it says that its physical existence and the human interpretation of what that existence means are two entirely different things. This exposes a fundamental straw man that bogs down any meaningful dialogue between the humanities and the sciences. Science is fond of perpetuating a notion that any claim of relativistic interpretation that might be based on environment, training, or ideological steeping is a claim for "cultural constructionism" that challenges even basic laws of physics (like gravity or thermodynamics). Such absurd claims have never been made by any postmodernist critic that I've ever read. Yet, to conflate cultural relativism with a notion of denying the basic facts of existence that we all must contend with is a favorite tactic of scientists hoping to discredit work outside of their sphere of knowledge.

Naturally, many scientists would make the same charge against a rhetorician who was hoping to discover how a concept like "loop quantum gravity" was being explained to non-physicists. But what scientists of this school fail to note is that the domain of science is one thing; the domain of public access to scientific ideas is quite another. Sokal's hoax was an act of rhetoric; it was designed to expose the pretensions and intellectual trendiness of cultural criticism. Yet the irony is that this act of rhetoric was also designed to show that rhetoric (or, at any rate, postmodern analysis) is not a factor in the "reality" of the world and that there is no connection between what something is and how we explain what something is. His stunt was a success in that it demonstrated how rhetorical framing can and does influence the way that we think about and interpret the natural world, even in the form of a hoax. It is, however, a failure to understand the very human process of casting the world in an image that is accessible to us, which carries with it its own interpretive limitations. Does the postmodernist claim that cognition is relative to the words we use to describe natural phenomena make the world less "real"? Does Sokal really want us to believe that the audience, and therefore the language used with that audience, really doesn't matter, or that the motives of the person reporting the information are irrelevant? On the contrary, cultural criticism demonstrates how reality is complicated by human sensory, intellectual, and linguistic abilities and limitations.

My main motive for bringing up this incident is to illustrate how, in fact, rhetoric *must* play a key role in the dissemination of scientific ideas and the maintenance of the scientific power dynamic. Sokal, though congratulating himself for his own mischievous creativity in exposing the "nonsense" and "gibberish" of postmodern theory, was unable to recognize how much his bad joke actually reinforced the rather postmodern claim that science, like all other disciplines, is a rhetorical activity. His only motive could be that he felt the "rigors" of disciplines from the faux left were wanting, that the apparent ease with which he could get a paper published in a leading cultural criticism journal (*Social Text* is published through Duke University Press, hence Stanley Fish springing into action) revealed how soft and desperate the humanities had become. This can only mean that he found the scholarship threatening somehow since, for all intents and purposes, such a journal only has interest in science insofar as it relates to the ideological and social influences it wields. He was, in other words, illustrating as meaningful exactly that which he was hoping to expose as meaningless; he was trumpeting the standards of scientific practice as an ideological force and he had established himself as self-appointed guardian of the doctrines that science holds in reverence. This is hardly an objective, detached response. Since other fields were not using the same standard (and in this much he is correct), the fields allegedly ignoring "academic rigors" were receiving what he had determined was too much credence and attention, and he felt it was his duty to construct a rendering of the field that would be easily ridiculed. This is unquestionably a rhetorical move, and one that textually reinforces the principles of Sokal's discipline: Only those fields of study that use a scientific methodology should be taken seriously; only those areas of inquiry that reinforce the laws of science should be considered; only those questions that match a scientific purpose should be asked. These are the proclamations of a man who is defending an intellectual way of life (he referred to cultural studies as "anti-intellectual," an egregious example of the pot calling the kettle black), and any methodology or path of inquiry that does not meet his particular academic orientation must be problematic. Had Sokal read and understood more cultural criticism or rhetorical theory, he probably would have seen that, from a theoretical point of view, his stunt had backfired.

From a public relations point of view, however, it enjoyed a disproportionate degree of success. Many people who have no knowledge whatsoever of either theoretical physics or of postmodernism (save for the ridiculously dismissive errata circling the phrase "anything goes") seemed validated in their neo-conservative impulse to be suspicious of the strange goings-on in academe. I wonder how pleased Sokal was with the prospect of Rush Limbaugh heartily endorsing his hoax. Many other journalistic pundits popped up their heads at the opportunity to comment on this issue, seeing it as a brilliant strategy for exposing the jargon-wielding, name-dropping disciples of postmodernism, further confounding the misinformation the public receives on such issues. All scholarship—science too—has its extremists, and any scientist who claims that the kooks in their respective disciplines are weeded out through the peer-review process is being naïve. Postmodernism has led to some remarkably self-aggrandizing "scholarship," but it is hardly alone, and to dismiss serious postmodern contributions to our understanding of language operations seems far more "anti-intellectual" than sifting through the quagmire to get to the valuable ideas. And in the end, postmodernist ideas have far more positive intellectual contributions to offer than not. Science a la Sokal, then, reflects a political orientation as much as an academic one; it forwards a set of values that it sees as central to the integrity of the discipline. It has a professional ethic. This realization should serve to validate the notion that while the contents of the natural sciences may not be entirely "socially constructed" (i.e., gravity exists, animals evolve, matter is made of particles, etc. with or without our observation and/or description of these phenomena), the process of articulating these ideas to others certainly is. Science is a human invention, and all human inventions involve language and communication. All language and communication is subject to rhetorical activity, either as a way to create identification with or to persuade other human agents.

The resistance the scientific community has erected to the notion that its various fields of study can be subject to outside analysis—especially from other fields of study—should demonstrate its desire for autonomy, a privilege of the ruling class. Many scientists view external interpretation of their activities as meddlesome and impossible. But this view overlooks the reason that many cultural critics and rhetoricians are interested in science in the first place: to determine the impact science has on our lives as a communicative apparatus that can be put into action. The academic examination of science has what is in my view a happy side-effect—it allows for interdisciplinarity. This is something that has been lacking for a long time, and the objections to interdisciplinary studies have less to do with the necessity of sequestering separate fields of knowledge for any practical purpose and more to do with the internal politics and territorial infrastructures of schools and universities. Scholars and teachers from such disparate areas of interest as history, sociology, psychology, communication, medicine, philosophy, political studies, literary studies, and rhetoric have commented on the role of science in society. Kenan Malik, mentioned earlier as the research psychologist at Sussex University, has written on the limitations of the scientific method in the human sciences, but also on the culture that scientists inhabit, in his book *Man, Beast, and Zombie: What Science Can and Cannot Tell Us about Human Nature.* A key point he makes early on in the book is that the movement of humanism was defined much differently during the Renaissance than it was in the Enlightenment, and it was certainly far different from how it is viewed today. His observation helps trace the intellectual history of the humanities (the early manifestation of which both rhetoric and the sciences were central), originally conceived not as a disciplinary division, but as a philosophical one where humanity was central to all important intellectual questions. In that early model, philosophy, science, rhetoric, mathematics, art, writing and languages, and music held equal sway in the intellectual progression of the person striving for education and knowledge. This is important because it marks for modernity a sharp departure from this model, one that has gone from a sense of the interrelatedness of knowledge to one where knowledge is strictly partitioned into various specialties and sub-specialties. There is, in fact, a palpable suspicion by most in academe (and in the general population) that anyone who professes to know something about a wide range of topics is a "dabbler," someone who does not have anything useful or provocative to say about anything because she only knows a little bit about everything. Any casual examination of the medical profession should be enough to confirm this attitude. One can no longer go to a general practitioner for anything but a referral to a specialist. For medicine, this may be a good thing, though it does tend to reflect the other deviation from early humanism. Whereas the Renaissance humanist saw mankind as "part of the natural order" (Malik 3) and, therefore, as a whole person in the complexity of the cosmos, specialized medicine tends to treat maladies not in a "holistic" sense of treating the whole person, but rather as a network of interrelated parts that are often treated as if they were not related at all. Whether this is appropriate or not is beside the point; it is more interesting to note that it informs a particular orientation regarding the medical hegemony, such that any exploration of "alternative" forms of medical treatment are met with disdain, and in some case, outright hostility by those who practice "traditional," "Western" medicine. Again, we see the emergence of a-scientific motives informing a particular reaction against alternative medicine. It may be that many "treatments" are purely placebos—does ginko biloba make us smarter? does enicea make our nervous system healthier?—but it may also be that the resistance by the medical community is motivated by a sense of professional protectionism. If such remedies do work (or, more importantly, people *believe* that they work), the reliance on doctors will be diminished. And we rely on doctors for everything from general preventative care to mental health. Drug companies provide incentives for pushing particular drugs. How do they figure into this equation? As I have reminded the reader regularly, many issues that have a scientific basis are appropriated by other agencies that have objectives that are not so unadulterated as those of the pure research scientist. Malik gives us much to think about in regard to the relationship between science and the other factors that contribute to how science affects our lives and our thinking. He asks why humanism, as a central tenet for knowledge, has taken such a beating in the last century. He acknowledges that, like any historical path, the reasons for the shift from humanism to a more ostensibly objective form of knowledge is a complex and intricate combination of lost faith, increased technology, and a need for a surrogate statement of certainty in an increasingly secular world. His summary of these developments gives us a useful picture of the present state of the humanities in our culture:

> The humanist tradition created a language through which to understand humanness by linking Man's inner world to his outer world, by viewing human beings as agents through which could be transformed nature, society and Man itself. By the end of the twentieth century, however, there was both a general disillusionment with the prospect of transforming the outer world, and a melancholy about the condition

> of inner Man. The result was an increasingly narrow, self-centered view of humanness and an increasingly degraded one. The language of humanism no longer seemed adequate, and yet no new language seemed available to replace it. (7)

Perhaps because of their transformative natures, science and technology helped supplant a humanistic view of the world—without which, we were in a state of nature in which we were as vulnerable to the laws of natural selection as any other animal. Yet, we carried into our civilization the notion that survival, albeit of a more socialized kind, was the most important goal. Survival, for twenty-first century Man, means much more than the simple propagation of the species; it means power and social protectionism against the forces of society that could keep us in positions of economic and class servitude. In the United States, humanism is generally viewed as dangerous because its only apparent function is to challenge the bottom-line definition of human purpose, to divert energy away from the pursuit of *status quo* economic survivalism and to channel it into areas that are seen as ultimately damaging to an Establishment. Humanism is viewed by many as a fruitless attempt to create a Utopian society through self-improvement and altruistic intellectual pursuits, ignoring the basically selfish nature of humanity. This outlook may not be new; pragmatic vocational pursuits have long been touted by the middle class as the only way to maintain its position in the economic hierarchy (especially in education; see Eagleton's *Literary Theory, An Introduction*, 2nd ed., 1996), but it takes on new vigor when all the other institutions that gave us meaning have been eroded over the centuries, leaving only one, corporatized version of institutional identity.

Malik's thesis is very similar to mine in some important respects. He claims, for instance, that "science is as much part of culture as is art. Science does not stand above the hurly-burly of everyday life. It is part of the hurly-burly, helping create it, and drawing sustenance from it" (7). This frequently overlooked fact is what has inspired so many scholars to reopen the chapter on science as a "social construct" that has so long been repressed—social constructionism not in the Sokalian sense where the laws of science are questioned, but in the rhetorical sense that science and society are reciprocally operating entities demonstrating an intersection of ideologies that influence each other in many unexpected (and counter-intuitive) ways. Early on in his book,

Malik makes and reinforces the scholarly distinction that separates the scientific discourse camps into those who hold a "social constructionist" view and those, like Richard Rorty, who deny any correspondence between scientific fact and objective reality. Malik sees both of these views as misguided, or at least unfortunately expressed, because they do more harm to the domain of science than to merely expose its limitations. His more judicious view is that

> [s]cience certainly gives us access to a reality that exists independently of human beings. In this sense science is different from other forms of knowledge, such as politics or literature. But the scientific process does not stand apart from the culture it inhabits. The questions scientists ask about the world, and the interpretations they place on their data, are often shaped by cultural attitudes, needs and possibilities. (9)

Malik is especially concerned with the limitations of the "human" sciences, and for good reason. As a psychologist himself, he recognizes that there is an inherent scientific methodological problem with an agent that is both the creator of the method and the object of study, for it is clear that people, no matter how well-trained or seemingly emotionless, are going to have some pretty strong preconception about the nature of their own species. While it is relatively easy to treat a new strain of bacteria with the proper scientific detachment, humans are far more complex, and any conclusions drawn about them, especially negative ones, cut close to the bone. Because of this, Malik concludes that "[t]he science of Man, therefore, cannot be understood wholly apart from the culture in which that science is produced" (9). While this is not, of course, a new idea, it is unusual to hear it from one of science's own (the relative status of psychology in the sciences notwithstanding, it is often more likely to hear a psychologist rigorously defend the scientific "purity" of his or her profession to its detractors in the hard sciences). Thomas Kuhn, himself a physicist, was ostracized by many in the scientific community for the impious pronouncements he made in *The Structure of Scientific Revolutions,* so it is clear that one of the cultural factors that has an impact on the scientist is to avoid blasphemy against the doctrine of scientific objectivity. One of the central problems with the human sciences, according to Malik, is that there is a fundamental paradox in the message that science tells us about the place of humanity in the natural world (as opposed to the earlier paradigm of Man as distinct from the materialism that governs the rest of nature):

> On the one hand, science has taught us to perceive nature in largely mechanistic terms, a process that has driven out magic and mysticism and "disenchanted" the natural world. On the other, we view humans as beings possessing consciousness and agency, qualities difficult to express in physical terms. We are happy to view human bodies as machines; but what we value about our fellow humans is that they do not act as machines—as robots or zombies—but as *people.* (11)

Moreover, Malik suggests, "[o]ur very success in understanding nature has generated deep problems for understanding human nature" (11).

This last observation is perhaps the most important one for considering the place in which we find ourselves regarding both the role of science and the foundations of what it means to be human. It is little wonder that so many have been so resistant to relinquishing older models of Man's centrality in the universe. Considering ourselves as just another product of nature at once simplifies and complicates things considerably; it simplifies the manner in which we view our material existence while complicating the less tangible aspects of what it means to be human. The human sciences struggle to reconcile this seemingly mutually exclusive dichotomy because it is difficult to defend the position that we are both physical and incorporeal, that we have what has been traditionally articulated—ever since Descartes, at least—as the "mind/body problem" or the "dualism of Man." This has long been a philosophical problem in the West that, it is interesting to note, has eluded many philosophies of Eastern origin. For many Eastern philosophies, mind and body are simply one and the same—humanity cannot separate that which we think from that which we are. In this respect, such philosophies have not been tainted by the materialistic reductionism of Western science—but many of the cultures in which these philosophies flourish have not fully reaped the benefits of science either. It seems, in some important respects, that we cannot have our cake and eat it too; we cannot hold to a strictly material explanation of the world with us in it. This conundrum may help explain why science and the humanities have for so long been at odds with one another—

they seem like such radically different perspectives on the nature of humanity that they can never overcome the dualism that seems to separate them (for another interesting, and remarkably progressive, perspective on this, see Betrand Russell's collection of 1953 lectures, *The Impact of Science on Society*).

A long-standing perception has been that science threatens to condense our most difficult and significant philosophical questions about the nature and origin of humanity to a series of dry technical queries that merely produce data but overlook the "soul" of the human organism. Both of the most recent major influences in psychology—evolutionary psychology and cognitive science—have gained a substantial following in the scientific community and, to a lesser extent, among the general population. Whereas evolutionary psychology argues for a Darwinian explanation of the origin of the modern human mind, cognitive science posits that the mind, like any other anatomical system in the human body, is a machine and can thus be "mapped," studied, and, one must conclude, *repaired* like a machine. According to Malik, there are three "principles" or underlying doctrines to the cognitive science of the mind: the brain functions like a computer; the brain is composed of separate "modules" like an array of inter-working processors; and these modules are innate in *Homo sapiens* (271). There are many other sources that boil down cognitive science, but Malik's is the most accessible and retains the primary theoretical descriptors. What is interesting is that all three of these principles see the mind in mechanistic terms—even if they afford us the dignity of having complex machines for brains in the model of the computer. The problem is that the metaphor of the computer parallels the state-of-the-art of human technology; is it possible that we are simply superimposing the metaphorical template onto the brain and seeing patterns that fit this template? And if so, doesn't this fall short of a scientific analysis?

We have seen how metaphors can dictate and therefore limit the way we understand certain scientific ideas, and this is another example. But it isn't my purpose here to challenge the theory underpinning cognitive science so much as it is to point out this trend in scientific (and humanistic, for that matter) thinking and to show how cognitive science fits in the historical narrative of scientific development. As soon as machines dominated the human landscape and the material philosophy dominated human thinking, our search for answers to burning questions tended to take form in these kinds of terms. It is reasonable to assume that the more machines are a part of our everyday existence, the more likely it is that they will influence our search for metaphorical connections in that direction. The metaphorical shift from a human-centered society to a technology-centered one is a disturbing prospect because it is an arrangement that favors the thing over the living. This might be easily compounded by a scientific school that says, in effect, humans *are* machines. In the strictest sense of the word, this is hard to deny; our anatomical bodies are electro-mechanical in nature, and this is more than a convenient metaphorical description. Bones, muscles, tendons, ligaments, circulatory systems, nervous systems, digestive tracts, respiratory systems, and the like are all mechanical—they use the laws of physics to keep the organism alive. Why shouldn't we then conclude that the brain is *literally* a "computer"?

As interesting as this question may be, the answer is relatively straightforward. A human being, even beyond our assumptions about the superiority of humanity or our self-assigned spot at the top of the evolutionary mountain, is more than the sum of his or her parts. The human body, as wondrous as it is, is nothing more than a mass of unanimated matter without the brain, the most complicated, intricate, and poorly understood system in the human body. And bringing something to "life" in the human sense is far more than making the machine walk and talk and perform tasks like a robot; there is something about the brain that breathes not only existence into a human being, but substance as well. Part of that substance, ironically, is a need for certainty, a drive so strong in some of us that we are willing to reduce the human animal to a set of "parts" that can be disassembled and analyzed like any mechanistic system. This is a failure of science. Malik, as a psychologist, recognizes that cognitivism, while at least going beyond behaviorism in its favorable depiction of human activity (for a behaviorist, any animal, humans included, merely responds passively to stimuli, not through any will of its own), still views humans as a remarkably predictable lot, requiring only careful dissection of the integral systems to understand the workings of the mind, a perspective he sees as limited and limiting. More importantly from the standpoint of a rhetorical examination, how does the favoring of a school of thought in the psychological community orient a way of thinking about ourselves that that may very well fall short of describing the whole person? How is the language appropriated and by whom, and what are the ripple-effects of this way of viewing humans?

If the human brain is the product of evolution, which it most certainly has to be, what is the problem with basing a psychological school of thought around this fact? This is a complicated question, and one that David Buller wrestles with in *Adaptive Minds.* The problem, he suggests, is not the idea that the brain is the result of evolutionary processes; it is the conclusion that evolutionary psychologists draw from this foundational premise. It is the practice of evolutionary psychology to look at our pre-historical ancestors for answers to the modern brain, and the conclusions they draw from this are more philosophically problematic than technically flawed. Looking at Pleistocene Man, for example, leads evolutionary psychologists to believe (and I use this word deliberately) that there is a universal human nature (the "modules" are, for all intents and purposes, a re-articulation of the idea of "instincts" updated with the more technologically sophisticated metaphors that I have just described) that all modern humans share. Rather, Buller suggests, it seems more likely that there are "many neurological structures that are relatively specialized and that resemble mental organs. But these structures . . . are not biological adaptations; they are flexible responses to the experiential conditions of an individual's life" (13–14).

I linger here for several reasons, all of which have less to do with who is right and who is wrong and more to do with the flurry of intellectual activity such debates create both in the scientific community and for general audiences. While most of the workaday population has little knowledge of or interest in evolutionary psychology or its principle arguments, I submit that knowledge of this area of the human condition should be part of any scientific or humanistic educational program because it effects everyone whether they are interested or not. Yet, unless a student is a psych major, chances are he or she will not be exposed to this debate in any significant way, and even there, chances are the coverage is superficial at best. At the same time, however, Buller insists that "Evolutionary Psychologists have been very successful in conveying their ideas to a broader public" and feels that his book should serve as a way to "convey to the same broad public the other side of the story" (16). Why is knowledge of this debate—or any other in science, for that matter—important? Because we have become a nation of technicians instead of thinkers. In a senior-level course I recently taught on popular culture and scientific discourse, my science students informed me that they were rarely, if ever, exposed to the history of science, the philosophy of science, or the debates within science between scientists, much less to the rhetoric of science. Why is this? In courses where the coverage of material is more important than the process of assessing that material, we should not be surprised that metacognitive processing of information is thrown enthusiastically aside or simply ignored. The distinction between training technicians and educating thinkers is significant when we consider that it is the latter process that encourages progression in scientific fields and reflection on content over mere compliance to the accepted hegemony that dictates many of the issues and procedures that affect our lives.

It is here important to emphasize that part of the issue informing the pejorative use of rhetoric is the public tendency to oversimplify the "truth/lie" dichotomy. In *The Liar's Tale: A History of Falsehood,* Jeremy Campbell reminds us that the reductionistic binary that separates truth from falsity is not only in error, but also that the thoroughly unclear and inconsistent distinction between the true and the false has a long, rich cultural history. Those doing much of the speaking in our own era, however, assume unreflectively that the dividing line between the domain of truth and the domain of untruth is clear and, more significantly, internalized by the average human being—even as they compromise this principle in the very expression of this view. In other words, all one need know is "right from wrong" in order to discern truth from falsity. "Truth" is, however, at best an elusive concept; while we can cite many simple examples of "truthful" statements (the sky is blue today, the spoon will fall if dropped, etc.), these truths are a matter of definition and categorization. The sky is "blue" because that is the word we use to describe the hue that we have collectively agreed is "bluish." We may, however, disagree about what shade of blue the sky is. Is it "powder blue"? Blue-green? Royal Blue? Interpretive responses to external realities that are reliant on definition (and language generally) always complicate the true/false binary, and when we begin to discuss the nature of abstractions, the distinction between truth and falsity can become very complex very quickly. For example, to suggest that "God is good" is a truthful statement depends very heavily upon the speaker's understanding of God and the nature of goodness, both of which are highly subjective and dependant upon the speaker's conceptualization that may be unique to himself, his group, or his cultural environment. Differing cultural and personal definitions of God and how he/she/it manifests goodness can vary widely, and expressing this in truthful terms has the effect of furthering an internal logic that may not be clear or deemed truthful by other parties. One might wonder, when we begin to consider the nature of truth and falsity in this way, how it is we are able to communicate at all. One possible answer is that truth is not a necessary condition of meaning, and it is perhaps less important to *know* what is true than to *believe something to be* true. For meaning, all one need is a set of established syntactic and semantic rules, an agreed-upon playing field on which the competition, contradiction, and coordination of words can be played out. This is not the same as saying that "there are no truthful statements" (which, if plausible, would be an internally contradictory statement) because truth is a value of the people who are engaging in a communicatory exchange. Put simply, what may be true for one group is not necessarily true for another. Is this rampant relativism? Some might think so, but it is perhaps more useful to suggest that the Absolute Truths that we so obsessively seek are, in fact, linguistically unattainable. Other cultures have seen the linguistic limitations of exclaiming the Truth: One example is Hinduism, which has long recognized that language is incapable of revealing Truth—to the extent that to utter the Truth is to simultaneously make it no longer Truthful.

Note here the distinction between capital T truth and lower-case t truth. Lower-case truths are situational, even personal. They often reflect more about the state of mind of the agent making the utterance than they do about the immutable nature of the truth. They are also temporally situated; what may be true now may not necessarily be true in the future. Truth in this sense is predicated on both perception and stability, and pragmatically speaking, such truths are transitional and, often, relative. Capital T Truths can be traced back at least as far as Plato, and are immutable, pure, and incorruptible. They do not, in fact, exist in our worldly realm at all, at least so far as Plato was concerned. This is why Plato was so apparently scornful of rhetoric: He felt that rhetoricians (in particular, the Sophists) were opportunists who taught people how to disguise the Truth with language and persuasion. Language, again, emerges as the problematizing force in matters of absolute Truth (at least insofar as we can trust Socrates voice in the dialogues as representative of Platonic ideas). Whereas Plato imagined a realm in which the worldly flaws and corruption of a physical existence were supplanted by perfect forms, the corporeal domain of human activity was saturated with language, and therefore, could not be trusted to reveal Truth with any certainty.[3] To underscore this prevailing sense of uncertainty, Campbell cites many thinkers from our philosophical past, but none is more important than Friedrich Nietzsche, who questioned the plausibility of "objectivity" in general, which he considered "a false concept of a genus and an antithesis *within* the subjective" (qtd. in Nelson et al. 7). For Nietzsche, the human species had no "organ" for discerning Truth, but it did have a natural instinct for falsehood. "Truth," as an abstraction taken from the subjectivity of normal human activities, was a necessarily manufactured fiction that we are not equipped to actually find. On the other hand, a natural aptitude for falsehood is an important survival mechanism for many species, and human beings have simply cultivated it in unexpectedly innovative and sophisticated ways. As rhetorician George A. Kennedy has noted, "[. . .] in daily life, many human speech acts are not consciously intentional; they are automatic reactions to situations, culturally (rather than genetically) imprinted in the brain or rising from the subconscious" (25). Our innate propensity for appropriate (if not truthful) responses to situations is something that is nourished by an instinctual foundation to survive, interact, protect, and socialize. Civilization gives us as many new ways to develop this propensity as there are situations that require response. Even in something like formalized speech, Kennedy insists that "the real intent of the speaker may be open to question; prejudice, insecurity, or hostility may unwittingly be conveyed, and a rhetorical tactic may have a different or the opposite effect from what is intended" (25). Objectivity, as laudable a pursuit as it may be in theory, is an ideal, a condition that we can imagine (like perfection, or absolute beauty, or complete freedom) but we can never achieve in practice.

To state this same idea with a somewhat different emphasis, Nietzsche was careful to make the distinction between Truth and a belief system that only professed to contain the Truth. Ken Gemes notes that Nietzsche coordinated the question of Truth around the pragmatics of survival, an observation echoed by Kennedy, who also provides a number of examples of creatures in nature that use deception as a form of self-preservation. One need not look very far to find this. Camouflage, a consistently used method of disguise in nature, can be seen in plants and animals alike. Many birds imitate the calls of rival species to fool them to distraction and away from their nests or food sources. Some snakes sport the colors of poisonous snakes even though they themselves are not toxic. Deception, it seems, is common in nature, much more so than the moral imperative to be honest. In the human species, doctrinal Truth as it is articulated in many institutions (especially religious) is, for Nietzsche, perhaps the most insidious of all deceptions; in this form, it is not a basic lie which is being promulgated but a lie masquerading as the Truth. Beliefs, then, while perhaps justified in the perpetuation and survival of a group has, for Nietzsche, nothing to do with the Truth. Gemes summarizes Nietzsche's position this way:

> In this light Nietzsche's rejection of the importance of truth is not so startling. After all, who but an ascetic fanatic would choose to have true but perhaps life-destroying beliefs? Nietzsche, like many philosophers of science, claims there is no clear connection between truth and various pragmatic virtues. Once we separate the question of pragmatic virtues from the question of truth the property of truth loses its importance. Indeed, if pragmatic virtues are no guide to truth it would seem that truth is unobtainable—for how could we ever recognize it—and hence doubly unworthy of our interest. (53)

Truth a la Nietzsche is a ritualized fiction, a condition manufactured for institutions and the individuals who control them to maintain their own power base. In terms of the conceptual relationship to rhetoric, the pragmatic element he articulates is perhaps his most important observation, since he sees in it a fundamental contradiction between the function of pragmatism over the pursuit of Truth. This requires some very specialized ethical rules to account for the circumstances we deal with daily—so foreign are they to our notion of Truth that we must hold Truth as a stylized abstraction not appropriate for material existence. Ultimately, this realization leads to a notion of Truth that has been reduced, on the one hand, to a question of belief, and on the other, to a question of actualization. Practically speaking, where matters of Truth are concerned, Americans tend to think in survivalist mode at all times: either in simplistically binary forms or in absolute abstractions, and neither position accounts for the complexity of meaning as it actually resides in daily living.

Philosophical inquiries about Truth are far too involved to summarize, confirm, or challenge here. There is a rich history of grappling with the concept of Truth in the philosophical tradition, and such a tradition cannot be dealt with reasonably in a single book chapter. My interest here is, rather, the conflation of nonsense and rhetoric that seems to be part of the public understanding of both terms as they relate to simplistic preconceptions about the nature of Truth/truth. Harry G. Frankfurt has made an interesting attempt to philosophically and etymologically trace the "history" of bullshit. Bullshit is an apparently elusive term, and Frankfurt concludes that bullshit differs from outright lying and can be used to describe those discursive moments "whenever circumstances require someone to talk without knowing what he is talking about" (63). According to Frankfurt, this is a situation that arises more in the modern world than in the past, since opportunities to speak in these instances are greater than they used to be in public life. But perhaps Frankfurt's most interesting contribution to the discussion is his assertion that the public at large feels it is "the responsibility of a citizen in a democracy to have opinions about everything, or at least everything that pertains to the conduct of his country's affairs" (64). This position is hard to deny; opinions seem to be at the heart of the very liberty we in America strive to embrace, as is evidenced through the common statement, "everyone has a right to his or her opinion." Opinions, in this light, are not just simple assertions about a given topic—they are a "right" that must be both possessed and guarded. Opinions, in this sense, become part of the composite of one' identity, something that is a point of ownership and reveals a sense of character about the person uttering the opinion. It is small wonder, given the current capitalistic, materialist, property-based notion of what our country is, that opinions should be abstracted to meet the threshold of our own ideological expectations about how important an opinion is to our concept of self.

Almost without question and without exception, science is taught to students and explained to the laity as a system of inquiry that has conquered its natural human propensity for self-deceit and the contamination of extraneous influences. It alone has managed to set aside the bias and personal desire that besets the rest of humanity and every other form of intellectual exploration. What is most interesting is that it took so long for this claim of objectivity to be challenged. While Nietzsche may have questioned this supercilious notion as early as the 1870s, it was hardly a widely accepted view. Even today, scientists-in-training are told that their field is detached, unfettered and disinterested (in the original sense). The Alan Sokals of the scientific community are quick to caricature any detractors as fringe lunatics who question the actual foundations of reality. Yet many rhetoricians, philosophers, and cultural critics have made strides in constructing a pedagogical foundation that captures the authentic meaning of "critical thinking" by suggesting that the way people are trained in and informed about science is steeped in tradition, dominant hegemony, and simple closed-minded conservatism.

The failure of science to recognize its own hubris seems, then, a cognitive failure; so preoccupied with the pursuit of certainty, science is portrayed as having only one possible path to the truth. With such a perception, imagining the ways that people really create knowledge and coordinate information seem difficult to grasp. Science cannot run interference with itself by thinking about how it works, only that it does. But as Hans-Georg Gadamer emphasized, much like Fleck, the individual approach to any given problem is only one facet of the equation; there is, according to him, a communitarian dimension to interpretation (Nelson et al. 9). We never, in other words, form an interpretation alone. We rely on past knowledge, new perspectives, old perspectives, and multiple experiences. Again, however, science is more often portrayed as a solitary search for truth than it is as a collaborative effort that depends on its own epistemological corpus of assumptions. Science is resistant—perhaps because it sees itself as objective—to Richard Rorty's notion that philosophy is a persuasive contribution to "the conversation of mankind" (qtd. in Nelson et al. 9). It may be necessary to go one step further to explain even this degree of certainty. Whereas rhetoric is often defined through the domain of persuasion, academic pursuits often have less to do with "convincing" others and more to do with "making sense." Both objectives are persuasive, though we tend to speak of "persuasion" as using language (in all it manifestations, including scientific) to promote a change of heart—ultimately, in fact, to move an audience to action. But how often does this really happen? More often, what we write or say is designed to garner a certain amount of agreement, to add to "the conversation," but it rarely creates converts. "Persuasion," in this sense, reflects an additive, not conversional quality. Students in the sciences and humanities alike get the impression that "persuasion" is a process of transposing one way of thinking onto another, and this is a by-product of the positivistic pursuit of truth because it assumes that there is only one possible answer to any question or one possible solution to any problem. If we as educators emphasize the importance of certainty, objectivity, and absolutism in knowledge, we should not be surprised that we perpetuate this critical error in judgment. Academic discourse has made inroads to challenge the age-old pursuits of non-relativistic truth as misguided, but it has yet to trickle down in any significant way to the Academy at large. Nelson et al. point out that for Rorty, as for others,

> inquiry that avoids trying to mirror nature escapes "seeing the attainment of truth as a matter of necessity," whether logical or empirical. The rejection of necessitarian truth and coercive argument is now shared by many philosophers, among them John Searle, Hilary Putnam, and Nelson Goodman. Even Robert Nozick, in many ways a polar opposite of Rorty, has written eloquently on behalf of less dictatorial and more persuasive styles of discourse. (9)

The last century has seen, then, some important revisions in the status quo of academic pursuits of truth and has produced thinkers interested in a more inclusive vision of a community of knowledge-making. All quarters of academia lag behind in learning this lesson, however, largely because the voice in the wilderness is a humanistic one. This statement is not intended as a fatalistic utterance regarding the future of the humanities but, rather, an acknowledgement that humanistic pursuits, traditionally at odds with science and technical fields, are simply not given the degree of credence they deserve. The effect within the humanistic disciplines, in other words, has been largely one of preaching to the converted, and this is what makes a more inclusive rhetoric of inquiry (to coin Nelson et al.'s phrase) so much more important. Rhetoric in the last half of the twentieth century has been far from cliquish; it has happily shared resources and knowledge with whatever field could aid it in the quest for a better understanding of the discursive human condition, even if this outreach effort has been rarely reciprocated.

11 The Education "Crisis"

Science and technology issues often manifest themselves in another oft-cited but underappreciated quarter of American social concerns: education. While this topic has come up throughout this book, it is important give it a bit more attention in order to determine its impact on the general state of learning in this country, especially as it is related to math, science, and technology. It is also important to determine not only how or how well science and math are being taught, but how the scientistic orientation influences educational ideas and attempts at reform.[1] For this, I will take an example of scientistic thinking from an area in which I have much experience, namely, the teaching of writing. James Berlin, who was mentioned in the introduction to this book, wrote in his seminal essay entitled "Rhetoric and Ideology in the Writing Class" that there are three main schools of thought regarding how writing should be taught, and they all have ideological implications. By this he means that theories about writing are driven by, as a basis of their very articulation, ideologically charged assumptions and objectives that implicitly ask "What exists? What is good? and What is possible?" (53). Because of these inherent ideological questions, Berlin insists that a "rhetoric can never be innocent, can never be a disinterested arbiter of the ideological claims of others because it is always already serving certain ideological claims" (54).

The first rhetoric ("rhetoric" in this context may be seen as roughly synonymous with a "school of thought" or, perhaps slightly more accurately, as a "theory of discourse") is the cognitivist rhetoric, a discursive, and decidedly scientistic, orientation characterized by an assumption that "the structures of the mind correspond in perfect harmony with the structures of the material world, the minds of the audience, and the units of language" (58). Expressionism, the second rhetorical orientation, is in many ways the antithesis to cognitivism and in its earliest form was a response to the elitist rhetoric of aristocratic and bourgeois culture following World War I, which insisted that writing was a "gift" that only certain people had access to. Expressionism challenged this exclusionary idea by democratizing writing and claiming that it was an art that both could and should be taught. In later years, especially during the 1960s when educational reform was gathering momentum, expressionism in rhetorical theory "argued for the inherent goodness of the individual, a goodness distorted by excessive contact with others in groups and institutions" (61). In the dialectic that Berlin has constructed, the synthesis takes the form of social-epistemic rhetoric, a discursive orientation that acknowledges not only its own rhetorical presence, but recognizes that "the real is located in a relationship that involves the dialectical interaction of the observer, the discourse community (social group) in which the observer is functioning, and the material conditions of existence" (66).

Given this very brief overview, I would submit that the current call to "return to the basics" of grammar, punctuation, spelling, and organization is at its heart a thoroughly positivistic notion that is closely tied to a cognitivist rhetoric. In an earlier chapter, I discussed the nature of cognitive psychology and its dismissal of the relationship between thought and language, especially as the cognitivist mode of thinking is explained by Steven Pinker. A "skills" oriented pedagogy follows a similar line of thinking, seeing the human mind and its ability to learn in mechanistic, anatomical terms. Cognitivist rhetoric, like cognitive psychology, is, in other words, an attempt to turn writing into a materialistic "science." People from all segments of society think they know what "good" writing is, and most further insist that they know how best to teach it. Of course, the vast majority of those who make absolutist claims about the nature of writing have no experience in how to teach it, for if they had, they would quickly realize that an essay can be grammatically and structurally perfect while still being lifeless, dull, and ultimately, meaningless. The call to "teach grammar" is, despite what its proponents may intuitively feel is correct, a pedagogical idleness. It is, in fact, quite easy to teach a set of "rules" that the student must focus on at the expense of inquiry, discovery, and relevance. Much more difficult is teaching a student to develop style, voice, and purpose, not to mention the regimented process necessary to truly improve one's craft. To teach "grammar" at the expense of these other qualities would be like teaching only musical scales and hoping that a child can then learn to compose a sonata.

Such skill and drill techniques reveal an educational heritage that is known in composition circles as "current traditionalism". It is a pedagogical practice that is characterized by some key assumptions, all of which stem from a philosophical viewpoint that is largely empirical and scientistic in nature. The three main postulations that inform this pedagogical approach are: 1) our external environment is perceived and understood through the senses alone; 2) like science, knowledge is arrived at through induction (reasoning from specific observations to general categories; and 3) that knowledge is independent of the agent, that it is "out there" waiting simply to be discovered (Williams 43). Historically, the instructional practices that inform current-traditional pedagogical methods can be traced back to the Greeks and Romans, who, as cultures relatively new to the idea of writing as a mode of eloquence, expression, and persuasion, attempted to carefully categorize what they felt made for effective rhetoric. Aristotle, so accustomed to classifying and cataloging, did the same with language. Cicero followed Aristotle's lead in this way, as did Quintilian, and later, Erasmus, De Quincey and Hugh Blair, but all of these famous teachers emphasized that a slavish over-reliance on rules ran counter to the internalization of knowledge that was the real basis for eloquence and effective rhetoric.[2] Medieval and Renaissance teachers drew heavily upon these sources as guides to the teaching of writing, and many of the practices endorsed by Aristotle, Cicero, and Quintilian survived not only these periods, but through the Industrial Revolution, were transferred to America through the British system, and cling doggedly until this very day. For example, the Greek notion of *progymnasmata*—a system of incrementally sophisticated writing exercises designed to improve student mastery of form and language—can still be seen in the so-called "rhetorical modes," which use assignment strategies such as definition, narration, process analysis, example, etc. as a means of providing students with practice in various rhetorical approaches. The ideas of invention, style, memory, expression, and delivery can be traced back to Aristotle and Cicero. Many of the "rules of grammar" that we abide by unquestioningly are actually rules of *Latin* grammar that have been transposed onto English. In short, many of the current-traditional practices of composition are quite ancient indeed, originally designed for a different time, place, culture, teacher, and student.

The education "crisis" and its subsequent call for retrogressive correctives have prompted a stark political debate, and a distressingly binary one at that. I will cite one representative advocate for the current-traditional model because her assessment of the problems in composition courses is typical of the errors that many in her political camp make. Heather MacDonald, a journalist with a degree in jurisprudence who has never, so far as her credentials indicate, taught a single composition course, confidently claims that "composition teachers have absorbed the worst strains in both popular and academic culture. The result is an indigestible stew of sixties liberationist zeal, seventies deconstructionist nihilism, and eighties multicultural proselytizing" (88). She deplores the "growth" model that has supplanted, in her mind, the more functional and utilitarian "transmission" model because she sees a loss of instructor authority in the shift to the new paradigm. She further laments the "Marxist" strain of thinking that has apparently corrupted composition studies, and she ridicules the notion that "coercing students to speak properly conditions them to accept the coercion of capitalism" (89). Such statements reveal anti-intellectualism at its most egregious, suggesting that a) Marxist theories (or any others that don't meet with the predetermined capitalistic ideological approval) are on their face illegitimate, and b) questions about the authority of teacherly practice is somehow another blow to the "standards" that such critics advocate.

While the top scholars in the field of comp studies have certainly adopted political perspectives along lines loosely similar to those described by MacDonald when trying to explore new pedagogical methods, she is woefully ill-informed both about who actually teaches freshman composition and what the study of composition pedagogy is all about. Composition is perhaps one of the largest service courses in academe, and one of her glaring oversights is that those who typically teach composition on the college level do so with little to no formal training in it—often, in fact, without degrees in English, rhetoric, or communication but in "related" fields like psychology or philosophy. Many of these same teachers, in fact, are hearty advocates of the current-traditional model that MacDonald embraces, and the modern version of this pedagogical approach uses the stiflingly tedious "skill and drill" or "imitation" method (among others) to further alienate students from a procedure of writing development that they already despise. Berlin notes that "cognitive rhetoric might be considered the heir apparent of current-traditional rhetoric" (57) and is a product of the new American university system that appeared during the late nineteenth century. Interestingly, this new system was characterized by an unswerving attention to the perceived benefits of science and economics: "[The new university system's] role was to provide a center for experts engaging in 'scientific' research designed to establish a body of knowledge that would rationalize all features of production, making it more efficient, more manageable, and, of course, more profitable" (57). Moreover, science and economics became absorbed into all facets of the university structure, including writing instruction. The nostalgic past that MacDonald pines for, in other words, is nothing more than the failed objectification of humanistic studies in more subjective fields like writing. It was to this model that scholars in the sixties and seventies were reacting.

For example, one alternative to the current-traditional model utilized by more progressive compositionists (though by no means the only one) is what MacDonald refers to as the "process school," a blanket term used to describe any variety of writing methodologies that emphasize the development of successive drafts and authorial *hexis* (from the Latin for "habit") in the formulation of a final product. She disapproves of this as well, declaring (without any apparent professional, educational, or experiential basis for the declaration) that "the drawbacks of the process school cancel out its contributions. In elevating process, it has driven out standards" (90). Convinced that this is a conspiracy to favor the "self-esteem" of students over their progress, she thinks she knows how students are being evaluated—not for "results" but for "growth." More to my point, she suggests that the "principles" of good writing are a static set of criteria that are "an objective standard of coherence and correctness" (90), a claim that reveals her misinformation about the dynamics of a composition course (and about the nature of writing in general); not only are most standards of this sort already in place, but they are a bare minimum for success in any adequately designed writing course. Furthermore, she erroneously equates "objectivity" with "standards," suggesting that academic activities that resist objectification are somehow not "rigorous"—a principle of assessment that is clearly based in the sciences, again reinforcing the idea that only scientific criteria are valid for the purpose of evaluation. In a well-conceived composition course (and there are many of them) "growth" is rewarded as something achieved beyond the minimalist "objective" requirements MacDonald endorses, and only the most rigid of professors would ignore improvement as a factor in learning (how, one might ask, can "growth" be considered irrelevant to the learning process? What *is* learning if not intellectual and personal "growth"?). What she advocates, in other words, is a mastery of material, a cognitivist model that assumes any worthwhile course has as its objective a body of content that can be readily quantified for easy assessment. It is a thoroughly materialist orientation, taking as given the idea that not only are the goals of learning ultimately materialistic, but so too are the means for achieving those goals.

Self-appointed educational pundits like MacDonald are a menace to progressive teaching not only because they have no basis in experience for the claims they are making, but also (and mostly) because the methodology they *do* advocate is at its core one that assumes the same analytic, mechanical structure that one might see in a mathematical equation. Such methods are ineffective because they stifle the development process of students serious about improving their writing, asking them to embrace an immature strategy long after its usefulness as a heuristic has expired, like asking a math student to learn quadratic equations and suggesting that they are appropriate for all mathematical problems. The "five paragraph essay" is one such boilerplate essay form, and one that many teachers fail to deviate from even when the formula was mastered by their students back in the seventh grade. The stubborn adherence to such recipes does not acknowledge that the best writing—far from being the predictably forced prose that always has a "thesis statement" in the first paragraph and the numbingly predictable "three main ideas" that follow—in fact embraces the unanticipated, reveals new thoughts, and forms new connections. Writing, a la MacDonald, is seen as writing for some profession, not as an act of intellectual discovery in it own right, not as a tool for academic inquiry and communication. It is an unswerving method like the procedure of drawing blood, not to be departed from lest the reader might actually have to concentrate on *what* is being said, not whether the author should have used a semi-colon or where she should have placed the topic sentence. Under the limitations of the current-traditional method, writing teachers and the students they teach become at worst police, at best technicians, but never craftsmen and certainly not artists.

Herein lies the distinction between the humanistic project and the scientific one; where the former celebrates subjective modes of expression to uncover fresh insights and to posit new possible perspectives, the latter insists upon an inevitability of expression that renders us all the same, a homogenous army of the banal and unimaginative. The cognitivist school reflects a linear trajectory, when most good writing in fact demands a more panoramic view. But the issue becomes one of what we are having students writing *for,* and if that purpose is the ability to write a grammatically unencumbered email or memo, then the cognitivist approach may work in this extremely limited capacity. If, however, we teach writing as a heuristic, as a mode of learning in and of itself, as a vehicle through which to forward new thoughts and connections, then this methodology simply does not work. Moreover, cognitivist "standards" contribute to the stifling sense of normalcy and mediocrity, making students even more lethargic about their education since it amounts to nothing more than the internalization of a set of "skills" that can later be bartered for professional earnings and security.

MacDonald as much as admits that the form of writing she is referring to is writing in particular professional disciplines.[3] But such declarations fail to account for the very real pedagogical value of freshman composition as a general introduction to academic writing, not as courses designed to be discipline-specific. They are, by and large, designed to provide an introduction to academic and personal writing that can be further enhanced and developed in the major, what is known as "Writing Across the Curriculum" (WAC) or "Writing in the Disciplines" (WID). This movement toward discipline-specific writing instruction has seen both success and failure: success in that it has made many in academia aware that a one-size-fits-all approach to writing instruction is not effective, that all courses using writing must shoulder the burden of writing instruction, and failure in that writing instruction is still largely seen as the domain of the English Department. Without getting into politics of why this is the case, my main interest in this institutional development has to do with the scientistic expectations of pedagogy, assessment, and methodology. Not only are English Departments mainly responsible for teaching writing at all levels (there are exceptions, of course, but they are rare), but they are also expected to document the relative success of their program in "data" that are numbers-based and easily digestible for an administration that responds most favorably to this one-dimensional rhetorical form. The problem with "data" such as these is that they mean reducing the writing process and product to a few discrete categories of "skills" for "outcomes assessment." Students are ranked by "scales" or sometimes "scores" that determine much of their academic fate it terms of writing. What this means is that those who teach writing and administer writing programs are being asked to force a humanistic peg into a ersatz scientific hole, even though the results of such data are, in the end, almost entirely useless for revealing how students write or how well students advance their craft.

Yet, this is exactly the sort of program that vocationally-oriented labor-producers within the cognitivist school advocate. The proponents of this school would have our children learn in the Medieval tradition of the scribe, mastering the form but not the substance, spending hours in pointless exercises that have as much to do with the improvement of writing as memorizing the color spectrum *ad nauseum* does with learning how to paint. They would have us believe that teaching writing is as simple as handing out a few grammar worksheets—so that we can "document" with "scientific accuracy" whether they learned the function of a correlative conjunction in a sentence. Self-appointed education watchdogs are more the reason for the failure of education today than the solution to its problems. Given the complexity and importance of educational questions to a democratic society, it is the most callous and idle sort of cultural analysis that concludes that we must go back to what we did fifty years ago or that we must marginalize, belittle, or ignore altogether decades of work done and ideas forwarded by people who have years of training in the field and experience in the classroom.

One of MacDonald's statements, for example, reveals how little she values thinking as part of the educational process: "[H]owever much a person may be enjoying the play of ideas in his own mind, that counts for nothing if he is unable to get a job. It also counts for nothing in *getting* a job" (MacDonald 96). Such a clear, cold, and unapologetic utilitarianism sees education as a meal ticket, a passport to the intellectually paralyzing, cubicle-partitioned world of the nine-to-five grind. Under this mode of thinking, a person's identity *is* his or her job, not just one component of it. Small wonder, then, that students get the message that humanistic courses like philosophy, literature, history, and the arts are a "waste of time." Furthermore, the logic is flawed; unless those doing the hiring for any given position are unrepentantly one-dimensional, I suspect that "the play of ideas" will contribute greatly to both getting and keeping a job. What makes one candidate more appealing than another during the employment screening process? In most cases, it is the expression and implementation of innovative ideas, evidence that an employee is able, to use a favorite corporatese phrase, "think outside the box." When an employer is swamped by hundreds of applicants with degrees in Business, there must be something that allows some resumes to rise to the top over others. And it has been thoroughly documented that humanities majors often make the most promising career employees because, go figure, they are not only imaginative and innovative, they are also good at solving problems—the kinds of problems that confront us everyday, not some hypothetical scenario in a textbook.[4].These abilities would be absent in graduates who have been taught merely a set of skills that simply make them exploitable gears in the great corporate mechanism.

It's hard to believe that any educated person would subscribe to the fantastic notion that *ideas* are a liability in the job market—unless, of course, that person were educated using a strictly cognitivist orientation and views employees as mere labor resources. The message is that, in the "real world" (96), striving for more than mediocrity will have you applying for food stamps. Ironically, the endorsement of these measures are rationalized under the banner of "standards" when the standards being advocated are only the most rudimentary kind. Teaching grammar is not difficult, nor is giving spelling tests, nor are any of the other cognitivist methods often championed by education "reformers"—who are more often reactionaries with no new ideas for how to address contemporary student needs but instead advocate a nostalgic return to the bye-gone days when everyone (they would have us believe) was literate and loved the instructional methods that made them so. The difficult components of teaching in the writing classroom are getting students to establish a voice, develop a style, maintain an argument, connect with others' ideas, negotiate a sea of texts, and construct a tone of readability that is driven by a genuine need to say something that is important to the writer. These more subjective areas of learning are extremely difficult to teach because they *are* subjective, and they require what Erasmus saw as an internalization of knowledge that could be used in service to other intellectual activities like writing. A method that embraces Erasmus's philosophy requires a broad training in all the humanities; but when we favor the scientistic over the humanistic in writing, we should not be surprised when students have no motivation to improve. For them, under a cognitivist system, there is nothing important to write *about*.

Ultimately, the reforms that the cognitivists see as imperative lead to an increased student alienation from their own education. Scores, grades, standards, documentable skills, credentialing, and academic averages may in fact be necessary administrative evils, but they have unfortunately become ends in themselves, the *reason* for getting and giving an education. Most students are only motivated by such numbers because they have been trained (either intentionally or not) to view them as the only important objective. They see their education as just a series of administrative hoops through which to jump, not an on-going intellectual process that requires independent thought, personal exploration, and the contemplation of "the best that has been written and said." I submit that this is an academic boondoggle of our own making, that we have overemphasized the significance of the grade point average at the expense of the most important reason of all to get an education: to exercise and augment that most prized human asset, the mind. The message we often send students is this: "Don't ask *that* question; answer *this* one." This is due in part to a fundamental inability to understand the value of an education, to see what a cultural endowment it is to have it available to us, to recognize that, even as recently as a fifty years ago, an education was by no means a foregone conclusion for many Americans—especially education on the higher levels. We have pilfered our privilege, and we now see it as our birthright as consumers to be able to "buy" an education. As much of a waste as this is, it is exacerbated by the insistence on a bottom-line, the repulsive idea that an education is a commodity, like a new car or sofa, that can be bought in installments. Finally, we put the last nail in the coffin of education when we assume that it can only be designed, delivered, and assessed using a cognitivist methodology of social-scientific numbers; students, under this system, become mere subjects in an on-going study, the results of which determine the funding of school districts, the security of jobs, and the public relations image more than what students are actually learning and for what purpose. Students are indifferent because they are often treated with indifference. They are unmotivated because often the system is unenthusiastic; teachers and administrators are jaded by decades of number-crunching and dead-end reforms. Students feel alienated because they are treated more like rats in a laboratory experiment than like human beings with minds, interests, and real, honest-to-goodness questions (and contrary to the popular characterization, students are not "dumber" than they once were; they have simply not been given a compelling reason to be smart). All many students really want—whether they realize it or not—is to be taken seriously, and to be given the opportunity to use their own, natural inquisitiveness as a motive for learning and personal growth. Some may see this as naively idealistic; I see it as common sense.

Another contributor to the academic apathy that students and adults alike seem to harbor in this country is that, historically, culture thrived upon the primary condition of a leisure class—a group on the hierarchical scale of social production that had others do the farming, building, repairing, and menial labor for them. This condition freed up the upper classes for other, more academic, scientific, and artistic pursuits. While in many ways little has changed in this regard—the "service class" still does the bulk of physical labor for the middle and upper classes and have the most remote access to educational ways out of their station in life—the middle class certainly has not taken advantage of this freedom because it is too busy "earning" to pay for the materials it owns to do much else. When members of the middle class do have leisure, they do not, in general, use it to enhance themselves intellectually, but rather, seek out activities that are expressly designed to avoid intellectual effort: sports, recreation, and entertainment, primarily. They therefore seek out education for only the most serviceable ends—to get a job to earn the cash to buy the stuff. This is what the middle class in America values, and it is what is rewarded.[5] It should come as no surprise, therefore, that intellectual energy has been depleted to the point where nearly all activities pursued by the middle class feed back into the marketplace in some way to maintain the all-important production/consumption cycle. Given this imbalance of academic wholeness, the middle class is afraid when they are faced with problems, and turn to one of the few places they trust to seek answers: science. And people trust science because it has tangible effects, is demonstrable in apparently certain terms, and, most importantly, impacts individuals directly. In a materialist, literal society such as 21st century America, material descriptions of and solutions to educational problems seem like the only rational choice. Without a body of knowledge like science toward which to turn, people might very well find themselves lost, uncertain, and ultimately, paralyzed.

The problem with this is complicated, since while on one level it seems to be a perfectly practical approach to a materialistic world, on another it fails to recognize the source of individual contentment (I avoid the word "happiness" because such a condition is a fiction, suggesting as it does a life-long consistency in emotional well-being that no one could possibly achieve). Contentment, a state of personal satisfaction and fulfillment that makes one's life as rewarding and trouble-free as possible, must also be *personally* defined and pursued, and it cannot be superimposed simply onto the goods people acquire. In other words, one is not the sum total of his or her possessions; it is a mistake to assume that the mere accumulation of things will result in personal satisfaction and fulfillment. As trite as it may sound, money cannot buy "happiness" because to assume that it can is a denial of the self. One sacrifices one's own development and personal enrichment for one's possessions, using his or her body and mind as merely a conduit for monetary income and outgo. The flow through one's bank account becomes the defining characteristic, with all physical and mental energy devoted to one's "balance," an ironic financial term that in fact reflects a marked *imbalance* in the whole person. In short, it may very well be that it is not *how much* one buys, but *what* one buys that counts, but this can never be the end sought in the quest for personal wholeness.

Bertrand Russell, in his 1912 book, *The Problems of Philosophy,* explains that the erroneous assumption by people "under the influence of science or of practical affairs" that philosophy is little better than "useless trifling, hair-splitting distinctions, and controversies on matters concerning which knowledge is impossible" is in part responsible for the deterioration of the modern intellect in the general population. He says that such a view of philosophy "appears to result, partly from a wrong conception of the ends of life, partly from a wrong conception of the kind of goods philosophy strives to achieve" (117). As an analytical philosopher, he embraces the value of the sciences and thinks people, being as they are directly affected by science even if they are wholly ignorant of it, should try to learn and understand it. He is no mere materialist, however, and feels that the value of philosophy has gotten poor press by those who have only a materialistic terministic screen through which to filter the world:

> But further, if we are not to fail in our endeavor to determine the value of philosophy, we must first free our minds from the prejudices of what are wrongly called "practical" men. The "practical" man, as this word is often used, is one who recognizes only material needs,

> who realizes that men must have food for the body, but is oblivious of the necessity of providing food for the mind. If all men were well off, if poverty and disease had been reduced to their lowest possible point, there would still remain much to be done to produce a valuable society; and even in the existing world the goods of the mind are at least as important as the goods of the body. It is exclusive among the goods of the mind that the value of philosophy is to be found; and only those who are not indifferent to these goods can be persuaded that the study of philosophy is not a waste of time. (117)

Such a description articulates well the condition of the U.S. middle class nearly 100 years later because it shows how thoroughly entrenched is the idea that only the practical has value. Yet, far from being testimony against the importance of practical matters, Russell's comments clearly demonstrate the ease with which the emphasis on practicality defines a view that holds practicality in high regard exclusively. The resulting complacency and isolation that befall people who cling stubbornly to the sole worth of the practical has a ripple effect on our most profound social problems. If for no other reason, we become sequestered in our concerns, self-absorbed to the point of narcissism, allowing others to worry about the state of society and to do something about the problems we face because we have no "practical" access to them. One must also have the all-important "incentive" to be moved to action—or even thoughtful contemplation—and there simply isn't time. There are mortgages to pay, car payments to make, and educations to finance. In the words of Rodney Crowell, "I don't have the means to power/I don't have the extra hour."

Perhaps our anti-intellectualism is a symptom of the downward slide of American culture so many people decry today. Perhaps "decadence" is the pinnacle of civilization; when people reach a certain point of acquisition, earning potential, and sheer ennui, they begin to indulge in a conspicuous consumption that becomes an end in itself. The worst thing that can happen to people in middle-class America today is boredom, it seems, and the greatest crisis is one that requires people to do some hard thinking. But who can be bothered with thinking abstractly about the problems that plague us today? We have no shortage of "opinions," but as someone who teaches young adults on a daily basis about the problems that effect us most, I can accurately report that, for any given "issue" that my students discuss, they parrot perhaps three primary arguments (at most—two is more common) as if by rote, arguments that have been prepackaged for consumer use by the media, parents, other teachers, and friends. This suggests that many students today are not trained to think in terms that will help them face and understand the big problems, and they may not be particularly inclined to make the effort in the first place, especially if there are prefabricated opinions available that they can acquire. "They" are not doing the thinking; it has already been done for them and can be possessed just like their Verizon cell phones and their Abercrombie and Fitch sweatshirts (this may help explain why so many students are so *needy*). The great educational challenge facing teachers today, in short, is getting students to think beyond their inherited stock responses—a symptom of the materialist mindset.

"Social Science" as an educational methodology has again saved the day. Students, and members of the large and influential middle class generally, do not have to make up their own minds on an issue; they need only mimic one of two possible, ready-to-use choices. Whichever one sounds best is probably right, and even if it isn't, who cares? The only reason they need to know such irrelevant stuff is because some dopey English prof made them write a paper on it. But where are the prepackaged ideas manufactured, and where can I get mine? As the Chair of Social Science at City College, CUNY, Steven Goldberg points out, social science has become the locus of all popular ideological concerns, using science (or, more accurately, quasi-science) to promote the political. In his essay, "The Erosion of the Social Sciences," Goldberg worries about how the social sciences, as a field that invariably spills over into mainstream culture, uses its influence to posit some decidedly unscientific propositions. Worse, he sees it as a tool of social control, one that permits certain ways of viewing societies and prohibits others. For example, he speaks from experience about how his book, entitled *The Inevitability of Patriarchy* was received with distaste and outright hostility because it made the unconscionable claim that patriarchal dominance in cultures is a product of biology, not cultural preference. The outrage voiced about this claim, he suggests, is the result of a shift from the scientific rigors of the hard sciences to the social sciences that lead to an epiphany, apparently reached in the 1960s, that "objectivity" is a myth, and therefore, the goal of social science was more a matter of promoting social change than using science to examine social phenomena:

> Then, at the height of the politicalization of the sixties, increasing numbers of sociologists began espousing the view that objectivity in the social sciences was a myth. And, since objectivity was a myth, the social sciences were not accountable to the logic of science [as if science is the only "logical" epistemology] and could, in effect, drop the "science" from social science and incorporate ideological dispositions into the "social." The belief that there could be no "value-free sociology" became an excuse to infuse theoretical framework with political wish. (99)

From these counter-culture beginnings, according to Goldberg, social science evolved (or mutated, depending on one's perspective) into the ideological monster that it now is. It is clear that political and ideological pressures restrain, redirect, and even demonize certain unpopular hypotheses (like those of Goldberg's), and it is equally clear that social science has been used to further particular political motives, especially in fields like education and social work. The question is, can there be such a thing as "social science" or will it always morph into social control? What is the definitive purpose of conducting social science if not to use the information gathered to change society?

Goldberg seems to think that a "pure" social science is possible, but if the "revolution" in the social sciences during the 1960s is any indicator, it is unlikely that the scientific integrity, objectivity, and detachment needed to raise social sciences to the status of the hard sciences will survive for very long in the face of the overwhelming temptation to use social science as a means of imposing political will onto social structures. Despite the likelihood that social sciences are, by their very conceptualization, prone to methodological corruption and misapplication, Goldberg insists that

> [s]cientific logic is infinitely our greatest defense against the bias of individual wish and ideology. It requires the scientist to do everything possible to refute his conclusions. To be sure, the scientist is human, and occasionally psychic need or ideological impulse trumps his knowledge that, in science, an ability to

> resist the demands of such need or impulse is what separates the children from the grown-ups. (99)

The message is that scientists can control their impulses and needs (unless a "trumping" "psychic need" arises, whatever that might be and however it might be avoided) and he or she is obligated to follow that doctrine (unless ethics—an ideological impulse?—intervenes).

Rhetorically, Goldberg has carefully guarded his claims about "pure" social science and what it has recently morphed into in a way that, it seems, invalidates the field of social sciences conceptually (a field, I should emphasize, that is the parent of current educational training). If human activity, group behavior, and the general ebb and flow of society or sub-groups within society can be quantitatively and objectively studied, what, one might wonder, can be done with the information *except* the application of ideologically charged and politically motivated agendas? If, for example, we determine conclusively that capital punishment does in fact deter crimes for which that punishment is a possible consequence, to whom do we give the "discovery" and what should the receiving party do with it? It is not like discovering a new and inexhaustible energy source that might be used to develop alternative technologies and help eradicate problems related to the depletion of non-renewable fuels. As a "social discovery," it by its very nature must have social applications, and hence, social consequences, all of which are dictated by social custom, acceptability, and good, old-fashioned principles—in other words, political interests. Goldberg's language—that scientific logic is our "greatest defense" against ideology—cynically presupposes that ideology is by itself an invasive, foreign, undesirable, and unnatural thing. Perhaps more significantly, he suggests a purity in social science that is, in the final analysis, simply knowledge for the sake of knowledge. It would be impossible to adopt a social discovery to a societal problem without invoking ideological overtones; therefore, one might wonder what the point of doing sociology could in fact be.

Goldberg couches all of this in his proposition that "if one has faith, as I do, that truth always wins out in the long run—that the emotions that wrap themselves around the nucleus of incorrect theoretical explanation or empirical claim always sooner or later attach themselves to something else—then one could be sanguine" (98). So, for Goldberg, as for any other scientist, the motive is the pursuit of "truth"—in this case, a "social truth." He defends this position by also defending cultural relativism as a viable terministic perspective through which to understand other cultures objectively. He calls cultural relativism "the quintessentially objective epistemological perspective and heuristic lens for the study of society" and asserts that "[o]bjectification of this sort is the path to scientific truth" (106). Unfortunately, the pejorative use of the term "relativism" has precluded acceptance of this notion on grounds that, if there is in fact a "truth," we must concede that such a truth (by definition) is inflexible, a premise that contradicts the very notion of cultural relativism. All truths become situational to the extent that calling them truths has little meaning. Goldberg, therefore, (who pedantically lectures the reader in the same essay about the evils of logical fallacy) can't at once claim to be seeking "truth" (if by truth he means an unyielding natural law) and then decide that such truths are also "relative" to the cultural context in which they arise.

Of course, Goldberg never bothers to define what he means by "truth" because that would be too much like doing philosophy, and philosophy, as Bertrand Russell has suggested, has no place in the mind of the practical man. Yet, it is the basis on which he rests his whole argument. Herein lies the consistent blindness on the part of science, especially of the social variety: The assumption that by *doing* science or *being* a scientist one can avoid the human pitfalls of ideology, bias, subjectivity, and language. In other words, though seeking the truth is a noble pursuit, we might have to concede the point that, while our minds are a wondrous apparatus for asking questions about absolute truth, they may in fact be inadequate to the task of actually finding it (to invoke the "natural law" of the universe argument a la Isaac Newton would be to ignore that quantum physics and string theory have already challenged the completeness of such laws. The question then becomes one of not whether such laws are "true," but whether they are "the whole and complete" and "unchanging" truth). A treatise on the philosophical implications of truth would far exceed the scope of this chapter. However, some discussion of our literalism in regard to such heady concepts as "truth" might help clarify why we find ourselves in a somewhat incompatible intellectual position.

Ralph Waldo Emerson once said that "there is a foolish consistency that is the hobgoblin of small minds." That sentiment should perhaps be upgraded (to use a technological metaphor) to the current era to read, "there is a foolish *literalness* that is the hobgoblin of an intellectually lazy society." Kenneth Burke once noted a "naïve verbal realism" that plagued American society (and humanity in general, though it is particularly prevalent in our loosely matter-of-fact culture), one which prevented people from seeing just how firmly situated their beliefs and convictions—even their reality—was in their use of language. He dedicated a good deal of his philosophical and rhetorical works to the idea that language shapes our world, that it is a means of "symbolic action," functioning not just as a convenient way to assign a one-to-one sign to signifier correlation, but as a "shorthand" to a surfeit of other associations, personal, social, cultural, and religious certainties, and to our assumptions about "what naturally goes with what." In the U.S., however, we tend to think of language only in its most utilitarian terms, finding rhetorical and philosophical analysis of it a confusing quagmire of metadiscourse that serves more to muddy than to clarify. But part of the point is that language is, at its core, a flawed means of expression; it cannot capture absolutely the centrality of any truth because it is by its nature both ambiguous and fluctuating. We therefore rely on our uncanny ability to use and comprehend, more or less, the symbolic nature of language, always invoking the tropes of metaphor, metonymy, synecdoche, and irony (what Burke calls the "four master tropes) to compensate for language's innate imprecision. Without these symbolic literary devices, we would only be capable of the sorts of communication that can be taught to orangutans: simple noun/verb constructions that *must* correlate not only to something in the world, but also to something in direct proximity to the user. The capacity of humans that may truly separate us from the animal world is our ability to abstract language from its immediate context and to formulate sentences and ideas that have never before (and will never again) exist. This amazing faculty means that we necessarily pack our words with "givens" that enable us to communicate with dispatch and alacrity, even though we sometimes sacrifice accuracy, even "truth," in the process.[6]

The pursuit of "truth," under these conditions, becomes more of a slogan than an actuality, especially under the myriad social situations we are confronted with every day. How, then, is it possible for the social scientist to claim that she seeks the truth using the scientific method when we know that social truths not only differ from society to society, but also fluctuate within our own society, shifting with the democratic notion of what are favorable social responses for any given time or place? And, again, assuming that such a truth can be discovered, how, then, do we go about putting it to some practical application? An affinity for science has in many ways contributed to the fiction that truth is attainable under any circumstances, so long as one is willing turn off his or her bias switch, adhere strictly to the method, and assume that whatever hypothesis being forwarded is wrong until it can be proven otherwise. But since most of us do not practice science, we can leave this superhuman ability to the scientists themselves, since they seem to be the only ones capable of achieving this ostensibly enviable state of mind. The general public is weak and lazy, but scientists are strong and rigorous; they are the modern-day clerics who are above the internal wrangle of ideological temptation and can be trusted to uncover the truth and use this knowledge appropriately and rationally.

Have I "proven" any of this? The question itself denotes a tendency to invoke a scientific standard where such a standard may not be appropriate. Calls for "proof" suggest an unyielding truth that under all circumstances will remain certain and stable. But the "real" world, especially that world occupied and manipulated by humans, is never this tidy. There are always exceptions, and it is a failure of the contemporary mind to assume that it is impossible to intellectually hold onto competing, even contradictory, ideas at once. We do it daily when we support ideas in theory that we will not invoke in practice, like the moral notion that killing is wrong, except in the form of capital punishment, self-defense, or going to war (even of one of a "preemptive" nature). The difference between humanistic explorations of the "human condition" and scientific ones is how we use and what we use as evidence. For the scientist, evidence is, as Kenneth Burke puts it, the compilation of data about "objects in motion." For the humanist, evidence is less about data, and more about observing "subjects in action." If the distinction seems a tricky philosophical hair-split, consider this: By treating human beings as "objects in motion," the social scientist has reduced the activity of the human animal to a set of physical laws that can be quantified, catalogued, and predicted, just like the physicist does when he details the behavior of balls rolling down inclined planes. If he hits upon a pattern that seems consistent, observationally and statistically, he can deduce that all humans, given the same conditions, will react in the same way. With this knowledge, he can predict certain outcomes and even act on the knowledge that the outcome is predetermined. The only way he can act with any consequence is in the form of social policy, such as those we see in education or the prison system. If a humanist looks at humans as subjects in motion, she has acknowledged the idea that people, once they are indoctrinated into a social environment, can act in a variety of ways depending on what factors contribute to their symbolic understanding of what that environment means and how their language helps to construct that meaning.

The social scientist, then, wears the accoutrements of the scientist, but in reality cannot possibly function as one. The "erosion" of social science Goldberg bemoans is a fiction because social science can't exist as a "science" in the first place, at least not one that can expect the same predictive accuracies that we would see in the hard sciences—human activity simply isn't suitable for understanding as objects in motion using the scientific method in the same way one would be able to predict the chemical reaction that results from combining combustible substances. We learn far more about humans by studying philosophy, history, rhetoric, art, and literature than we ever can by classifying, naming, and attempting to predict human behavior in any stable and rigid way. Such attempts cannot help but lead to political and cultural control, so my response to Goldberg is, "What do you expect?" Knowing full well that this "declaration" may alienate my sociologist cousins in the academy, I should point out that most social scientists themselves recognize the "softness" characteristic of the science of their profession. The distinction between "hard" and "soft" science, however, has always been puzzling; does "soft" imply that nothing in social science is ever conclusive, and if so, why is it so important to assign it the moniker of "science" at all?

One possible answer is that the success of science—both as an epistemological system of knowledge and as an accepted force of intellectual authority in Western cultures—is something that many fields not as privileged as science may covet. It is not unusual, for example, for under-established disciplines to attach themselves to recognized intellectual traditions. Composition Studies has, perhaps, been the most egregious offender of this in the past. I recall how, for years, the Conference on College Composition and Communication would offer panels with titles like, "Heideggerian Analysis of Current-Traditional Composition Pedagogy" and "The Phenomenological Turn in First-Year Composition Studies," which usually involved some serious stretches (and outright misreading) of the philosophical principles described to the process of teaching writing to college freshmen. Before we rush to the MacDonaldian conclusion that Composition Studies is not a feasible nor bona fide discipline, we should note that the topics of these types of panels, rather, indicate a disciplinary insecurity that Comp Studies has since outgrown, one that fostered the need to associate with disciplines that commanded the rigor and respect that had eluded Comp Studies for many years. It was seen as a necessary capitulation to an academically elitist orientation, since only certain ways of knowing were taken seriously. It was a desperate attempt to maintain viability when many, more reactionary academics (and traditional disciplines) questioned its legitimacy. Social science, while enjoying a much longer history of acceptance, has done more than adopt topics or principles that "belong" to other disciplines; it has adopted the methodology of the hard sciences so that it can *call* itself a science. (It is interesting to note, incidentally, that Comp Studies likewise borrows heavily from social science itself—like most other educational fields—as a way to shoulder the positivistic burden of evidence that so many academic areas and their governing administrations insist on having.)

What all of this may suggest is that we have, in contemporary culture (particularly contemporary academic culture) one and only one apparent standard of evidence or argument for any intellectual claim or theory: a scientific one. In educational and academic circles, for instance, there exists a troublesome preoccupation with "outcomes assessment," a process which involves many different kinds of "measurement instruments" but which has only one universal language: numbers. On any typical student evaluation, for example, instructors no longer ask (or are allowed to ask) for *written* comments on the success or failure of a particular course, its instruction, the texts, or the material, but to use what is referred to as an "SIR" (Student Instructional Report) that looks like the answer sheet for the classic Scantron standardized test. All answers to statements such as "Instructor made appropriate use of technology in the classroom" are then ranked on a scale (usually from one to five, one representing "strongly disagree" and five representing "strongly agree"). The evaluation is then averaged into a "mean" score between one and five, and the overall success of the course is based on this mean. An instructor who receives a mean score of 4.1, therefore, has averaged around a four for all responses (which correlates symbolically with "agree"). These evaluations are used for everything from tenure decisions to course assignment to promotion, and consistently low scores can, of course, threaten a professor's job. But are they accurate? In general, the best that such an evaluation can hope to show is that the respondent agrees or disagrees—along a very finite spectrum—with the statements made by the evaluation's designers; there are no opportunities to explain certain responses nor to formulate one's own assessment of a given course. Responses are limited to those which the evaluation will allow, thus "bureaucratizing" any possible deviation from a set of conditions that the *evaluator* (not the professor or the students) find important and acceptable. It should come as no surprise, therefore, that students expect (like the good consumers they are) a limited set of choices in their education. Prepackaged options mean that one need not ever bother to think outside of the available constraints. Just pick a response and fill in the circle.

We are so used to these instruments that we rarely question their overall effectiveness, but the loosely social scientific purpose behind such evaluations has less to do with precision or "conclusiveness" and more to do with convenience and an effort to appear "objective." It functions, that is, as shorthand for what would otherwise be much more complete, interesting, and useful types of feedback (like essays, short responses, or even student interviews) but without the depth or specificity of these methods. Given the rather perfunctory structure of the evaluations, one must wonder what, indeed, they *are* capable of showing, regardless of their understood limitations. Moreover, because we are so accustomed to using what *looks* like a scientific process in doing everything from evaluating college courses to taking "personality tests," as a generic population, we have largely ceased to question their validity. We generate massive files of data and numbers only to use these figures for highly dubious purposes. Why, for example, does a therapist use an "emotional maturity test" when first meeting with a client? Is it because he trusts completely the results, or is it because it saves him an awful lot of time trying to figure out the problem through discussion and analysis, even if the "score" is questionable? What is "emotional maturity" anyway? Is it like an I.Q. test for feelings? And like an I.Q. test, which only measures one kind of intelligence, are there other "ways" of feeling? It's as if we have been so thoroughly saturated with the notion that science can solve even the most complicated human problem that we insist on relegating to it every conceivable problem or "issue" that confronts us. We no longer even ask the question of whether or not another way of addressing the problem might be more effective; we assume that if it looks like science, walks, talks, and acts like science, it must have what it takes to solve a given problem.

The rhetorical purpose behind the professional allegiance to science is both simple and clear: In a society that favors the products and privilege of science, in a society that is flagrantly materialistic and literal, it is advantageous to adopt the features of science's methodology and to emulate the characteristics of it success. Professions that use a scientific approach have a greater degree of credibility with the general public, certainly, even if that same credibility is not extended by those in the hard sciences. It is important for people like Goldberg to insist that they seek truthful answers using a method that has been closely linked to truth and reality in our social consciousness because it validates what they do. Once the results are extended to practice—in the form of standard "tests" and "assessment instruments," among other things—the predictability factor *seems* to increase. However, lurking behind the assumption that because one used a quasi-scientific methodology to reach one's conclusion—and therefore the conclusions are themselves universally useful—is the grim specter of insignificance. The once primary discipline of logic, for example, has failed to generate relevance these days largely because people are no longer formally trained to accept the conditions under which logic must operate. In order for logic to thrive, it must have practitioners; people do not intuitively understand the rules that govern the structure, and when they don't play by the rules, the system is irrelevant. The same can be said of the social sciences: In order for scientific conclusions to have merit, people must behave in ways that can be placed within the categories that the social sciences create. If the category doesn't exist, and enough people demonstrate the behavior, activity, or impulse, then perhaps a new category will be constructed and a kind of pseudo-algorithm can be devised. But unlike an algorithm, people behave unpredictably, confounding the system and belying its reliability. Is social science, then, the best way to produce knowledge about human activity? Human nature? Our social condition?

But what of the state of education in the sciences themselves? Scientists and educational pundits are fond of reminding us that aptitude in science is not at an acceptable level—certainly not at the level that we nostalgically enjoyed during the height of the science education movement in America, the post-war years between 1945 and 1970.[7] As is the case in many other subjects, science education is seen as a faltering institution that, because of funding cutbacks, under-prepared students, and under-trained teachers, cannot keep pace globally with our European and Asian counterparts. This declaration is based largely on performance on standardized tests which, given the nature of scientific material that prompts either correct or incorrect responses, indicates *something,* though it is not clear what that something is. Questions about the validity of these claims aside, it is important to note how this becomes a national issue that is perceived to be of some import. Like the "literacy crisis," many assert that a similar emergency exists in science education. Take this article from the March 16, 2004 issue of *Business Week Online,* alarmingly entitled "America's Failure in Science Education," for example:

> For anyone concerned about strengthening America's long-term leadership in science and technology, the nation's schools are an obvious place to start. But brace yourself for what you'll find. The depressing reality is that when it comes to educating the next generation in these subjects, America is no longer a world contender. In fact, U.S. students have fallen far behind their competitors in much of Western Europe and in advanced Asian nations like Japan and South Korea. (Symonds)

This is a "depressing reality" *because* we are not a "world contender," a fate unthinkable to competitive-minded Americans who feel they must dominate in all global activities. The attitude is in many ways a holdover from the Cold War days when technological superiority was seen not just as a basis for bragging rights, but as a means of assuring national security and world domination. Today, however, the argument for bolstering the training in science is almost strictly an economic one. For Americans, the idea of falling behind economically is tantamount to a collapse of the entire nation's infrastructure. From the same article:

> This trend has disturbing implications not just for the future of American technological leadership but for the broader economy. Already, "we have developed a shortage of highly skilled workers and a surplus of lesser-skilled workers," warned Federal Reserve Board Chairman Alan Greenspan in a Mar. 12 ad-

> dress at Boston College. And the problem is worsening. "[We're] graduating too few skilled workers to address the apparent imbalance between the supply of such workers and the burgeoning demand for them," Greenspan added. (Symonds)

The rhetorical threat of an economic downfall carries with it so many adjacent associations that almost no further explanation is required. At the mere suggestion of an imploding economy, we envision a loss of our already garish standard of living, the dissolution of jobs, and a general state of suffering comparable to the Great Depression. Add to this our latent sense of xenophobia, and we have the makings of a bona fide crisis:

> Until now, America has compensated for its failure to adequately educate the next generation by importing brainpower. In 2000, a stunning 38% of U.S. jobs requiring a PhD in science or technology were filled by people who were born abroad, up from 24% in 1990, according to the NSB. Similarly, doctoral positions at the nation's leading universities are often filled with foreign students. (Symonds)

Oddly, though science training is apparently on a downslide, we still have droves of foreign-born students attending our universities to be educated in the sciences. The implicit message here is that secondary education is the real culprit, that our institutions of higher education are still among the best in the world, but that we are allowing them to be used to train students who will then take our own knowledge home and somehow use it against us economically. Even more alarming are the consequences, and reactionary warnings help complete the sense of gravity surrounding this issue:

> The consequences of this could be enormous. Because the quality of a nation's workforce has such a huge influence on productivity, effective school reform could easily stimulate the economy more than conventional strategies, such as the Bush tax cuts. Consider what would happen if the U.S. could raise the performance of its high school students on math and science to the levels of Western Europe within a decade. According

> to Eric A. Hunushek, a senior fellow at the Hoover Institution at Stanford University, U.S. gross domestic product growth would then be 4% higher than otherwise by 2025 and 10% higher in 30 years. (Symonds)

I don't wish to suggest that there are not some serious problems with education in the sciences in this country; my point has more to do with the severity attributed to this problem over other failures in education. While people have long hysterically lamented the condition of "literacy" in this country, they have done so with only the most rudimentary understanding of what literacy entails and only because, again, it made us "less competitive" with other nations economically, compromising our own workforce to the extent that employees (so the usual scenario goes) are unable to produce something as basic as an interoffice memo. Likewise, the failure to adequately educate our youth in the sciences is not seen as a failure of learning per se, but only as a shortcoming that will create the incidental economic, technological, and even military repercussions of under-qualified science students. The conclusion drawn from this failure is glibly articulated at the end of the article:

> As America sleeps, other nations that have long since recognized the critical importance of science and technology education to their futures are moving ahead. The U.S. has grasped this lesson in many Olympic sports, where strong national programs have been established to ensure that America has world-class athletes. Unless the nation applies the same approach to science education, it stands to lose far more than a few gold medals. It could ultimately squander its leadership of the world economy. (Symonds)

The low performance of science education, then, is not viewed as a collapse of knowledge in the sense of learning about and understanding more fully the world and our place in it; it is seen as a letdown of our duties to the means of production, as a utilitarian shortcoming that robs us of our rightful place at the technological head of the international community. To compare education in science with Olympic athletes, in my mind, is reductionistic, implying as it does that our emphasis on sports overshadows our efforts in educating our youth. While this may in fact be true, it is unclear how a similar approach can be usefully extended to science education—a case of comparing apples and oranges in terms of the needs that drive both types of accomplishments. Like the "literacy crisis," the "science crisis" is seen as a problem that marks the erosion of standards that puts us at a political and diplomatic disadvantage because it compromises our ability to steamroll over other nations with our economic and technological supremacy (a "crisis" that will undoubtedly prompt some kind of "war"—like a "War on Ignorance"). But strictly economic motivations are not the educational objective of scientists or science teachers, at least not as a primary condition of learning the sciences. The National Science Teachers Association sees its mission as one that is designed to

- Model excellence
- Embrace and model diversity through equity, respect, and opportunity for all
- Provide and expand professional development to support standards-based science education
- Serve as the voice for excellence and innovation in science teaching and learning, curriculum and instruction, and assessment
- Promote interest and support for science education, collaboratively and proactively throughout society
- Exemplify a dynamic organization that values and practices self renewal ("NSTA's Guiding Principles")

Nowhere in this list of goals is there any mention of an objective that seeks to bolster the national economy and make our country scientifically and technologically competitive with other nations. While this may, in fact, be an implicit motivation, it is clear that the NSTA is far more interested in teaching science because it is more conducive to an active and informed democracy, just like schooling in any other realm of learning. A functional democracy must have a broadly educated populous for it to work properly and intelligently. Absent this, people base decisions on feelings and impressions, not on a working knowledge of the culture and society in which we live. Wisdom becomes an abstraction for the idealistically naïve. "Intelligent" people in the "real world" think only in terms of how to play the financial game. That's what has currency and meaning. Moreover, the economic motivations for increased levels of performance in the sciences do not appear, through this statement at least, to be the motivations of the scientific community. Practically speaking, all educational incentives these days tend to be fiscal in nature, but it is a failure of insight to insist that education in science, or any other subject, must necessarily have an accompanying financial payoff. Again, the capitalistic orientation so prevalent in this culture penetrates every component of it, such that everything is "about the economy." Gone (assuming it ever existed) is the notion that one might actually learn something for the sheer fascination of it or as a way to exercise the mind in ways that make it truly an instrument of power.

We likewise appear to have cavalierly abandoned the commonsense idea that a "hands-on" approach will not only have the most realistic results for students who wish to learn how science is actually conducted, it will also create a learning environment that students respond positively to. Standardized tests, designed to assess knowledge of facts and basic skills, show only one possible level of learning, one that requires young people to have, perhaps unrealistically, a high threshold for tedium. While repetition and tedium are certainly a part of any area of research, having a clear understanding of the purpose and a long-term objective that is relevant to the activity certainly has more appeal and more learning power than the simple internalization of facts. Howard Gardner, professor at Harvard's School of Education, notes that we are better off

> resist[ing] the tendency to focus on increasing our students' scores on these tests. These tests don't measure whether students can think scientifically or mathematically, they just measure a kind of lowest common denominator of facts and skills. So getting students to do well on them doesn't necessarily mean much in the real world. It doesn't even mean that students will have successful careers in science and technology.

Alfie Kohn, another educator who has spoken out against standardized testing, offers what he refers to as "five fatal flaws" in the standardized testing mania that has been part of the exit requirements of high school students for decades. The most significant problem, in terms of sacrificing intellectual problem-solving over a merely memorized body of facts, numbers, and names, is the distinction between "ego-based motivation" and "task-based motivation." Julia Harris, of the Eisenhower National Clearinghouse for Mathematics and Science Education (ENC), explains Kohn's position this way:

> There is a vast difference between focusing on performance and focusing on learning. The latter approach is aimed at getting students to focus on what they're doing, while the former mainly makes them worry about how well they're doing it. Kohn stresses that a results-oriented kind of instruction—in which the outcome is a product of some sort rather than a process—encourages students to think that their success or failure is due to their intelligence (or lack thereof). "When you over-emphasize achievement, kids are thrown for a loop when they don't do well. They will pick easier tasks when given the chance." ("Author Takes on High-Stakes Tests")

The tendency to "pick easier tasks" is a problem educators at every level experience, but it is not simply a matter of laziness on the part of the student. When students are gauged by performance over ideas and an effort to make sense of them, they will invariably choose the easier task because their performance will be ranked higher. The ultimate result of an "ego-based" approach is one that stifles genuine thought and reflection in favor of a "good grade." When grades are rewarded, that's what any typical student will put her energies into pursuing.

This debate is framed rhetorically as one that pits "standards" against progress, "accountability" against actual learning. It gives the public the impression that education is a partisan issue, that the more conservative contingent is interested in raising the basic principles of education whereas the liberal educators are trying to systematically sabotage the value of a traditional education in this country. This binary representation, like so many other issues that confront us today, is both an oversimplification and an effort to discredit genuine educational reform. It promotes an unrealistic sense of the real impact of educational standards, suggesting that the problem is not one that effects the students so much as it effects the nation. Public school administration is perhaps one of the most complicated infrastructures to establish, maintain, and evaluate. It deals with what really happens when we fail to educate, but more interestingly, the process by which this is articulated almost always assumes that we know what a "good education" is. In terms of how any failure to promote the sciences is perceived by the general population, it is generally held that decreased standards in science means decreased human resources to compete globally. Training in science, then, becomes a matter of national priority—it goes beyond simply driving the standards that we expect from education; it becomes synonymous with what it means to have an education that is redeemable in the marketplace. Just as science is a commodity in the general sense of how it is instrumental in driving the economy, so too does this orientation spill over into education, giving the person on the street the simple impression that all we need to do to fix the educational crisis in science and mathematics (or any other subject) is to force schools into stringent curricula and test their students at increasingly frequent intervals to make sure our teachers are "doing their jobs." Such an approach is a thinly-veiled effort to promote certain nationalistic objectives—traditional values (whatever those might be), the so-called "basics" of literacy in both the sciences and the humanities, the importance of having a vibrant and largely homogenous, indistinctive workforce, and the idea that all assessment criteria can be condensed into easily quantifiable data—in the name of protecting our children's "future"—which, given this short-sighted agenda, will simply amount to a future overflowing with more of the same.

It is an effective rhetorical slant to establish since, on its face, the promotion of such standards seems so common-sensical. It is difficult to argue against the need for increased scores on standardized tests in the sciences since this seems so fundamental to our economic success. It is equally difficult to argue against "values" and a strong "work ethic." But what makes this perspective such an egregious case of rhetorical sleight-of-hand is that none of the assumptions underlying them have much merit. It is not at all clear, for example, that our values have degenerated into narcissism, hedonism, and decadence, as so many have implied. It is more likely that our "values" have simply shifted to something that doesn't resemble our traditional, romantic outlook of how America, and the people who inhabit it, should be. We are experiencing a dialectical swing that began as early as the 1950s, but it is a social and historical trajectory that has not yet righted itself into some sort of synthesis. The problem with the clarion call to return to "the good old days" is one that fails to recognize that our cultural proximity to the principles of those days is much more remote. Education, perhaps more than any other social institution, must move with the momentum of historical change. And since science and technology is so central to the shift I am describing, they must also be constantly monitored to determine their proper roles in our schools and in our lives. School administrations are quick, for example, to promote technology in the classroom, though they often do so while attempting to maintain the same pedagogical approaches that we used decades ago. Certainly, new technology necessitates new methods, and the representatives of such methods are still our most valuable resources: the people who bring all their experience, training, and knowledge into the classroom and try new things not because it is "trendy" to do so, but because there is a genuine need to reevaluate how best to educate our youth in the 21st century.

Moreover, the "education crisis" is, if not a complete fiction, certainly hyperbole developed to force changes in education in certain preordained and conservative, usually politically-motivated directions—directions that often have less to do with concern for what our children are actually learning and more to do with a nostalgic notion of what the American past actually was. This is not to say that problems do not exist in our schools, but it is important to determine from whence the advocates of certain simplistic educational measures ideologically originate. Applying a healthy dose of rhetorical scrutiny to the claims of the self-proclaimed educational pundits who prophesy alarmist warnings about the state of our society will help us determine whether there is really a literacy crisis or whether science education is actually faltering to the detriment of our nation. Rather than accept at face value that which we are told about our schools and our children, it is important to ask some difficult questions: what does it really mean to "be educated"?; what is the role of science, math, and technology in our society, and what does it mean to possess proficiency in these subject areas?; how important is standardized testing when assessing this proficiency?; *can* standardized testing assess proficiency in meaningful ways in these areas?; who is advocating what educational reforms, and what other ideological motives might be driving the language they use to describe the state of our schools? These are difficult questions because they involve so many separate social, cultural, historical, and ideological factors that providing a conclusive answer is, at best, elusive, and at worst, impossible. However, if we fail to ask the questions at all, we do so to the peril of the very education we fear is in decline.

12 Conclusion

These sentences contain the culture of nations; these are the corner-stone of schools; these are the fountain-head of literatures. A discipline it is in logic, arithmetic, taste, symmetry, poetry, language, rhetoric, ontology, morals, or practical wisdom. There was never such a range of speculation.

—Ralph Waldo Emerson, "Plato; or, The Philosopher"

I have argued in this book that the relevance of science, technology, empiricism, positivism, and a preoccupation with facts and data in this country necessitates both an understanding of science and a language with which to describe and assess its activities, whether these activities are genuine or distorted for a-scientific purposes. We have become a society that is at once enamored with science and suspicious of it, but regardless of our feelings about the abilities and relevance of science, we are dependent upon it. Our reliance on technology alone guarantees that we must depend on scientists, engineers, and technicians to keep the complex infrastructure of our world running, and without people who possess that knowledge, we are paralyzed. Medicine, for example, like technology, is fast becoming so central to our lives that it is hard to imagine what life would be like without a specialist to consult whenever we have an ache or pain—a condition unlike the past, when our doctors were general practitioners and we sought them out only when we were truly ill or injured. Two of the most important areas of scientific influence and advancement, then, in twenty-first century America are computers and medicine, creating enormous opportunities for exploitation and new marketing frontiers.

The relationship between those who produce new scientific and technological knowledge and those who stand to benefit from it is a tenuous one. This is due in large part to the lack of education in science, but just as importantly, in the way that science is often manipulated for extraneous purposes by those who are neither scientists nor interested in the scientific enterprise for its knowledge value, except insofar as that knowledge can be reduced to a barterable commodity. In a free-market, capitalistic society such as our own, everything becomes commodified in this way—in the pursuit of "pure capitalism" where *all* things are divided up and owned—and the greatest end to which science and technology has been put is in the production of new markets, opportunities for profit ranging anywhere from the selling of the latest gadget to the industrialization of mental health to the selling of rivers, forests, airspace, and even air itself. These markets are not, by themselves, malevolent forces deliberately designed to oppress the weak or separate us from our money, but they do have these effects when acquired by certain agencies. If America is actually controlled, as we so often hear, by the Mega Corporations that have the money to direct public opinion through consumerism, then science simply becomes one more resource through which such forces can operate profitably. The things we value as a society have become so literalized that the very word "value" means the monetary worth of a product, a device, a service, or a person. To refer to people as having "net worth" is itself the ultimate dehumanization, a reduction of people to the objectified accumulation not of who they are, but what they own, what they can barter for profit, what their "function" in the larger economic machine is, or what the value of their estate happens to be. People, through this system, become mere conduits through which the production and consumption of material goods flow, losing all sense of identity aside from their possessions and "assets"—a revealing distortion of that word that suggests the benefits inherent in a person are what can be gained when all material ownership is liquidated into hard cash. Science, like everything else, is not invulnerable to commodification, and may in fact be more susceptible than other knowledge systems because it deals primarily *with* materialism, whether it be in the form of finding new fuel sources, creating new drugs, or producing a faster computer processor.

I don't want people to mistake what I'm saying as advocating some idyllic and utopian notion of a better society through the denial of material comforts or the casting off of modern conveniences; the products of material culture have many benefits, and I would rather live in a time when the menial tasks of daily living are relegated to machines that have a far greater threshold for the tedium of repetition than human beings should be made to endure. I would rather live in a time when polio, small pox, and leprosy are diseases of a remote era. I like that the world is more accessible through advances in technology, communication, and transportation. But what have we done with the liberation afforded us by modern living? Have we, as a whole society, seized the opportunity to improve our civilization, to use our leisure to create or to think? In many ways, our goals for living are the same as they were 150 years ago: Grow up, get a job, get married, have kids, and repeat the cycle. There is, of course, nothing wrong with this, but if improved living conditions do anything, they should push us to develop the whole of human society, to construct institutions that encourage and even reward new ideas and innovative communal opportunities, to protect the basic social principles on which this country is based, and to foster the development of art and philosophy in everyone. Instead, we march aimlessly, driven by a programmed sequence of consumer duties and fueled by an ethic of quantity and acquisition. Science has given us the gift of freedom, and we have squandered it irresponsibly in pursuit of greed and self-interest.

Science and technology have their revenge effects, of course, not the least of which is that with increased control over our environment comes an increased obligation to protect that which we control. Here as well we have had an uneven track record. Any advancement in environmental protectionism has been unraveled by subsequent administrations that want to return to a *laissez faire* mentality that is both outdated and dangerous. If corporate America is propelled strictly by a profit motive, all other, more principled motives fall by the wayside. In a democracy that values a strict majority opinion (as opposed to a more responsible democratic process), the efforts to instill and maintain consumerism assure that such short-sightedness will be met with misinformed approval. Whenever money becomes a central focus, monetary arguments gain increased currency. If consumers feel that costs for protecting the environment, the worker, or the security of jobs will somehow be passed on to them for the goods they feel are indispensable parts of their lives, public opinion will always vote for hands-off policies. They will readily embrace the economic benefits of NAFTA, or decry the evils of unionization (except when it comes to their own job). If taxes will be raised to make government intervention logistically possible, people would rather take their chances with corporate self-regulation, which we know from experience would be laughable if only the consequences weren't so severe. Science is caught in the middle—in between the crushing plies of corporate motive and consumer desire, between the often incompatible principles of social responsibility and professional survival.

Since this is a highly complicated problem, the answers are equally complex and ongoing. The answer requires commitment, time, imagination, innovation, and, of course, money. An education that values all forms of knowledge and training in how to reconcile this knowledge with the situations that tangibly confront us is perhaps the best corrective for the downward spiral I've just described, but educational "values" have taken on an equally materialist meaning. They've had to because, like science, educational institutions have to survive in a market environment just like everyone and everything else. The best rhetorical possibility for change is to make an argument (in any space where that argument needs to be made) that reintroduces the value (in the most encompassing sense of that word) of non-scientistic epistemological systems that have been fundamental contributors to the greatest civilizations in history while avoiding capitulation to the pressures of a strictly material orientation in the process. If that seems like a tall order, it's because it is. The only meaningful locus for small-scale change that I've been able to muster is in the classroom, because students "get" this idea, once it's pointed out to them—and they don't really like the notion of having their identities seized in such a cavalier way for the benefit of the few. They see the empowerment of knowledge, and at times the quest for it even *replaces* the motive for mere acquisition, because it becomes something that they have not only earned, but something that has become part of their new identity. If such knowledge is used with integrity and ongoing vigilance, within a generation or two, perhaps a truly new and responsible society will begin to emerge, one that values humanity in all its many facets, faces, ways of knowing, thinking, believing, talking, and acting.

While this clearly means training in science and technology, we must not do so at the expense of equally important areas of knowledge—ones that include thorough education in history, politics, communication, the arts, the fine arts, philosophy, and rhetoric. It means a commitment to a general, liberal arts education. It means that we must stop thinking of education in terms of its stock-market return value and start thinking about it as a necessary component of a working democracy. It means we must instill in students, parents, and administrators the value of broad-based learning, and stop submitting blindly to the notion that immediate relevance is the only measure of an academic subject's value. It means preparing students to think at an early age, to encourage ideas, questions, community projects, independent inquiry, and above all, compassion. It means we must depoliticize the curriculum in the sense that it supplies students with the intellectual equipment to assess different ideas, not use the curriculum as a means to reflect and implant a particular ideological orientation or a specific political agenda. It means we must recognize that different knowledge structures require different criteria and evaluation heuristics and not attempt to quantify the unquantifiable.

I have suggested that central to this project is an emphasis on rhetoric. This is not to say that training in rhetoric is a panacea for all social maladies; it is, in fact, possible that a poor education in rhetoric could have as many detrimental effects as positive ones. Reckless instruction in rhetoric might teach students how language is manipulated, thus demonstrating how language can be used for strictly self-serving ends. Hence, it is important to not only teach *rhetorica utens,* but also *rhetorica docens*—both the how of effective language use and the evaluative, heuristic nature of rhetorical studies. Above all, it must not be taught without careful attention to the other key humanistic disciplines, especially philosophy and history. Philosophy, as a means for teaching students how to think well, also provides an ethical basis that may have more genuine applicability than what religion can offer students. Philosophy allows students to ask the important questions without fear of reprisal, or worse, ostracism from their peers for being weird or different or impious. Dialogue is so often shut down before it even begins because people are uncomfortable with important questions, especially those that might challenge received doctrine from other quarters. Alas, philosophy is one of those expendable programs of study at most American universities. Except for the largest institutions, philosophy departments, if they exist at all, are made up of one or two faculty members, and students rarely get access to anything but the most introductory courses. Philosophy, as a discipline, needs to change this—to *do* philosophy instead of just teach the ideas of dead philosophers, to make a case for a higher relevance, and to show how what philosophy offers is an insight into the understanding and implementation of ideas that point toward a better world.

History is, perhaps, even more central to a complete liberal arts education than philosophy or rhetoric, and I am frequently saddened by just how little history my students know. But part of the reason historical education has failed with so many students is that the arguments for why it is important do not resonate with students today. Even if statements like "we must know where we've been to know where we're going" and "those who do not know history are doomed to repeat it" have a superficial truthfulness to them, they have a very thin penetration factor because they seem abstract, remote as they are from any young person's own experiences or sense of reality. We also make the mistake of teaching exclusively through the "great events," "great figures," and "great dates" approach which relies too heavily on memorization and ignores, even resists, the importance of dramatizing history, of breathing life back into the people of the past. Most importantly, we must not fool ourselves into believing history is a science, nor should we want it to be. If we teach history as a sequence of wars, generals, and treaties, it is little wonder that students find it boring and irrelevant; if, on the other hand, we tap into their innate love for narrative, and we talk about the "real" people of history, we have a drama that speaks to students and a lineage that begins to make sense. The study of literature is perhaps one of the best ways to achieve this. While it doesn't always deal with "real" people, it certainly deals with real ideologies, real conditions, and real philosophies, reflecting the mindset of an era and providing us with an opportunity to dramatize the very principles and trajectories we hope students will get out of history. Then, students have a basis upon which to think, a reason to want to understand our world and themselves, and a system of values that becomes truly worth remembering, practicing, and protecting.

The broader American population—those either out of school or the product of a limited educational experience—must be diligent in keeping themselves informed. This brings us full circle to the need to understand science and its influence on modern life, policy, and decision-making. I submit that understanding science means understanding all the other, often neglected, areas of knowledge—the humanities, the fine arts, and the spiritual disciplines and practices. Without these, any knowledge of science becomes barren and meaningless; it is only through a thorough knowledge of our own being as human agents thinking and acting in the world that science's potential is unlocked. To understand science, one must also understand the nature of language. Language, in the end, is the common denominator—no human activity is conducted without it. To understand the nature of language, one must know how it is used, how it affects our ways of knowing, and how it influences, repels, delights, enrages, distorts, clarifies, controls, and confuses. To deny ourselves a deeper knowledge of how language works dooms us to be manipulated by those who *do* know how it works. In our interpretive role as human language-users, it is important to remember that words are only as good as the person using them. If we know the authentic purposes behind scientific discourse in all its diverse forms, we are one step closer to understanding our own purpose as language using animals.

Notes

Notes to the Introduction

[1] I will outline these rhetorics more completely in chap. 11.

[2] A notable (and highly recommended) recent exception to this absence is Cary Nelson and Stephen Watts' book, *Office Hours: Activism and Change in the Academy* (Routledge 2004). The focus of this book, however, is more on the problems of labor relations, graduate studies, and university politics than on informing the public about the general practices of academia. It is targeted, in other words, at people who are already part of the academic culture.

[3] In April of 2006, for example, newly-elected (Democratic!) New Jersey governor John Corzine proposed that aid to higher education in the state be cut by an unprecedented 167 million dollars, provoking immediate panic in college administrators across the state with almost instant repercussions for the lowest and most exploited members of the university hierarchy: adjuncts and instructors.

Notes to Chapter One

[1] For a more detailed discussion of this, see Tietge, *Flash Effect: Science and the Rhetorical Origins of Cold War America*. Athens, Ohio: Ohio University Press, 2002.

[2] Prometheus, you may recall, was the Greek mythological figure, a Titan, who stole fire from the gods for use by humans. Zeus, king of the Greek god world, infuriated by the insubordination of Prometheus, punishes him by chaining him to a rock where a giant bird appears every day to eat his liver (his liver, magically, regrows every time, making the punishment an eternal one—that crimes against gods should carry sentences that last an eternity is a literary tradition that is best illustrated by Dante). The notion that there is a domain that should be reserved strictly for our deities is at least as old as literature, and Shelley borrowed this idea to demonstrate what perils await us if we transgress against the sanctified rights of beings that are superior to us. Science is the "fire" in Shelley's novel, and hers is a warning against the injudicious use of science to tamper with the "natural order" of the universe.

[3] Kenneth Burke has also observed that the connection between science, religion, and magic is an important (if counterintuitive) one insofar as they are, as Clarke posits, nearly indistinguishable at a certain point in their development. It is the characteristic of wonder that ties these practices together, for it is clear that the awe we experience spiritually, the amazement we feel when shown magic, and the fascination that heralds us when new science is demonstrated for us all derive from the same sense of human curiosity and need to know new things.

[4] Common language usage is always useful for bearing out such claims. I called the plumber yesterday, and the receptionist (now known as the office coordinator) informed me that the plumber (which she referred to as the "on-site technician") would be at my house (single-family protective structure?) within two hours. The professionalization incurred through the technological labeling of common vocations lends an element of credentialing to the job. The rhetorical impact is that we trust those with technical-sounding job titles because we assume that the title was earned through specialized training, though there is often no reason whatsoever to make this assumption.

[5] Insurance companies create yet another wrinkle in this rhetorical dynamic since the prescription of drugs is often reliant upon whether a patient's insurance will approve the cost of drug treatment. Doctors and patients alike are rightfully suspicious of allowing what are essentially businessmen make key medical decisions about appropriate treatment. Again, economics and science intersect, usually not in a way favorable to the person seeking medical treatment. For a detailed discussion of the economic ramifications of health coverage, see Sherman Folland et al., *The Economics of Health and Health Care,* 4th ed. New York: Prenitce Hall, 2003.

[6] Also useful as a study of the rhetorical implications of health insurance and medicine generally (especially how media forces exploit a growing national hypochondria) is Judy Z. Segal's, *Health and the Rhetoric of Medicine,* Carbondale: Southern Illinois UP, 2005.

[7] This truism has, actually, been challenged by certain cultural critics in recent years. Some critics, like James Twitchell and Judith Williamson, argue that advertising is less of an economic phenomenon and more of a cultural one. Coca-Cola, for example, is such a mainstay in American self-identity that advertising ceases to be about product recognition for the purpose of purchase, and becomes a system of identification that reminds consumers of their product-based birthright. No one would stop buying Coke if advertisements for it were suddenly discontinued, at least not right away, because the product is part of the American cultural landscape, a nationalistic identifier that is part of our collective heritage. Why, then, would the Coca-Cola company continue to shovel millions of dollars into the advertising machine? Because to fail to do so would be to deny this very heritage. The immediate goal of advertising (in this case, at least) may not be so much to "sell" a product as

it is to maintain an image. Ultimately, however, the image and the sales are closely tied—it just becomes a more complex cultural relationship than the more straight-forward approach of lesser-known products or of products that do not enjoy the historical status of Coca-Cola. See James Twitchell, *Twenty Ads that Shook the World,* Three Rivers Press, 2001; and Judith Williamson, *Decoding Advertisements: Ideas in Progress,* Marion Boyers Pub, 1994. Naomi Klein, likewise, argues that corporations have more or less abandoned the strategy of touting product performance in favor of the "image" approach—one that emphasizes the hyped benefits of adhering to a "life-style" that is associated with a product or through recognition of ubiquitous corporate icons. See Klein's *No Logo,* New York: Picador Press, 2002.

[8] That *The Wall Street Journal* has become such a mainstream news source (not to mention a source of apparent scientific legitimacy) in this country is itself an interesting development. We have so married in our minds the notion that economics drives everything that we trust the authority of a publication whose primary purpose is economic analysis. "Smart" people read *The Journal,* and if they are smart with money, they must be smart about everything else. I am reminded of that frequent and specious "rhetorical" question, "if you're so smart, why aren't you rich?"

Notes to Chapter Two

[1] Edward Tenner again offers useful insight into this phenomenon by noting that the way the media (and therefore, the public) prioritizes environmental problems and events means that our attention is redirected towards sensational events rather than the chronic issues that really have the most impact on our environment. We have a panicky impulse to look for—and respond to—the "big disasters" like hurricanes, floods, and earthquakes, while the little, daily encroachments chip away at the quality of our air, forests, plains, rivers, and oceans. For a more detailed discussion of this, see Tenner's *Why Things Bite Back: Technology and the Revenge of Unintended Consequences.*

[2] In a recent "Nova" episode called "Dimming the Sun," the process of global warming is complicated by another phenomenon known as "global cooling," which, currently, is in a very precarious balance. It appears that the accelerated global warming produced by human activity is masked by a global cooling that is the result of other human activities, suggesting that the effects of global warming are far more serious and immediate than originally believed. For a more detailed account of this phenomenon, see the PBS website, http://www.pbs.org/wgbh/nova/sun/.

[3] Also of interest is the apparent human-centeredness of any natural disaster. This relates directly to the age-old (religious) assumption that we are the stars of the worldly stage, that everything that happens is the result of

either reward or punishment for human conduct. President George W. Bush, following the Katrina hurricane in the North American Gulf of Mexico, attributed the carnage that the storm produced in Louisiana, Mississippi, and Alabama to an act of God, implying that there was a supernatural "reason" for the storm. This, despite the fact that the region had been expecting a catastrophic hurricane for years and that it happened in the late summer—hurricane season—may explain the ridiculously inept response from FEMA, the Federal Emergency Management Agency; if one holds to a doctrine of predestination in matters that are really the domain of natural science, "prevention" becomes a low priority. Some rabbis, in addition, have recklessly speculated that Katrina was punishment for the Bush administration's support of the forced evacuation of Gaza! Consider this story from WorldNetDaily, an online "newspaper":

> Rabbi Joseph Garlitzky, head of the international Chabad Lubavitch movement's Tel Aviv synagogue, recounted for WND a pulpit speech he gave this past Sabbath:
>
> "We don't have prophets who can tell us exactly what are God's ways, but when we see something so enormous as Katrina, I would say [President] Bush and [Secretary of State Condoleezza] Rice need to make an accounting of their actions, because something was done wrong by America in a big way. And here there are many obvious connections between the storm and the Gaza evacuation, which came right on top of each other. No one has permission to take away one inch of the land of Israel from the Jewish people." (Klein)

This is a textbook example of the *post hoc ergo propter hoc* fallacy (Latin for "after this, therefore, because of this"). Moreover, it is human hubris and taking the effects of natural occurrences personally that lead to poor judgment and high death tolls during such tempests as Katrina, not the vengeful wrath of God punishing leaders for their international policy. This is an obvious danger because it reflects a theological preferentiality, an argument that God is on "our" side, not "yours." Palestinians can, of course, make the same argument as those of their Israeli enemies, and such attitudes foster increasing mistrust and perpetuate the very hatred that is the true cause of suffering in that region. Katrina was a coincidental disaster, but the Jewish tradition of the vengeful god using the forces of nature to punish the enemies of Israel runs strong. One must wonder, however, why God didn't simply exact his wrath on the Palestinians themselves. It would seem a far more efficient measure.

Notes to Chapter Three

[1] The issue of legitimate knowledge is, in fact, a larger one in science studies and particularly for the rhetoric of science. Science has, in many ways, appropriated all knowledge as being subject to its standard, a practice which becomes problematic for areas of knowing that are not easily quantifiable or rely on more subjective criteria in order to construct knowledge claims. This will be discussed in more detail in subsequent chapters.

[2] For an eye-opening examination of how the myth of The Liberal Media is actually perpetuated by GOP manipulation of the media itself, see Sheldon Rampton and John Staubers' *Banana Republicans: How the Right Wing is Turning America into a One Party State,* Tarcher/Penguin, 2004. See also Eric Alterman's *What Liberal Media? The Truth About Bias and the News,* Basic Books, 2003.

[3] Naturalist David Attenborough has much the same effect. His *Trials of Life* series was a triumph of nature photography to which he lent an air of British sophistication without sacrificing audience identification in the process. He was an excellent professor with the most amazing of classrooms: all of nature. The combination of technical superiority and Attenborough's enthusiasm and expertise was an immensely effective one. His series should be viewed as the paragon for representations of natural science to the general public.

[4] See Tietge, *Flash Effect: Science and the Rhetorical Origins of Cold War America.* Athens, Ohio: Ohio University Press, 2002.

[5] See Edward Tenner's *Our Own Devices: How Technology Remakes Humanity.* New York: Vintage Books, 2004.

[6] During the 2006 Winter Olympic Games, figure skaters were vocal about the quantification of their art in the form of "objective" scoring criteria, something which in the past had been much more "subjective." Their complaint was that the art of figure skating had been reduced to mere technique, and this had a detrimental effect on the content of the routines because, if a skater knew that a perfectly executed triple axel was needed in order to bring the score to a mathematical victory, then the skater would perform it even if it did not fit compositionally with the rest of the routine. Skaters rightfully argued that a routine was more than the sum of its parts, that a skilled and beautiful skater would compose a routine based on more subjective criteria that made the performance a holistic experience, not just a series of "tricks" designed to push up the objective score. Here again we see an example of scientific criteria and objective standards being cavalierly applied to an activity that requires a more subjective judgment. The objective criteria were defended, of course, in the interest of fairness, but what suffered ultimately was the skating itself.

[7] This claim, incidentally, is itself not valid. One of the problems with unification theories like Quantum Mechanics and String Theory is that they

are not, in fact, scientifically verifiable through traditional experimentation. Rather, they are only mathematically verifiable, which means that scientists must demonstrate their theoretical existence using a language system, not physical testing. It this respect, such theories are much like logic in the sense that they can only be legitimated with systems for which only humans provide the rules.

[8] On a history and theory listserv to which I subscribe, one recurring theme is the issue of whether or not the study of history is a "science." While some challenge this idea, the vast majority of professional historians insist that history is (and should be) a science. Here are some examples of posts that have appeared on this listserv thread:

> So, is there a science of history or not? If that question means, are there 'laws of history'?, and that is a sensible requirement, then as we have suspected, the answer is in the negative, perhaps, and therefore the claim for jurisdiction by science fails. Large unruly crowds will be storming the 'history lab' demanding a refund, and the decision must pass to neutral judges, confronted with lawyers for each side, if that might suffice. To be a trifle melodramatic, we can then negotiate the terms of surrender by scientists, who can be taken forthwith to a place of geisteswissenschaft by Diltheyan types in academic gowns. (H-Net, "History and Theory," March 21, 2006)

> Would that metrics could encompass the scope and scale of history that has so many unknowns and unquantifiable variables. Please note that Secretary of Defense Donald Rumsfeld played the US-Iraq war by the numbers. (H-Net, "History and Theory," March 14, 2006)

> Mathematics is a concept that suggests rigor of process, not just numbers. My next door neighbor at Penn State was a mathematical philosopher who specialized in semiotic theory. She approached her work in the same way a Boolean logician would approach such work—only at a much more sophisticated level. We can certainly hypothesize that the course of events followed a certain path, define what we would expect to find on that path, and even assign probabilities in a quasi-numeric sense. In fact, we do this all the time. We just are not trained to clearly state what we are doing. And that has allowed the extremists of the "literary

> turn" to chop out the guts of history. (H-Net, "History and Theory," March 14, 2006)

> Suppose we have a view of why Charlemagne's empire collapsed. It is certainly possible to make predictions as to what we should find if that view is correct. To the degree that that view is verified by previously unknown data we can ascertain a quasi-mathematical probability that the stated view is accurate. As someone who has labored long in the vineyards of quantification, this is hardly a statement of triumphalism—there are just an enormous number of variables to take into account—but it is certainly a way of framing our work which promises great utility and rigor. (H-Net, "History and Theory," March 14, 2006)

"Rigor," incidentally, is the domain of science exclusively, if we are to believe the proponents of the "science is knowledge" school. Apparently, science is the only sphere of knowledge that practices intellectual "rigor," and it is on this basis that science coveters want to argue for their fields as scientific. To use another epistemic stance would be, evidently, to sacrifice "rigor."

> Those who argue that history is not and cannot be a science usually base their arguments on 'laws' (e.g. Newtonian). They seem to think that the essence of science is the discovery of laws, and argue 'a priori' that there can never be any laws of human history-ergo: no scientific history. However, the core idea for all sciences, is the concept of verification; i.e. that knowledge claims are expressed with proposition which indicate what invariable elements have to appear for the proposition to be true. Further, any competent qualified researcher can reproduce those elements and get the same results. Without verifiable propositions, there can be no science. For example, recently a South Korean biologist made claims about cloning experiments that could not be reproduced (verified) by other biologist. Eventually the South Korean had to admit that his knowledge claims were not true. Those who advocate scientific history seek no more—simply, criteria for knowledge claims that other historians can verify. (H-Net, "History and Theory," March 7, 2006)

How does one scientifically verify the truth value of a claim made in a text about, say, The Battle of Hastings, except to verify it through another text (anthropological relics can be seen as "texts" in this framework as well)?

While one may find consistency between the texts of the time about many—even most—matters, this is hardly the same as saying that they have been "proven" in the scientific sense. Since we cannot exhume the dead to ask them, all we have to rely on are a string of texts that leads to some probable conclusions. But this is not science—it's scholarship, and the rigor practiced by careful scholars predates methods used by scientists by many centuries. Why, suddenly, are these methodologies either a) dismissed as unreliable, or b) transformed into (labeled as) a "science" that retains most of the original practices of the scholarship it was meant to supplant?

Notes to Chapter Four

[1] Carl Sagan was also a leading planetary astronomer while he was alive, so it is perhaps an important component of being a popularizer of science to have reached laudable professional levels among one's peers before trying to educate the masses, though as we have seen with media figures like James Burke, this is clearly not a prerequisite. Nevertheless, the manner in which professional reputation aids in the scientific ethos of individual scientists may contribute to their "expert" status, suggesting that these individuals are consulted more for their knowledge in a specific area of science rather than the more generalized scientific teaching that characterizes Burke.

[2] A good example of this is the "Trinity Test," the codename given to the experimental explosion of the first atomic weapon (known as The Gadget) during the Manhattan Project. This name was assigned to the experiment by Robert Oppenheimer himself, an amazing scientific figure who commanded a wide range of intellectual talents beyond physics. "Trinity," of course, refers to the "three-personed God" of John Donne's sonnet and of Catholic doctrine generally, and Oppenheimer was also reported to have quoted the Bagihvad Gita after witnessing the explosion of Trinity: "I have become Death, destroyer of worlds." For a more detailed discussion, see Tietge, *Flash Effect,* Ohio University Press, 2002.

[3] For a detailed and highly illuminating discussion of the role of rationality in fundamentalism, see Karen Armstrong's *The Battle for God,* Ballentine Books, 2000. Armstrong argues that twentieth century fundamentalism "is a reaction against the scientific and secular culture that first appeared in the West, but which has since taken root in other parts of the world" (xiii). The danger, she further asserts, is the confusion between *logos* and *mythos*—the fundamentalist tendency to mistake the literary for the literal, a problem that is increasing among students who have very little literary or rhetorical training to make foundational symbolic distinctions between the function of a text and the "reality" of the world they inhabit. In other words, absent solid humanistic training, people have a bad habit of taking all texts literally. This trend will be discussed further in Chapter 11.

[4] Chomsky is almost as famous for his anarchistic leanings (and writings) as he is for his scholarly work in linguistics. While *Syntactic Structures* is still a mainstay of linguistic study in graduate programs (in which Chomsky founded the idea of generative grammar, which, incidentally, has profound implications for evolutionary psychology), a quick Internet search of Chomsky reveals that he has actually written more on classical left-wing politics than he ever has on linguistic theory.

[5] Wright provides another example of this linguistic tendency, though I don't think it was his intention. He writes:

> One might expect that, given enough time, beetle predators would up the ante, developing some clever way to neutralize the beetle's noxious spray. In fact, they have. Skunks and one species of mouse, the biologists James Gould (no relation) and William Keeton have written, "evolved specialized innate behavior patterns that cause the spray to be discharged harmlessly, and they can then eat the beetles." Evolutionary biologists call this form of positive feedback an "arms race." Richard Dawkins and John Tyler Bonner, among others, have noted that arms races favor the evolution of complexity. Yet Gould's two books on the evolution of complexity don't even mention the phenomenon. (Wright)

An "arms race" between insectivore and insect is, of course, an anthropomorphic description of the process, and, again, we have named something that reflects our connection with what something else in the environment does naturally. Such naming is, of course, metaphorical (and "natural" for our species), but it also reflects our tendency to invert a natural process in a way that shows our need to describe things in terms of the order we place on it.

[6] People often ask, "Why is there disease, starvation, and suffering if God loves us, if there is a rational scheme to His plan?" The Faithful usually reply, "He works in mysterious ways, and there is a purpose to it all, but we are not privy to it." A more satisfactory response (in one sense, at least) can only be provided by evolutionary theory. There is disease, starvation, deformity, and, hence, suffering, because nature is random and indifferent—it makes adjustments without a conscious directionality, and things therefore go wrong (from our perspective as species trying to survive). On the other hand, "going wrong" is relative; if disease (in the bacteriological, viral, or syndromatic sense) is viewed as the survival of one organism at the expense of the other, then one organism is simply surviving and the other perishing. This is how natural selection works. Even more importantly, nature is amoral, a condition that we usually mistake for "evil." Western civilization is filled with literary, philosophical, and theological examples of our fear of na-

ture as an malicious entity. The paradise of the Genesis myth is described as an oasis from the toil and pain of nature, and our original fall meant that we were thrust into this bad place forever. Nature, in this example, is antithetical to God, the domain of godlessness, the place where Satan himself dwells. However, the reality of nature is that it simply exists, doing what nature does indifferently to our presence or our future as a species. We are no better or worse off in this way than any other organism, but that is a humbling realization that many resist because, put simply, it means we aren't "special."

[7] The Intelligent Design/Creationist debate will be covered in more detail in Chap. 8. It is necessary to include it here as well because Gould's most influential contribution to popularizing science has to do with re-educating people on the misinformation driving this debate. Gould is an important popularizer of science, and his main incentive is to clear up the tenets of evolution where creationists have been the bane of his professional existence.

[8] Gould anticipates objections to this position early in the book and feebly counters it by saying he "most emphatically do[es] not argue that ethical people must validate their standards by overt appeals to religion . . . and we all know that atheists can live in the most firmly principled manner" (57). But, he says, he will "reiterate that religion has occupied the center of this magisterium in the tradition of most cultures" (58). That religion controlled moral boundaries "*has* been" the case, but is "not *necessarily*" the way it has to be seems to be the message but, if this is so, the magisterium of religion comes up very empty indeed. What, we might wonder, is Gould "allowing" religion to have sovereignty over?

[9] As perhaps a rather odd mental exercise, try to think of death in terms of all the eons during which you weren't alive. Since I was born in the mid-sixties, I was not around for infinitely longer than I have been around (billions of years, perhaps longer), and this, in some odd respect, must be what death is like. I was, of course, not conscious to know that I wasn't alive prior to my birth, and therefore, the prospect was not something I was equipped to worry about (those who believe in reincarnation have gotten around this prospect as well—both pre-existence and death are covered under this way of believing). It is only through the microscopic window of our brief existence that we get a taste for life in the first place, and the prospect of finally giving that up is painful and frightening on one level, but a condition we are far more "accustomed" to on another. It is a strange notion to consider that, at some very finite time, we suddenly became aware, conscious, thinking entities. This is, in fact, a "gift," but only in terms of its vast unlikelihood. It strikes me as important to consider this as an ethical question when we so cavalierly forfeit our own lives or ask others to forfeit theirs for "God, king, and country." It is far easier to sacrifice a precious life if one believes there is an afterlife full of rewards or that our memory will have some sort of lasting power. It is far more difficult to throw away the only chance at existence we

have if we know this *is* our only chance at living. Perhaps we would reconsider the ease with which we kill our fellow species. As Clint Eastwood so poignantly put it in *Unforgiven*: "It's a hell of thing to kill a man, to take away everything he has and everything he's ever gonna have."

[10] Stanley Kubrick explores the implications of a machine that does make this evolutionary leap in the form of HAL 9000 in what many consider his cinematic masterpiece, *2001: A Space Odyssey*.

[11] It is a myth that animals are unable to anticipate future events; any trained dog belies this notion when expecting a reward for a trick performed or an order obeyed. My dog, a Shetland Sheepdog (or "Shelty") is highly intelligent and almost infinitely trainable. He can sense, for example, when my family is preparing to leave the house, and it bothers him because, as a sheepdog, he understands that he is about to lose his flock. He paces, circles, and, finally, enters his kennel (his own little cave) long before we have said anything to him or left the house. He senses this through certain unmistakable signs (we are all standing, talking to each other, laying down plans for our outing, whatever it may be), and he acts on those signals because he knows we are *about to* leave. This indicates, in the short-term, at least, his ability to foresee upcoming contingencies and to act appropriately (usually) upon his experience with this same scenario in the past.

[12] Pinker has not read these thinkers very well, however, as is evidenced by the erratum that Mark Johnson is a "linguist" (357). He is, in fact, a philosopher whose primary area of study is the philosophy of language. Whether this slip was intended or not, I find it interesting that Pinker assigns him the moniker "linguist," since "philosopher" is too overly refined for our present anti-intellectual population. "Linguist" lends itself to the assumed objectivity and certainty of the scientist, while philosopher conjures images of the effete academic with too much time on his hands to think about frivolities.

[13] While most people are probably aware of this, Alfred Kinsey was the biologist-turned-sexologist who released the so-called "Kinsey Report" in 1948 (the actual title of the book was *Sexual Behavior in the Human Male*), a hugely popular publication that aided in challenging many of the myths and taboos regarding sexual behavior, followed several years later by *Sexual Behavior of the Human Female* in 1953. The stir these books created is often cited as the origin of the "sexual revolution" of the 1960s.

Notes to Chapter Five

[1] In a recent "Nova" documentary, the topic was the search for the "Ghost Particle," neutrinos that were being emitted by the sun and could tell scientists much about the origins of the universe. In terms of this discussion, an interesting component of the debate was that the theoretical physicists assumed that the cosmologists were wrong about their theory of neutrinos

and had actually thwarted much of the research process under the banner of the Standard Model, an excepted series of facts and laws about the nature of particles. It turns out that the model was wrong and that the cosmologists were right, even though they had been working on experiments that reflected what one scientist referred to as a "socially unpopular result." The point is that scientists who rely on the authority of other established texts sometimes create a serious blind spot in their ability to see new possibilities, and this often relates directly to the ethos of earlier scientists who have established "certain" paradigms. This is ironic since many of these same scientists, ones who rely upon the consensus of their peers, would deny the "social constructivist" view of reality.

[2] Thomas Huckin notes that "[s]cientific journal articles vie for attention in the scientific marketplace, and a key measure of their success is the degree to which they are cited by other researchers" (24). Scientists, like the public, are influenced by the megalithic institution of the scientific establishment, giving credence to those have helped forged accepted knowledge. This is, of course, appropriate, but it is also rhetorical in the sense that any efforts to challenge the scientific *status quo* are met not with individual opposition, but with the force and power of disciplinary essentialism.

[3] The word "jargon," interestingly, is a Middle English word derived from the Middle French word for the "chattering of birds" (*OED*). It is used today in the pejorative to connote unintelligibility or deliberate obfuscation for the purpose of concealing an otherwise straightforward concept. The colloquial usage of the word is, of course, an unfair representation of the specialized language that may be necessary to deal with more advanced concepts than ordinary language will allow. Much jargon, however, does function as a linguistic and rhetorical obstacle to keep those in the know sequestered from the uninitiated.

[4] The Internet makes such journals far more accessible now than in the past. Until recently, professional journals would either be housed in a university library where access by the general public was limited or through an interlibrary loan system. In either case, anyone interested in finding out about the activity of the scientific community through these journals would have to make a concerted effort to track down information. The Internet now has many such journals readily available (often for a fee), so that non-specialists can get a better sense of how scientists communicate with one another. This creates problems of its own, as we shall see.

[5] Sokal and company, no doubt, would attribute this trend to the Leftist Postmodern Conspiracy against science and its accompanying jealousy toward the monopoly science has on the truth. However, the reality of this is much more complicated, and a detailed analysis of the cause of popular mistrust for anything complicated or pluralistic would make a fascinating study. Some contributors to the increasing doubt about the conclusions science is

able to draw with certainty (and about intellectualism as a whole) because of its inherent complexity might include the failure of education to promote and encourage open thinking in the general population, the diminishment of listening and concentration skills that have been one result of a tempestuous consumer culture, and the inability to learn the basic tenets of sound argumentation. Still, science has more currency as a knowledge base than its humanistic counterparts, largely because it is capable of reductionistic representation while simultaneously retaining the main thrust of the argument with which it hope to endow its audience, even if such a representation is in actuality a diminishment of the complexity of the claim being made.

Notes to Chapter Six

[1] See also Harding, Sandra G. *Feminism and Methodology.* Bloomington, IN: Indiana UP, 1987; *The Science Question in Feminism.* Ithaca, NY: Cornell UP, 1986; *Whose Science? Whose Knowledge? Thinking From Women's Lives.* Ithaca, NY: Cornell UP, 1991.

[2] A particularly appalling example of just how married we are to the idea of our consumer obligation was the advice given by the Bush administration following the attacks of September 11th, 2001. In order to curtail the deleterious effects of the evils of terrorism, it was suggested that the American populous maintain its "way of life." The best thing we could do for our country was to continue spending. This, it was argued, would show those who hoped to disrupt American resolve that their efforts were futile, that we would stubbornly continue our lifestyle of mass consumption. The advice given to the patriotic citizen in order to thwart further attempts against our country (both George W. Bush and Rudolph Giuliani said as much), in other words, was "go shopping."

[3] *Star Trek* and its various spin-offs are the conceptual pioneers in this area. The vast array of alien beings we encounter on such shows as *Star Trek: The Next Generation*, for example, are often amplifications of single human characteristics. The Klingons are the alien representation of our warlike selves taken to a cultural extreme; the Ferengi reflect the greed and materialism of a society obsessed with the acquisition of wealth and material goods; The Vulcans are noble, if sometimes insensitive, reflections of our rational, "logical" side; even the Borg, it has been argued, are composites of the automaton-like consumer, scaling up "assimilation" and consumption to a level of irresistible force, absorbing all resources (and cultures, and "technological distinctiveness") that it encounters. They are the great cosmic colonizers, and they function as a political commentary on the mindlessness of mass consumption and the indifferent evils of a self-perceived imperial destiny. Note too the symbolic effect of the prosthetics: Klingons are harsh-looking with dangerous, ridged foreheads that look like they could be used as weapons;

the Ferengi are hideously ugly with their huge ears and small, pointy teeth, suggesting the superficial unattractiveness of so shallow a species, as if, in the recognition of their own physical and emotional ugliness, they pursue the acquisition of things to compensate for their lack of character, soul, or wisdom; the Vulcans are clean, well-groomed, and uniform in style and appearance, suggesting a minimalism appropriate for a species so driven by logic, one which would wish to avoid the vain superfluities of fashion or style; the Borg, most interestingly, are equipped with a wide variety of technological attachments—mechanical hands and arms, computerized eyes, tools designed for specific applications—much like the cell-phones, laptops, Blackberries, and other technological ubiquities that we ourselves use. An over-reliance on integrated technology, the message seems to be, has the potential for making us mindless and susceptible to central control. In the end, we rarely meet anything or anyone in *Star Trek* that we don't already clearly recognize as part of ourselves.

[4] As another example, in a related story on July 2nd of 2004 titled, "Noxious Alien Weed Flourishes in the South," the Associated Press introduces the problem of an invasive Brazilian weed in Texas this way:

> One biologist compares the persistent green weed to 'The Blob,' the title character in the 1950s sci-fi classic flick that grows and grows and consumes everything in its path. Other scientists describe the plant as looking like little heads of lettuce or squished green grapes. Then they use terms like noxious, invasive and just plain scary. Even the species name sounds sinister: *salvinia molesta*.

[5] For a detailed discussion of the relationship between science fiction and our cultural fascination with Mars, see Robert Markley's *Dying Planet: Mars in Science and the Imagination*. Durham, NC: Duke UP, 2005.

Notes to Chapter Seven

[1] The Japanese applied similar thinking to submarines and fighter aircraft, and speed and stealth were also highly regarded design preferences. Japan was the first to produce a single-man submarine, which was produced for its low profile and high maneuverability. There was also a "kamikaze" version of this, essentially a manned, submersible torpedo that could be laid on target with great precision (being, perhaps, the first truly "smart" torpedo, unfortunately for its pilot). This, too, is a design option which reflects the Imperial attitude of individual sacrifice for the glory of the Emperor. The famous Mitsubishi "Zero" fighter, while a fine aircraft in nearly all respects, had an Achilles heel that was exploited by the Allies. In an effort to save weight and materials, it had no armor to protect the pilot. The pilot had to

rely on the enhanced maneuverability and speed this choice afforded the aircraft. It also, for the same reasons, had no self-sealing fuel tanks, which meant a direct hit on a fuel tank almost inevitably led to a victory for the Allied marksman. Again, the pilot, while certainly valuable as a trained flyer, was expendable in the face of the greater good of the Empire, which was unwilling (and unable) to commit the resources and expense of these measures to their aircraft design for the benefit of a single man.

[2] This same imagistic/ideological representation can be seen in nearly all objects that have some cultural influence and/or function. Sally Gregory Kohlstedt has noted a similar process of meaning-construction in not just a single object, but in entire museums. Kohlstedt argues that the museum administrators of the past who arranged objects for public display were "intent on attaining mastery over the objects that constituted their museums" and in reflecting the principles and ideological details of the culture from which they arise (586). More importantly, museum arrangement in Western societies is a decidedly empirical, even positivistic enterprise that determines rationalistic classification when making decisions about both particular displays and the layout of museums as a whole. She concludes that "living museums . . . require that our histories tease out the ways that science becomes expressed in particular times and places and encourage multiple narratives that, if not objective or complete, carry the capacity to be read as simultaneously understandable" (601). Science, it seems, is not only manifest in the design of objects of technology, but in the arrangement of objects for public display, a cultural text that indicates both a reverence for the aesthetic qualities of a logically organized set of artifacts and reinforces the cultural orientation of the time and place of that arrangement. This is a fascinating example of how the lens of science can color the cultural refraction of objects from a different time and place, a process that invites much further scholarship in the area of scientific cultural studies.

[3] See also David J. Buller's *Adapting Minds: Evolutionary Psychology and the Persistent Quest for Human Nature.* Cambridge, MA: MIT Press, 2005 and Steven Pinker's *How the Mind Works.* New York: W.W. Norton and Co., 1997.

[4] One might assume from the last part of this statement that Burke is responding to the positivistic trends in analytic philosophy which he sees as inferior to—or, at least, less useful than—the more literary philosophical methods of Plato, Heidegger, or Nietzsche.

[5] However, in the January 15, 2006 edition of the *New York Times,* assumptions about the integrity of scientists is challenged. In a feature entitled "Trial and Error," David Dobbs reports that

> John Ioannidis, an epidemiologist, recently concluded that most articles published by biomedical journals are flat-out wrong. The sources of error, he found, are numerous:

> the small size of many studies, for instance, often leads to mistakes, as does the fact that emerging disciplines, which lately abound, may employ standards and methods that are still evolving. Finally, there is bias, which Ioannidis says he believes to be ubiquitous. Bias can take the form of a broadly held but dubious assumption, a partisan position in a longstanding debate (e.g., whether depression is mostly biological or environmental) or (especially slippery) a belief in a hypothesis that can blind a scientist to evidence contradicting it. These factors, Ioannidis argues, weigh especially heavily these days and together make it less than likely that any given published finding is true. (Dobbs)

Such questioning of scientific professionalism is becoming more common. *The Chronicle of Higher Education* reported on a study in which over one-third of some three thousand research scientists interviewed admitted to "fudging" data for a variety of reasons, most of which had to do with job stability and/or tenure (see "Scientific Misbehavior is Rampant, Study of 3000 Researchers Find" in the June 17, 2005 issue, pg. 11A). In Dobb's article, the suggestion is that peer-review is perhaps to blame or the lapses in ethics, but regardless of the cause, these episodes demonstrate the very real professional pressures that beset scientists just as they do the rest of professional world.

[6] For a more in-depth discussion on scientific reporting processes, see also Bruno Latour, *Science in Action.* Cambridge, MA: Harvard UP, 1987.

[7] The emphasis on visualization is also a function of the journalistic venue in which this idea appears. *Scientific American,* like all popular sources of scientific information, tends to package its features in visuals. While on the one hand this may be construed as a function of the contemporary preference for receiving information through imagistic means, it has also proven to be a relatively stable device for this publication. In issues that I reviewed for another project, most of which came from the early 1950s, visualization (especially in the form of graphs, tables, and artistic renderings) was still a large part of the format. The technology for delivering those visuals has certainly improved, thus making it easier and more cost-effective, but it is not a new phenomenon.

[8] In at least one 2000 study, bullying and self-esteem were seen as irrelevant associations. Jane L. Ireland concluded in that study that "[t]here were no significant difference among bully categories in self-esteem scores" (184). This suggests that the connection between bullying and self-esteem had either been manufactured or that new research had downplayed the relationship between these characteristics (see "Social self-esteem and self-reported bullying behaviour among adult prisoners," *Aggressive Behaviour,* 28.3 [2000]: 184–197). In a study in the *Scandinavian Journal of Psychology,* the "conclusion" was similarly inconclusive:

> Bullying was positively correlated with self-concept scores. However, this was true only of boys. According to cross-tabulations, there were significantly more bullies among children with learning difficulties (LD) than would have been expected by chance. Victimization, on the other hand, was not related to LD. LD children's proposed victim status was in some degree supported by cluster analysis: a group of LD children emerged, who not only scored high on bullying, but also tended to be victimized by others. In addition, two groups of bullies appeared: one whose members could be interpreted as socially unskilled and another as socially skilled. This finding is in line with recent theoretical reasoning, which calls into question the idea of bullies as a unified group, lacking in social skills. (Kaukiainen et al. 269)

Finally, in another study, this time in the journal *Physiology and Behavior,* the author uses a behavioralist methodology in the study of rats and pigs to focus on the emotional toll bullying takes on the victim, concluding that "Victims of bullying are known to suffer from depression, anxiety, sociophobia, loss of self-esteem, psychosomatic diseases, and other behavioral symptoms" (Bjorkgvist 435). Considering the range of focal points (prisoners, school bullies, victims) and the lack of a consistent methodology (physiological studies of other animals, self-reports, and qualitative research), it should come as little surprise that certain myths are generated about the relationship between bullying and self-esteem and that the conclusions drawn by psychologists are less than conclusive. In other words, "bullying," as a nebulous term used to describe a variety of human behaviors in a multitude of settings, is not well-contained as a point of psychological research. There are likely to be as many conclusions drawn as there are studies conducted about the phenomenon of bullying. The question then becomes how complicit the psychological community itself is in the perpetuation of these "popular perceptions" about bullies.

[9] Also of interest for the credibility test here is that this book, along with several others published by Holstein, is "self-published"—meaning that it has gone through no editing or review process by a reputable publisher, much less by peers in the psychological community. It was published through AuthorHouse, which describes itself as "the leading self-publishing company in the world, [which] has helped authors achieve their book publishing goals with over 27,000 books in print." Furthermore, AuthorHouse provides virtually unlimited authorial freedom by allowing the author to "determine how many and when copies of your published book are printed and you select your own royalty schedule. When your book is finished, it's available for order at more than 25,000 retail outlets worldwide, on the Internet at Amazon.

com, BarnesandNoble.com, and through the AuthorHouse online publishing company book store" (AuthorHouse). Typically, the *number* of books a publisher publishes is not nearly as important as the quality of the titles and of the authors. Self-publication is becoming more common as publishing costs through technological advances is providing options that were not available using traditional publishing procedures. Significantly, it also means that public must be considerably more keenly aware of source credibility.

[10] Alcoholics Anonymous demonstrates the power of discursive control to forge symbolic acts of contrition and surrender to the larger group. Paul Antze notes that the success of AA is largely contingent upon the defining characteristics that it lends to alcohol, alcoholics, and the group:

> There is an impressive feat of persuasion implicit [in AA] as well. Somehow, although it takes a point of departure from an extremely narrow concern, AA has enlisted the allegiance of its members to an elaborate symbolic order. The group seems to perform the major social and psychological offices of religion with considerable success: it gives identity and a sense of purpose to members, it schools them in the general conduct of their lives, it binds them to the community. (151)

AA becomes, then, not simply an instrument for aiding the drinker in not drinking; it provides a symbolic attachment to a "higher power," the group, and the larger community. In doing so, it controls the discourse, allowing members to discuss their alcoholism only in predetermined and acceptable ways. It maintains the logic that part of changing the person is changing the language one uses to frame his or her problems. The same can be said of most pop-psych therapists.

[11] Note the use of the first name behind the authoritative title, a rhetorical move that creates a sense of personability and professionalism; "Dr. Laura" Schlessinger also uses this tactic. Her terminal degree, incidentally, is in physiology, not psychology, but the fields are close enough in spelling to hoodwink the uninitiated. She also has an impressive line of merchandise for sale, ranging from t-shirts to posters. One of her books, titled *The Proper Care and Feeding of Husbands,* would be banned by feminists groups if the word "wife" were substituted for "husband," but in the present state of our social prioritizing, men don't have the luxury of being offended. It's supposed to be funny. Her "companion book," *Woman Power: Transform Your Man, Your Marriage, Your Life* is equally insulting, embracing as it does the a concept at least as old as Chaucer's "Wife of Bath Tale," that what women really want is control, and the best way to gain it by changing your man (see http://www.drlaurashop.com/). This dangerous philosophy, I suspect, has caused far more harm in relationships than good. Unfortunately, Dr. Laura can get

away with this because the "masculinist" movement is laughed at by the public at large. That my spell-checker identified this word as misspelled should be an indicator of something.

Notes to Chapter Eight

[1]It is interesting to note that the Hebrew concept of God through the lens of the Old Testament is not at all how we have come to think of Him. He is, in scripture, the God of the Jews, suggesting that there are in fact other gods which believers "will not place before" the Jewish god. By the time of the New Testament, God is not really in the picture at all, but is replaced by a man who claims to have privileged access to Him. The conflation of these distinct gods, oddly, never seems to phase fundamentalists who, if they were to read the scripture of both testaments "literally," would have some serious questions about this obvious inconsistency.

[2]The Second Law of Thermodynamics is far more complex than the Creationist author wants to reveal. It says that "*in every neighborhood of an adiabatically isolated system there are states inaccessible from G.* (This formulation of the Second Law is known as the *Principle of Caratheodory.*)" (DePree and Axelrod 730, emphasis original).

Notes to Chapter Nine

[1] Anyone who has taught freshmen, as I have for my entire professional life, understands the difficulty people have with making sense of large bodies of information. The Internet, in this respect, has become a great equalizer, much to the chagrin of professors everywhere, because the phrase "all things being equal" is taken literally by students. My students find it very difficult to distinguish between credible sources and those that are merely "opinions" (a favorite word of theirs), sources written by anyone without support or evidentiary credence. This says nothing about sources that seem superficially credible, but in fact engage in subtle levels of rhetorical presentation and outright deceit to further a specific agenda (the White House homepage is a perfect example). This suggests that the general public, while no stranger to information, must have more advanced critical tools to unpack, decode, and ultimately assess the validity of the information they do receive.

[2] Rather than use Burke's pentadic system, which involves an analysis of the act, scene, agent, agency, and purpose and the alignment of the "ratios" that subsequently emerge to ultimately define the motive of a dramatistic environment, my modification uses a septadic process and adds context, purpose, information, and includes motive in the analysis itself rather than as the end goal.

[3] It is interesting to note that during the other great shuttle disaster, *Challenger,* the press routinely referred to the accident as an "explosion," when, in fact, the o-ring that failed on the engine nacelle caused the shuttle to fly apart in much the same way that Columbia is presumed to. For this disaster, we had the luxury of images (played on news broadcasts *ad nauseum*) that seemed to support this description. However, it is clear on close examination of the Challenger footage that although the o-ring failed, producing fire from the starboard engine, it did not cause the shuttle to explode. The insistence of the media to use the verb "explode" was an obvious way of dramatizing the incident, even when visual footage, if closely scrutinized, contradicted this description. This is a testament to the willingness of the public to accept certain explanations over others even in the presence of contradictory (in this case, visual) evidence. It further testifies to the media's power to imprint an image and attach a word in our minds and our willingness to accept it as "factual" or "truthful" when, of course, it is neither.

[4]For further discussion of the *Challenger* accident, see Diane Vaughan's comprehensive report, *The Challenger Launch Decision: Risky Technology, Culture, and Deviance at NASA.* Chicago: U of Chicago P, 1996.

[5]Another public "issue" (insofar as it can be viewed an issue for the general public) that might benefit from residual field analysis should be an almost entirely scientific question: whether or not Pluto should or should not be considered a planet. The latest verdict on this question is that, based on the astronomically-determined criteria needed to classify a celestial body a "planet," Pluto simply does not qualify. MSNBC recently reported that Pluto had been "demoted" to the status of "dwarf planet" (written by senior science writer Robert Roy Britt). The summary of the "event" was articulated this way by MSNBC: "Capping years of intense debate, astronomers resolved Thursday to demote Pluto in a wholesale redefinition of planethood that is being billed as a victory of scientific reasoning over historic and cultural influences. But the decision is already being hotly debated" ("Scientists Decide Pluto's No Longer a Planet"). What prompted the reemergence of this debate, apparently, was the question of whether or not astronomer Mike Brown's recent celestial discovery, 2003 UB313, could be considered a planet because it is in fact larger than Pluto (Roy, "Object Bigger than Pluto Discovered, Called 10th Planet"). Interestingly, the demotion of Pluto to a planetoid "dwarf planet" is, contrary to MSNBC's report, *all about* cultural influences: Because of the discovery, the question of what defines a planet cannot be considered a "victory of scientific reasoning" on the simple basis that it was not motivated by scientific purity, but by how much honor Brown was entitled to as the astronomer who had discovered the "tenth planet." If we examine the cultural, professional, and rhetorical purpose of this definitional question in particular, we see that it has more to do with the relative status of the *scientist* than with the planetary rank of Pluto. Finally, the level of public

interest surrounding this decision is unexpected; many people (when polled using the question "do *you* think Pluto should be classified as a planet?") responded with a resounding "yes" because, they argued, that's what they learned in school and were "used to." The idea of reducing the number of planets populating the solar system to eight instead of nine had the effect of challenging received wisdom, and on that basis, much of the general public wanted the classification to remain intact. But the ultimate question of what all this means for science is rather baffling. Certainly, scientists will find the definition useful for future discoveries, but none of this effects either the reality of Pluto's existence nor its innate status as an object orbiting our sun. In sum, Pluto's designation in the eyes of earthlings has changed its *actual* position in the celestial lineup not one whit as it orbits our sun blissfully unaware of how much attention it is receiving.

Notes to Chapter Ten

[1] Sokal's title's attempt at irony is less than subtle. Clearly, the boundary that was transgressed was the untouchable bulwark of science.

[2] I am not aware of this as an actual field of study; sociologists are, by their own definition, already scientists. Fish may have been attempting to simplify an academic distinction, or he may be aware of a subfield that is not generally acknowledged. There are, however, sociologists like Maurice N. Richter, Jr. who have made claims for the cultural dimension of scientific activity. This idea has been around a long time. Richter's book, *Science as a Cultural Process,* was published in 1972.

[3] Millennia later, the shift away from truth and toward meaning became complete, philosophically speaking, with the rise of Postmodern attention to the nature of interpretive complications, particularly in texts that had been traditionally viewed as sources of Truth and Beauty. It is no accident, therefore, that the shifting nature of language has become a much more centralized focus of inquiry among philosophers, rhetoricians, and literary critics in the last half century. In fact, for rhetoricians, language has always been the central component to understanding. What has changed for them is a departure from the ancient tradition of merely describing the effective use of language (*rhetorica utens*) toward a more hermeneutical treatment of the nature of language itself. Truth, in the Platonic sense, became less important for many language scholars because it was becoming increasingly evident (through the efforts of Derrida, Foucault, Lyotard, Gadamer, Barthes, etc.) that the correspondence between language and Truth was a tenuous one at best. In its starkest form, Postmodern thinking denied not only absolute Truth, but any stable meaning whatsoever; in its more utilitarian (and ultimately useful) form, it emphasized the inherent ambiguity of language and the multiple interpretations, associations, identifications, and meanings that could be derived from its use as a symbolic activity.

Notes to Chapter Eleven

[1] As evidence for the claim that a scientistic orientation dominates the field of education, usually at the expense of other epistemological approaches like qualitative research or simple judgment, experience, and expertise, consider this statement from the US Department of Education, released on January 25th, 2005:

> DEPARTMENT OF EDUCATION
>
> RIN 1890–ZA00
>
> Scientifically Based Evaluation Methods
>
> **AGENCY:** Department of Education. **ACTION:** Notice of final priority.
>
> **SUMMARY:** The Secretary of Education announces a priority that may be used for any appropriate programs in the Department of Education (Department) in FY 2005 and in later years. We take this action to focus Federal financial assistance on expanding the number of programs and projects Department-wide that are evaluated under rigorous scientifically based research methods in accordance with the Elementary and Secondary Education Act of 1965 (ESEA), as reauthorized by the No Child Left Behind Act of 2001 (NCLB). The definition of scientifically based research in section 9201(37) of NCLB includes other research designs in addition to the random assignment and quasi-experimental designs that are the subject of this priority. However, the Secretary considers random assignment and quasi-experimental designs to be the most rigorous methods to address the question of project effectiveness. While this action is of particular importance for programs authorized by NCLB, it is also an important tool for other programs and, for this reason, is being established for all Department programs. Establishing the priority on a Department-wide basis will permit any office to use the priority for a program for which it is appropriate. (1)

The message here is clear: School districts and researchers in education must employ a scientifically-based mode of evaluation or lose out on federal funds. "Quasi-experimental designs" are considered the "most rigorous," and there is little question that the drafters of this policy have no inhibitions about furthering the notion that only scientific data pass for acceptable knowledge.

All other forms, it is implied, "lack rigor" and are therefore not "real" knowledge. The ramifications of this policy are equally clear: All students will be evaluated on a quantitative basis, and therefore, all curricula will mirror how best to accomplish the goals set forth in a quantitative assessment. That is, if students are expected to perform well on quantitative tests, a quantitative pedagogy of "skill, drill, and kill" will dominate the classroom. All other forms of knowledge and learning—including creativity, critical thinking, independent inquiry and collaborative learning—are considered irrelevant and ineffective. This is the policy that will determine the nature of our democracy in years to come.

[2] For further discussion on the nature of early Greek and Roman writing practices and pedagogy, see James J. Murphy's *A Short History of Writing Instruction,* 2nd edition, Hermagoras Press, 2001.

[3] Ironically, MacDonald would probably fail her essay in many Freshman Composition courses because it is not properly supported or well-argued, but rather based on vague impressions and unsubstantiated assumptions. It is, however, grammatically sound.

[4] See, for example, Julie DeGalan and Stephen Lambert, *Great Jobs for English Majors,* NTC Publishing Corp., 1996. To site a few examples of jobs available for English graduates, students can seek careers in writing, editing, and publishing; advertising and public relations; business administration and management; technical writing; and, of course, teaching.

[5] In a recent, controversial decision at several school districts in New Jersey, for example, those students who finished high school as valedictorians were rewarded not with a scholarship fund, but with a new Ford Mustang.

[6] This not the same idea as Deconstruction, the much maligned "postmodern" analytical "system" attributed to Jacques Derrida that says, in essence, that all words merely point to other words, that a text has no stable meaning, and that whatever content a reader can extract from a text merely seeps into other texts, forever. This extreme form of literary dissection was an important milestone in the reconsideration of the nature of language, but it was overstated and has generally given way to more moderate views of texts with most serious students of language, literature, and rhetoric today.

[7] As an interesting side note, George W. Bush, in his 2006 State of the Union Address, made a rare nod towards the importance of education (No Child Left Behind doesn't count because the program is not about "education" but "accountability"), but only in terms of how it translated into productivity and competition with the world. He was interested in increased funding and support for "math and science" only, again suggesting that the nation's interest in education is in quantifiable subjects that can help aid the country's economic stability and global dominance. Conservative legislators are far less interested in humanistic learning and liberal arts education because the content of courses in these areas might actually push the public to think for itself.

Works Cited

Adams, John Charles. Rev. of *A Rhetoric of Science: Inventing Scientific Discourse,* by Lawrence J. Prelli. *Rhetoric Review* 9.2 (Spr. 1991): 369-72.

Alterman, Eric. *What Liberal Media? The Truth About Bias and the News.* NewYork: Basic Books, 2003.

Antze, Paul. "Symbolic Action in Alcoholics Anonymous." *Constructive Drinking.* Mary Douglas, ed. Cambridge: Cambridge UP, 1987.

Aristotle. *On Rhetoric: A Theory of Civic Discourse.* Trans. George A. Kennedy. Oxford: Oxford UP, 1991.

Armstrong, Karen. *The Battle for God.* New York: Ballantine Books, 2000.

"Asteroid Data Sheet." *Space.com.* 2007. 5 Dec. 2007 <http://www.space.com/scienceastronomy/solarsystem/asteroids-ez.html>.

AT&T. "A Brief History: The Bell System." 2007. 1 June 2007. <http://www.att.com/history/history3.html>

AuthorHouse. 2006. Author Solutions Inc. 7 Dec. 2007 <http://www.authorhouse.com/>.

Barrett, Stephen. "Endorsements Don't Guarantee Reliability." *Quackwatch.* 13 February 2003. 8 Dec. 2007 <http://www.quackwatch.org/01QuackeryRelatedTopics/endorsements.html>.

Barrett, Stephen, and William T. Jarvis. *The Health Robbers: A Close Look at Quackery in America.* New Jersey: Prometheus Books, 1993.

Baumeister, Roy F. "Exploding the Self-Esteem Myth." *Scientific American.* January, 2005. 84–96.

Bazerman, Charles. "Intertextuality: How Texts Rely on Other Texts." *What Writing Does and How It Does It: An Introduction to Analyzing Texts and Textual Practices.* Mahwah, NJ: Erlbaum, 2004. 83–97.

Becker, Barbara Holstein. "Books by Dr. Barbara." 2008. 2 May 2008 <http://www.enchantedself.com/books-by-dr-barbara-holstein.html>.

Beale, Walter H. *A Pragmatic Theory of Rhetoric.* Carbondale: Southern Illinois UP, 1987.

Berger, Peter L., and Thomas Luckmann. *The Social Construction of Reality.* New York: Anchor Books, 1967.

Berlin, James A. "Rhetoric and Ideology in the Writing Class." *Teaching Composition: Back-Ground Readings.* Ed. T. R. Johnson and Shirley Morahan. New York: Bedford/St. Martin's, 2002. 53–73.

—. *Rhetoric and Reality: Writing Instruction in American Colleges, 1900–1985.* Carbondale: Southern Illinois UP, 1987.

Berthoff, Ann E. ed. *Richards on Rhetoric: Selected Essays 1929–1974.* Oxford: Oxford UP, 1991.

Bizzell, Patricia, and Bruce Herzberg. *The Rhetorical Tradition: Readings from Classical Times to the Present.*" 2nd ed. Boston: Bedford/St. Martin's, 2001.

Bjorkgvist, Kaj. "Social Defeat as a Stessor in Humans." *Physiology and Behavior.* 73 (2001): 435–42.

Blakesley, David. *The Elements of Dramatism.* New York: Longman, 2002.

Blankenship, Jane, and Janette Kenner Muir. "On Imaging the Future: The Secular Search for 'Piety'." *Communication Quarterly* 35 (Winter 1987): 1-12.

Blettner, Maria, et al. "Mortality from Cancer and Other Causes among Airline Cabin Attendants In Germany, 1960–1997." *American Journal of Epidemiology.* 156.6 (2001): 556–65.

Bloor, David. *Knowledge and Social Imagery.* 2nd ed. Chicago: U of Chicago P, 1991.

Booth, Wayne C. *Modern Dogma and the Rhetoric of Assent.* Chicago: U of Chicago P, 1974.

Bratich, Jack Z. "Amassing the Multitude: Revisiting Early Audience Studies." *Communication Theory.* 15.3 (Aug. 2005): 242–65.

Bridis, Ted. "Columbia Probe Reveals Memo." Associated Press. 11 March 2003.

Britt, Robert Roy. "Object Bigger than Pluto Discovered, Called 10th Planet." *Space.com.* 29 July 2005. 8 Dec. 2007 <http://www.space.com/scienceastronomy/050729_new_planet.html>.

—. "Scientists Decide Pluto's No Longer a Planet." *MSNBC Online.* 24 Aug. 2007. 8 Dec. 2007 <http://www.msnbc.msn.com/id/14489259/>.

Brooker, Peter, ed. *Modernism/Postmodernism.* New York: Longman, 1992.

Brooks, Rodney A. *Flesh and Machines: How Robots Will Change Us.* New York: Pantheon Books, 2002.

Brown, Kathryn S. "Charisma, Content Make for Effective Scientific Presentations." *The Scientist.* 11.20 (13 Oct. 1997). 7 Dec. 2007 <http://www.the-scientist.com/1997/10/13/14/1>.

Buller, David J. *Adapting Minds: Evolutionary Psychology and the Persistent Quest for Human Nature.* Cambridge, MA: The MIT Press, 2005.

Burhop, E. H. S. "Scientists and Public Affairs." *Society and Science.* Ed. Maurice Goldsmith and Alan Macay. New York: Simon and Schuster, 1964.

Burton, Gideon. "The Forest of Rhetoric." *Silva Rhetoricae*. 2007. 6 Dec 2007 <http://humanities.byu.edu/rhetoric/silva.htm>.

Burke, Kenneth. *Counter-Statement*. 1931. 2nd ed. Berkeley: U of California P, 1968.

—. *Permanence and Change: An Anatomy of Purpose*. 1935. 3rd ed. Berkeley: U of California P, 1984.

—. *Attitudes Toward History*. 1937. 3rd ed. Berkeley: U of California P, 1984.

—. *The Philosophy of Literary Form*. 1941. 3rd ed. Berkeley: U of California P, 1973.

—. *A Grammar of Motives*. 1945. Berkeley: U of California P, 1969.

—. *A Rhetoric of Motives*. 1950 Rpt. Berkeley: U of California P, 1969.

—. *The Rhetoric of Religion: Studies in Logology*. 1961. Berkeley: U of California P, 1970.

—. *Language as Symbolic Action: Essays on Life, Literature, and Method*. Berkeley: U of California P, 1966.

"Bussard Ramjet." Davidszondy.com.2007. 8 Dec. 2007 <http://davidszondy.com/future/space/bussard.htm>.

Cameron, James, and Steven Quale. *Aliens of the Deep*. Imax film. 2005.

Campbell, Jeremy. *The Liar's Tail: A History of Falsehood*. New York: Norton, 2001.

Campbell, John Angus. "Charles Darwin: Rhetorician of Science." *The Rhetoric of the Human Sciences*. Ed. John S. Nelson, Allan Megill, and Donald N. McCloskey. Madison: U of Wisconsin P, 1987. 69–86.

Carlston, Donal E. "Turning Psychology on Itself: The Rhetoric of Psychology and the Psychology of Rhetoric." *The Rhetoric of the Human Sciences: Language and Argument in Scholarship and Public Affairs*. Ed. John S. Nelson, et al. Madison: U of Wisconsin P, 1987. 145-162

Cazenave, Anny. "How Fast Are the Ice Sheets Melting?" *Science*. 314.5803 (24 Nov. 2006): 1250-1252.

Cherry, R.D. "Ethos Versus Persona: Self-Representation in Written Discourse." *Written Communication*. 15.3 (July 1, 1998): 384–410.

Cherwitz, Richard ed. *Rhetoric and Philosophy*. Hillsdale, NJ: Erlbaum, 1990.

Churchman, Deborah. "Voices of the Academy: Academics' Responses to a Corporatised University." n.d. 2 May 2008 <http://web.archive.org/web/20040111144851/http://www.esib.org/commodification/documents/Churchman.pdf>

Cline, Austin. "Evolution and Creation in Public Schools." *About.com*. 2007. 7 Dec. 2007 <http://atheism.about.com/library/decisions/indexes/bldec_CreationismIndex.htm>.

Collins, Harry, and Trevor Pinch. *The Golem at Large: What You Should Know About Technology*. Cambridge: Cambridge UP, 1998.

Connolly, William E. *The Terms of Political Discourse.* 2nd ed. Princeton, NJ: Princeton UP, 1983.

"The Crafty Attacks on Evolution." *New York Times.* Editorial. 23 Jan. 2005. 7 Dec. 2007 <http://www.nytimes.com/2005/01/23/opinion/23sun1.html?pagewanted=1>.

Crusius, Timothy. *Kenneth Burke and the Conversation After Philosophy.* Carbondale: Southern Illinois UP, 1999.

Mann, Adrian. "Daedalus Arrives at Barnards Star." *This Is Rocket Science.* 2002. 8 Dec. 2007 <http://www.aemann.pwp.blueyonder.co.uk/prints/daedalusa.html>.

Dampier, William C., and Margaret Dampier. *Readings in the Literature of Science.* New York: Harper & Brothers, 1959.

Darrow, Karl K. "The Quantum Theory." *Scientific American.* 186.3 (March 1952): 47–54.

Darwin, Charles. Excepts from *The Descent of Man. Readings in the Literature of Science.* Ed. William C. Dampier and Margaret Dampier. New York: Harper & Bothers, 1959.

Davis, Erik. "Connect This! James Burke's *Connections2.*" 5 Dec 2007 <http://www.techgnosis.com/burke.html>.

"Deep Impact Scores Bull's-Eye." Associated Press. 4 July 2005. 2 May 2008 <http://www.wired.com/science/space/news/2005/07/68085>.

Defending the Teaching of Evolution in the Public Schools. National Center for Science Education. 2007. 8 Dec. 2007 <http://www.natcenscied.org/>.

Dennett, Daniel C. *Darwin's Dangerous Idea: Evolution and the Meaning of Life.* New York: Simon and Schuster, 1995.

—. "The Mythical Threat of Genetic Determinism." *The Chronicle of Higher Education.* 49.12 (31 Jan. 2003): B7-9.

"Declining By Degrees: Higher Education at Risk." PBS Home Video, 2005.

DeGalan, Julie, and Stephen Lambert. *Great Jobs for English Majors.* Lincolnwood, IL: NTC Publishing Group, 1996.

DePree, Christopher G., and Alan Axelrod, eds. *Van Nostrand's Concise Encyclopedia of Science.* Hoboken, NJ: John Wiley and Sons, Inc., 2003.

"Detailed Biography." *The Edison Papers.* Rutgers University. 20 August 2007. <http://edison.rutgers.edu/bio-long.htm>

Connections. "The Trigger Effect." The Discovery Science Channel. 12 March 2004. Mick Jackson, Writer. 1st aired BBC, 1978.

"Do You Believe Evolution to Be True?" *Creation Science Homepage.* 7 Dec. 2007 <http://emporium.turnpike.net/C/cs/quest.htm>.

Dobbins, James H. "Planning for Technology Transition." *Defense AT&L.* March/April 2004, 14–17.

Dobbs, David. "Trial and Error." *New York Times.* 15 January 2006. 7 Dec. 2007 <http://www.nytimes.com/2006/01/15/magazine/15wwln_idealab.html?_r=1&emc=eta1&oref=slogin>.

Dr. Phil.com. "About Dr. Phil." 2007. Peteski Productions Inc. 7 Dec. 2007 <http://www.drphil.com/shows/page/bio/>.

"Drake's Equation for Extraterrestrial Life." Homo Excelsior: The Trans Humanist Wiki. 13 Jan 2007. 7 Dec 2007 <http://www.homoexcelsior.com/Drake%27s_Equation>.

"Dimming the Sun." *Nova.* Dir. Duncan Copp. PBS. 18 Apr. 2006. WBGH Boston. Produced by David H. Koch. <http://www.pbs.org/wgbh/nova/sun/>.

Dyck, Joachim. "Rhetoric and Psychoanalysis." *Rhetoric Society Quarterly* 19.2 (Spr. 1989): 95-104.

"Earth." *Global Warming Kids Site.* United States Environmental Protection Agency. 23 Oct. 2006. 9 Sept 2007 <http://www.epa.gov/globalwarming/kids/greenhouse.html>.

Eaves, Lindon. "Spirit, Method, and Content in Science and Religion: The Theological Perspectives of a Geneticist." Zygon 24.2 (Jun. 01, 1989): 185+.

Eagleton, Terry. *Literary Theory: An Introduction.* 2nd ed. Minneapolis: U of Minnesota P, 1996

Ellul, Jacques. *Propaganda: The Formation of Men's Attitudes.* Trans. Konrad Keelen and Jean Lerner. New York: Knopf, 1968.

—. *The Technological Society.* Trans. John Wilkinson. Foreword Robert K. Merton. New York: Random House, 1964.

Enstrom, James E., and Geoffrey C. Kabat. "Environmental Tobacco Smoke and Tobacco Related Mortality in a Prospective Study of Californians, 1960–1998." *BMJ.* 326 (May 17, 2003): 1–10.

"The *Ethos* of Rhetoric." University of South Carolina Press. 10 March 2006. 6 Dec 2007 <http://www.sc.edu/uscpress/>. Path: Complete Catalog. .

Eubanks, Philip. "Poetics and Narrativity: How Texts Tell Stories." *What Writing Does and How It Does It.* Ed. Charles Bazerman and Paul Prior. Mahwah, NJ: Erlbaum, 2004.

Fahnestock, Jeanne. *Rhetorical Figures in Science.* Oxford: Oxford UP, 1999.

—. "Arguing in Different Forums: The Bering Crossover Controversy." *Landmark Essays on the Rhetoric of Science.* Ed. Randy Allen Harris. Mahwah, NJ: Hermagoras Press, 1997.

"fatman_littleboy." National Atomic Museum. 2003. 8 Dec. 2007 <http://www.atomicmuseum.com/tour/photos/fatman_littleboy.jpg>.

Feyerabend, Paul. *Knowledge Without Foundations: Two Lectures Delivered on the Nellie Heldt Lecture Fund.* Oberlin College, 1961.

—. *Against Method: Outline of an Anarchistic Theory of Knowledge.* Atlantic Highlands, NJ: Humanities Press, 1975.

—. Science in a Free Society. London: NLB, 1978.

—. *Three Dialogues on Knowledge.* Cambridge, MA: Blackwell Publishers, 1991.

Feyerabend, Paul, and Grover Maxwell, eds. *Mind, Matter, and Method: Essays in Philosophy and Science in Honor of Herbert Feigl.* Minneapolis: U of Minnesota P, 1966.

Fish, Stanley. "Professor Sokal's Bad Joke." *The New York Times.* 21 May 1996, p. A23. New York University Department of Physics. 8 Dec. 2007 <http://physics.nyu.edu/~as2/fish.html>.

Fleck, Ludwik. *Genesis and Development of a Scientific Fact.* Chicago: U of Chicago P, 1979.

Folland, Sherman, et al, *The Economics of Health and Health Care.* 4th ed. New York: Prentice-Hall, 2003.

Forbes, R .J., and E. J. Dijksterhuis. *A History of Science and Technology, Vol. 2: Nature Obeyed and Conquered, The Eighteenth and Nineteenth Centuries.* Baltimore, Penguin, 1963.

Foss, Sonja K. *Rhetorical Criticism: Exploration and Practice.* 2nd ed. Prospect Heights, IL: Waveland Press, Inc., 1996.

Foucault, Michel. *Discipline and Punish: The Birth of the Prison.* trans. Alan Sheridan. New York: Vintage Books, 1979.

—. *The Order of Things: An Archaeology of the Human Sciences.* New York: Vintage Books, 1973.

Frankfurt, Harry G. *On Bullshit.* Princeton, NJ: Princeton UP, 2005.

Frobish, Todd S. "Alter Rhetoric and Online Performance: Scientology, Ethos, and the World Wide Web." *American Communication Journal.* 4.1 (Fall 2000). 8 Dec. 2007 <http://acjournal.org/holdings/vol4/iss1/articles/frobish.htm>.

Fuller, Steve. "'Rhetoric of Science': A Doubly Vexed Expression." *Southern Communication Journal* 58.4 (Sum. 1993): 306-11.

Gaonkar, Dilip Parameshwar. "The Idea of Rhetoric in the Rhetoric of Science." *Southern Communication Journal* 58.4 (Sum. 1993): 258-95.

Gardner, Howard. "Low Scores Are No Disgrace." *The New York Times.* 2 March 1998. New Century School. 7 Dec. 2007 <http://danenet.wicip.org/ncs/forumgardner.htm>.

Glassner, Barry. *The Culture of Fear: Why Americans Are Afraid of the Wrong Things.* New York: Basic Books, 1999.

Glickman, Lawrence B. "Born to Shop? Consumer History and American History." *Consumer Society in American History: A Reader.* Ithaca, NY: Cornell UP, 1999. 1–16.

Goldberg, Steven. "The Erosion of the Social Sciences." *Dumbing Down: Essays on the Strip Mining of American Culture.* New York: Norton, 1996.

Gould, Steven Jay. "Nonmoral Nature." *A World of Ideas: Essential Readings for College Writers.* 7th ed. Ed. Lee A. Jacobus. New York: Bedford/St. Martins, 2006. 597–613.

—. *Rocks of Ages: Science and Religion in the Fullness of Life.* New York: Ballantine Books, 1999.

—. *The Structure of Evolutionary Theory.* Cambridge, MA: The Belknap Press of Harvard UP, 2002.

Grafton, Anthony. "Magic and Technology in Early Modern Europe." Dibner Library Lecture. 15 October, 2002. Washington DC: Smithsonian Institution Libraries, 2005.

Grassi, Ernesto. *Rhetoric as Philosophy: The Humanist Tradition.* Trans. John Michael Krois and Azizeh Azodi. Carbondale: Southern Illinois UP, 2001.

Greene, Brian. *The Elegant Universe: Superstrings, Hidden Dimensions, and the Quest for the Ultimate Theory.* New York: Vintage Books, 2000.

Greene, John C. "The Kuhnian Paradigm and the Darwinian Revolution in Natural History." *Perspectives in the History of Science and Technology.* Ed. Duane H. D. Roller. Norman: U of Oklahoma P, 1971. 3–25.

Gross, Alan G. "Does Rhetoric of Science Matter? The Case of the Floppy-Eared Rabbits." *College English* 53.8 (Dec. 1991): 933-41.

—. "Rhetoric of Science Is Epistemic Rhetoric." *Quarterly Journal of Speech* 76 (1990): 304-6.

—. "Rhetoric of Science without Constraints." *Rhetorica* 9.4 (Fall 1991): 283-99.

—. "What if We're Not Producing Knowledge? Critical Reflections on the Rhetorical Criticism of Science." *Southern Communication Journal* 58.4 (Sum. 1993): 301-5.

Gross, Alan G,. and Joseph E. Harmon. "What's Right About Scientific Writing." *The Scientist.* 13.24 (6 Dec 1999) 7 Dec 2007 <http://www.the-scientist.com/1999/12/06/20/1>.

Gross, Paul R,. and Norman Levitt. *Higher Superstition: The Academic Left and Its Quarrels With Science.* Baltimore: Johns Hopkins UP, 1994.

Grove, J. W. *In Defense of Science: Science, Technology, and Politics in Modern Society.* Toronto: U of Toronto P, 1989.

Harding, Sandra G. *Feminism and Methodology.* Bloomington, IN: Indiana UP, 1987.

—. *The Science Question in Feminism.* Ithaca, NY: Cornell UP, 1986.

—. *Whose Science? Whose Knowledge? Thinking From Women's Lives.* Ithaca, NY: Cornell UP, 1991.

Harris, Julia. Eisenhower National Clearinghouse for Mathematics and Science Education. n.d. 2 May 2008 <http://web.archive.org/web/20020210180925/http://www.enc.org/topics/assessment/testing/document.shtm?input=FOC-001576-index>.

Harris, Randy Allen, ed. *Landmark Essays on the Rhetoric of Science.* Mahwah, NJ: Hermagoras Press, 1997.

Haskins, William A. "Ethos and Pedagogical Communication: Suggestions for Enhancing Credibility in the Classroom." *Current Issues in Education.* 3.4 (2000). 6 Dec 2007 <http://cie.ed.asu.edu/volume3/number4/>.

Hawking, Stephen. *A Brief History of Time,* 10th Anniversary ed. New York: Bantam Books, 1996.

Hendrixson, Anne. "U.S. High School Social Studies Textbooks: Perpetuating the Idea of Over-Population." *Population and Development Program at Hampshire College.* 9 (Spring 2001). 23 August 2007 <http://popdev.hampshire.edu/projects/dt/pdfs/DifferenTakes_09.pdf>.

Holstein, Barbara Becker. "Practical Steps to Enchantment: Improving Your Self-Esteem." 2007. 7 Dec 2007 <http://www.more-selfesteem.com/improving_self_esteem_article.htm>.

Honner, John. "Science vs. Religion." *Commonweal* 121.16 (Sept. 23, 1994): 14+.

Horvat, Robert E. "The Science Education for Public Understanding Program: What's New with SEPUP." *Bulletin of Science, Technology, and Society.* 13.4 (1993): 208+.

Horwich, Paul, ed. *World Changes: Thomas Kuhn and the Nature of Science.* Cambridge, MA: The MIT Press, 1993.

"How the Fresh Air Machine Makes Your Air So Clean." *Fresh Air Machine.* 2006. 29 July 2007 <http://www.freshairmachine.com/air/howitworks.htm>.

Hübler, Mike. "The Drama of a Technological Society." *KB Journal* 1.2. Spring 2005. 1 June 2007 <http://kbjournal.org/node/60>.

Huckin, Thomas. "Content Analysis: What Texts Talk About." *What Writing Does and How It Does It: An Introduction to Analyzing Texts and Textual Practices.* Mahwah, NJ: Lawrence Erlbaum Associates, Publishers, 2004. 13–33.

Hunt, Everett. "The Rhetorical Mood of World War II." *Quarterly Journal of Speech* 29 (1943):1-5.

Hyde, Michael J. ed. *The Ethos of Rhetoric.* Columbia, SC: U of South Carolina P, 2004.

Ireland, Jane L. "Social Self-Esteem and Self-Reported Bullying Behaviour among Adult Prisoners." *Aggressive Behaviour.* 28.3 (2000): 184–97.

"James Burke." Royce Carlton, Inc. 2007. 11 Dec 2007 <http://www.roycecarlton.com/speakers/burke.html>.

Jansz, Jereon. "The Emotional Appeal of Violent Video Games for Adolescent Males." *Communication Theory* 15.3 (2005): 219-41.

Jenkins, Roger A., et al. "Environmental Tobacco Smoke in an Unrestricted Smoking Workplace: Area and Personal Exposure Monitoring." *Journal of Exposure Analysis and Environmental Epidemiology.* 11.5 (Sept./Oct. 2001): 369–80.

Jensen, Arthur R. "The Debunking of Scientific Fossils and Straw Persons." *Contemporary Educational Review.* 1.2 (1982). *The Debunker's Domain.* 8 Dec. 2007 <http://www.debunker.com/texts/jensen.html>.

Jones, W. T. *A History of Western Philosophy: Hobbes to Hume.* San Diego: Harcourt, 1969.

Journet, Debra. "Metaphor, Ambiguity, and Motive in Evolutionary Biology: W. D. Hamilton and the 'Gene's Point of View.'" *Written Communication.* 22.4 (October 1, 2005): 379–420.

Kalfus, Ken. "Last Night at the Planetarium." *Dumbing Down: Essays on the Strip Mining of American Culture.* Ed. Katherine Washburn and John F. Thornton. New York: Norton, 1996. 138–48.

Kaukiainen, Ari, et al. "Learning Difficulties, Social Intelligence, and Self-Concept: Connections to Bully-Victim Problems." *Scandinavian Journal of Psychology.* 43.3 (July 2002): 269–89.

Kennedy, George A. *Comparative Rhetoric: A Historical and Cross-Cultural Introduction.* Oxford: Oxford UP, 1998.

—. *A New History of Classical Rhetoric.* Princeton, NJ: Princeton UP, 1994.

Kimball, Roger. *Tenured Radicals: How Politics Has Corrupted Higher Education.* New York: Harper and Row, 1990.

Klein, Aaron. "Did God send Katrina as Judgment for Gaza?" *WorldNetDaily.com.* 7 Sept. 2005. 7 Dec. 2007 <http://www.worldnetdaily.com/news/article.asp?ARTICLE_ID=46178>.

Klein, Naomi. *No Logo.* New York: Picador Press, 2002

Klinkenborg, Verlyn. "Grasping the Depth of Time as a First Step in Understanding Evolution." *The New York Times.* 23 Aug. 2005. Op/Ed. Online. 7 Dec. 2007 <http://www.nytimes.com/2005/08/23/opinion/23tue3.html?ex=1125720000&en=91408d465c1b41bf&ei=5070&emc=eta1>.

Klotz, Irving M. "Postmodernist Rhetoric Does Not Change Fundamental Scientific Facts." *The Scientist.* 10.15 (July 22, 1992): 9–12.

Kohlstedt, Sally Gregory. "Thoughts in Things: Modernity, History, and North American Museums." *Isis.* 96.4 (December 2005): 585–602.

Kuhn, Thomas S. *The Structure of Scientific Revolutions.* 2nd ed. Chicago: U of Chicago P, 1970.

Kurdas, Chidem. "Accumulation and Technical Change: Marx Revisited." *Science and Society* 59.1 (Spring 1995): 52+.

Kurzman, Charles. "The Rhetoric of Science: Strategies for Logical Leaping." *Berkeley Journal of Sociology* 33 (1988): 131-58.

Kurzweil, Ray. *The Age of Spiritual Machines: When Computers Exceed Human Intelligence.* New York: Penguin, 1999.

Lakoff, George, and Mark Johnson. *Metaphors We Live By.* Chicago: U of Chicago P, 1980.

Latour, Bruno. *Science in Action.* Cambridge, MA: Harvard UP, 1987.

LeBlanc III, H. Paul. "Accountability and External Ethical Constraints in Academia." Conference Paper. 1996 Annual Meeting of the Illinois Speech and Theater Association. Oakbrook, IL. 19 Oct. 1996. UTSA Dept. of Communication. 30 July 2007 <http://communication.utsa.edu/leblanc/articles/art11.pdf>.

"Lee Smolin." *Edge: The Third Culture.* 2002. 7 Dec 2007 <http://www.edge.org/3rd_culture/bios/smolin.html>.

Livingston, David. "Space as a Popular National Goal." *Beyond Earth: The Future of Humans In Space.* Ed. Langston Morris and Kenneth Cox. Burlington, Ontario: Apogee Books, 2006.

Lyne, John, and Henry F. Howe. "'Puntuated Equilibria': Rhetorical Dynamics of a Scientific Controversy." *Landmark Essays on the Rhetoric of Science.* Mahwah, NJ: Hermagoras Press, 1997.

MacDonald, Heather. "Writing Down Together." *Dumbing Down: Essays on the Strip-Mining of American Culture.* New York: W.W. Norton, 1996. 88-96.

Malik, Kenan. *Man, Beast, and Zombie: What Science Can and Cannot Tell Us about Human Nature.* New Brunswick, NJ: Rutgers UP, 2002.

—. "Inventing allies in the sky." *New Statesman.* 19 Feb 2007. 5 Dec 2007 <http://www.newstatesman.com/200102190041.htm>.

Marano, Hana Estroff. "Big Bad Bully." *Psychology Today.* Sept/Oct. 1995. 7 Dec. 2007 <http://www.psychologytoday.com/articles/pto-19950901–000020.html>.

Margolis, Eric. "Tragedy on the Somme: A Second Balaclava." *World War I: Trenches on the Web.* 2002. 1 June 2007. <http://www.worldwar1.com/sfsomme.htm>

Markley, Robert. *Dying Planet: Mars in Science and the Imagination.* Durham, NC: Duke UP, 2005.

Martin, Harold. "The Aims of Harvard's General Education A." *College Composition and Communication.* 9 (1958): 87–90.

Martins, David S. "Compliance Rhetoric and the Impoverishment of Context." *Communication Theory.* 15.1 (Feb. 2005): 59–77.

McChesney, Robert W. *Corporate Media and the Threat to Democracy.* New York: Seven Stories Press, 1997.

—. *Rich Media,Poor Democracy: Communication Politics in Dubious Times.* New York: The New Press, 1999.

McCroskey, James C., and Thomas J. Young. "Ethos and Credibility: The Construct and Its Measurement after Three Decades." *Central States Speech Journal.* 32 (Spring 1981): 24–34.

McGuire, J. E., and Barbara Tuchanska. *Science Unfettered: A Philosophical Study in Socio-Historical Ontology.* Athens, OH: Ohio UP, 2000.

McGuire, J. E., and Trevor Melia. "The Rhetoric of the Radical Rhetoric of Science." *Rhetorica* 9.4 (Fall 1991): 301-16.

—. "Some Cautionary Strictures on the Writing of the Rhetoric of Science." *Rhetorica* 7.1 (Win. 1989): 206–15.

McMillen, Liz. "Scholars Who Study the Lab Say Their Work Has Been Distorted." *The Chronicle of Higher Education.* 28 June 1996. Section: Scholarship, A8.

Medhurst, Martin. "Religious Rhetoric and the *Ethos* of Democracy: A Case Study for the 2000 Presidential Campaign." *The Ethos of Rhetoric.* Columbia, SC: U of South Carolina P, 2004.

Miller, Carolyn. "Expertise and Agency: Transformations of *Ethos* in Human-Computer Interaction. *The Ethos of Rhetoric.* Ed. Michael J. Hyde. Columbia: U of South Carolina P, 197–218

Mirsky, Steve. "Check Those Figures. Hey, It's Playtime! Icons—You Can Too." *Scientific American.* Jan 2004. 18 Feb 2006 <http://www.sciam.com/article.cfm?articleID=000DE05C-EDF3-1FD3-A7EA83414B-7F012C>.

"The Model T." *The Henry Ford.* 1999. 8 Dec. 2007 <http://www.hfmgv.org/exhibits/showroom/1908/model.t.html>.

Monastersky, Richard. "The Number that's Devouring Science." *The Chronicle of Higher Education.* 14 October 2005. Section: Research, A12.

—. "Scientific Misbehavior Is Rampant, Study of 3000 Researchers Finds." *The Chronicle of Higher Education.* June 17, 2005: A11.

Monk, Ray. *Ludwig Wittgenstein: The Duty of Genius.* New York: Penguin, 1990.

Moore, James. "Science and Religion: Some Historical Perspectives, By John Hedley Brooke." *History of Science* 30.89 (Sept. 01, 1992): 311+.

Moorhouse, H. F. "The 'Work' Ethic and 'Leisure' Activity: The Hot Rod in Post-War America." *Consumer Society in American History: A Reader.* Ed. Lawrence B. Glickman. Ithaca, NY: Cornell UP, 1999.

"NASA Investigates Revolutionary Space Exploration Concepts." National Aeronautics and Space Administration. Press Release. 15 June 2005. 7 Dec 2007 <http://www.nasa.gov/home/hqnews/2005/jun/HQ_05_151_NIAC_Awards.html>.

"NSTA's Guiding Principles." National Science Teachers Association. 2007. 8 Dec. 2007 <http://www.nsta.org/about/overview.aspx>.

Nelkin, Dorothy. "The Science Wars: What Is at Stake?" *The Chronicle of Higher Education.* 26 July 1996. U of Washington. 23 Sept. 1996. 8 Dec. 2007 <http://weber.u.washington.edu/~jwalsh/sokal/nelkin.cg>.

—. "An Uneasy Relationship: The Tensions Between Medicine and the Media." *Research Policy.* 27.8 (Dec. 1998): 869–79.

Nelson, Cary and Stephen Watt. *Offices Hours: Activism and Change in the Academy.* New York: Routledge, 2004.

Nelson, John S. et al. "Rhetoric of Inquiry." *The Rhetoric of the Human Sciences: Language and Argument in Scholarship and Public Affairs.* Madison, WI: U of Wisconsin P, 1987.

Nye, David E. *America As Second Creation: Technology and Narratives of New Beginnings.* Cambridge, MA: MIT Press, 2004.

"Noxious Alien Weed Flourishes in the South." Associated Press. 6 July 2005. 2 May 2008 <http://web.archive.org/web/20050706020907/http://apnews.excite.com/article/20050702/D8B3IF481.html>.

O'Hear, Anthony. "Science and Religion." *The British Journal for the Philosophy of Science.* 44.3 (Sept. 01, 1993): 505+.

"Orion-class Nuclear Starship." *Encyclopedia Astronautica.* 2007. 8 Dec 2007 <http://www.astronautix.com/lvfam/orion.htm>.

Palmer, Tom. "The James Burke Fan Companion." Feb 2007. 5 Dec 2007 <http://www.palmersguide.com/jamesburke/jbc_newsletter_02.html>.

Park, Robert L. "Last Night at the Planetarium." *Dumbing Down: Essays on the Strip Mining of American Culture.* New York: Norton, 1996.

—. *Voodoo Science: The Road from Foolishness to Fraud.* Oxford: Oxford UP, 2000.

Penick, J. L., et al., eds. *The Politics of American Science: 1939 to the Present.* Chicago: Rand-McNally., 1965.

Perdue, Tim. "Drake's Equation Simulator." *Perdue, Inc.* n.d. 8 Dec. 2007 <http://www.perdue.net/drakesequation/>.

Perelman, Chaim. *The Realm of Rhetoric.* Notre Dame, IN: U of Notre Dame P, 1982.

Phillips, Adam. "A Mind Is a Terrible Thing to Measure." *The New York Times.* Editorials/Op-Ed. 26 Feb. 2006. 8 Dec. 2007 <http://www.nytimes.com/2006/02/26/opinion/26phillips.html>.

Pincus, Walter. "Funds for Atomic Bomb Research Cut from Spending Bill." *Washington Post.* Tuesday, November 23rd, 2004. A06.

Pinker, Steven. *The Language Instinct: How the Mind Creates Language.* New York: Harper-Collins, 1994.

—. *How the Mind Works.* New York: Norton, 1997.

Prelli, Lawrence J. "The Rhetorical Construction of Scientific Ethos." *Landmark Essays on the Rhetoric of Science.* Mahwah, NJ: Hermagoras Press, 1997.

—. "Rhetorical Logic and the Integration of Rhetoric and Science." *Communication Monographs* 57.4 (Dec., 1990): 315-21.

Quine, W. V. *Ontological Relativity and Other Essays.* New York: Columbia UP, 1969.

Rampton, Sheldon, and John Stauber. *Banana Republicans: How the Right Wing Is Turning America into a One-Party State.* New York: Tarcher/Penguin, 2004.

Robinson, B. A. "Public Beliefs about Evolution and Creation." *Religious Tolerance.org*. Ontario Consultants on Religious Tolerance. 7 May 2007. 8 Dec. 2007 <http://www.religioustolerance.org/ev_publi.htm>.

—. "Trends Among Christian Believers in America." *Religious Tolerance.org*. Ontario Consultants on Religious Tolerance. 18 Oct 2006. 5 Dec 2007 <http://www.religioustolerance.org/chr_tren.htm>.

Rennie, John. "15 Answers to Creationist Nonsense." *Scientific American.com*. July 2007. 8 Dec. 2007 <http://www.sciam.com/article.cfm?articleID=000D4FEC-7D5B-1D07-8E49809EC588EEDF>.

Restivo, Sal. *Science, Society, and Values: Toward a Sociology of Objectivity*. Bethlehem: Lehigh UP, 1994.

Rhodes, H. T. "Science As a Public Trust." *Arts and Sciences Newsletter*. 18.2 (Fall 1997): 1–6.

Richards, I. A. "Towards a Technique for Comparative Study." *Richards on Rhetoric: Selected Essays 1929–1974*. Ed. Ann E. Berthoff, Ann E. Oxford: Oxford UP, 1991. 38–52.

—. *The Philosophy of Rhetoric*. Oxford: Oxford UP, 1965.

Richter, Maurice N., Jr. *Science As a Cultural Process*. Cambridge, MA: Schenkman Pub. Co., 1972.

Robbins, Bruce. "On Being Hoaxed / Draft of article for *Tikkun*." New York University Department of Physics. 23 Sept. 1996. 8 Dec. 2007 <http://physics.nyu.edu/faculty/sokal/robbins_prelim.html>.

—. "Reality and Social Text." *In These Times*. 8 July 1996. *Sokal & Social Text*. jwalsh.net. 24 Nov. 1997. 8 Dec. 2007 < http://jwalsh.net/projects/sokal/articles/rbbns2itt.html>.

Robbins, Bruce, and Andrew Ross. "Editorial Response to Sokal." *Social Text*. 23 Sept. 1996. 23 Sept. 1996 <http://www.designsysy.com/socialtext/sokal.htm>.

Rogers, Thomas F. "Creating the First City on the Moon." *Beyond Earth: The Future of Humans In Space*. Ed. Langston Morris and Kenneth Cox. Burlington, Ontario: Apogee Books, 2006.

Rosen, Ruth. "A Physics Prof Drops a Bomb on the Faux Left." *Los Angeles Times*. 23 May 1996, p. A11.

Ruse, Michael. *The Evolution-Creation Struggle*. Cambridge, MA: Harvard UP, 2005.

Russell, Bertrand. *The Impact of Science on Society*. New York: Simon and Schuster, 1953.

—. *The Problems of Philosophy*. Oxford: Oxford UP, 1997.

Russo, Richard. *Straight Man*. New York: Vintage Books, 1997.

Sagan, Carl. *Cosmos*. New York: Random House, 1980.

—. *The Demon-Haunted World: Science as a Candle in the Dark*. New York: Random House, 1996.

Sartre, Jean-Paul. *Essays in Existentialism*. New York: Kensington, 1993.

Schoen, E. L. "The Roles of Predictions in Science and Religion." *International Journal for Philosophy of Religion.* 29.1 (Feb. 01, 1991): 1+.

"Scientific Misbehavior is Rampant, Study of 3000 Researchers Find." *Chronicle of Higher Education.* (17 June, 2005). 11A

Schopenhauer, Arthur. *Philosophical Writings.* Virginia Cutrufelli trans. New York: Continuum, 1998.

"Scientific American's Circulation Is Red Hot." *Scientific American.* 2007. 7 Dec 2007 <http://www.sciam.com/mediakit/print/index.cfm?section=circulation>.

Scott, Eugenie C. *Evolution vs. Creationism: An Introduction.* Berkeley: U of California P, 2005.

Segal, Judy Z. *Health and the Rhetoric of Medicine.* Carbondale: Southern Illinois UP, 2005.

Searle, John R. *Minds, Brains, and Science.* London: Penguin, 1984.

Sewell, Gilbert T. "The Postmodern Schoolhouse." *Dumbing Down: Essays on the Strip Mining of American Culture.* Ed. Katherine Washburn and John F. Thornton. New York: Norton, 1996. 57–67.

Sherry, John L. "Media Saturation and Entertainment-Education." *Communication Theory.* 12.2 (May 2002): 206–24.

Shulevitz, Judith. "When Cosmologies Colide." *The New York Times.* January 22, 2006.

Simanek, Donald. "What Is Science? What Is Pseudo-Science?" *Donald Simanek's Skeptical Documents and Links.* University of Pennsylvania Lock Haven. Nov. 2005. 8 Dec. 2007 <http://www.lhup.edu/~dsimanek/pseudo/scipseud.htm>.

Silva, Vesta T. "In the Beginning Was the Gene: The Hegemony of Genetic Thinking in Contemporary Culture." *Communication Theory.* 15.1 (Feb. 2005): 100–23.

Segal, Judy, and Alan Richardson. "Scientific *Ethos:* Authority, Authroship, and Trust in the Sciences." *Configurations.* 11.2 (Spring 2003): 137–44.

Segal, Robert. "Paralleling Religion and Science: The Project of Robert Horton." *Annals of Scholarship* 10.2 (1993): 177+.

Segre, Eduardo. "Religion versus Science?" *American Journal of Physics* 62.4 (Apr. 01, 1994): 296+.

Seife, Charles. *Alpha and Omega: The Search for the Beginning and the End of the Universe.* New York: Penguin, 2003.

Selzer, Jack. "Rhetorical Analysis: Understanding How Texts Persuade Readers." *What Writing Does and How It Does It.* Mahwah, NJ: Erlbaum, 2004.

Search for Extra-Terrestrial Intelligence Institute. 2007. 7 Dec 2007 <http://www.seti.org/>.

Sewall, Gilbert T. "The Postmodern Schoolhouse." *Dumbing Down: Essays on the Strip Mining of American Culture.* New York: Norton, 1996.

—. *Zero: The Biography of a Dangerous Idea.* New York: Penguin , 2000.

Simons, Herbert W., and Trevor Melia, eds. *The Legacy of Kenneth Burke.* Madison: U of Wisconsin P, 1989.

Shapin, Steven and Simon Schaffer. *Leviathan and the Air Pump: Hobbes, Boyle, and the Experimental Life.* Princeton, NJ: Princeton UP, 1985.

Shermer, Michael. "A Bounty of Science." *Scientific American.* (February 2004): 33.

—. "The Fossil Fallacy." *Scientific American.* (March 2005): 35.

Shulevitz, Judith. "When Cosmologies Collide." *The New York Times.* (Jan. 22nd, 2006): Editorial Essay section.

Smolin, Lee. "Atoms of Space and Time." *Scientific American.* (Jan. 2004): 67–75.

Soffer, Oren. "The Textual Pendulum." *Communication Theory.* 15.3 (Aug. 2005): 266–91.

Sokal, Alan, and Jean Bricmont. *Fashionable Nonsense: Postmodern Intellectuals' Abuse of Science.* New York: Picador, 1998.

Sokal, Alan "A Physicist experiments with Cultural Studies." *Lingua Franca.* May/June 1996, 62–64. New York University Department of Physics. 23 Sept. 1996. 8 Dec. 2007 <http://www.nyu.edu/ gsas/dept/physics/faculty/sokal/lingua_franca_v4.asci>.

—. "Sokal's Reply to *Social Text* Editorial." New York University Department of Physics. 23 Sept. 1996. 8 Dec. 2007 <http://www.nyu.edu/gsas/dept/physics/faculty/sokal/reply.htm>.

—. "Transgressing the Boundaries: Towards a Transformative Hermeneutics of Quantum Gravity." *Social Text.* Spring/Summer 1996. New York University Department of Physics. 23 Sept. 1996. 8 Dec. 2007 <http://www.nyu.edu/gsas/dept/physics /faculty/sokal/transgress_v2/transgress_v2.htm>.

"Space Flight." *Scientific American.* 186.5 (May 1952): 38.

Sparrow, Bartholomew H. *Uncertain Guardians: The News Media as a Political Institution.* Baltimore: Johns Hopkins UP, 1999.

Stocking S. H., and L. W. Holstein. "Constructing and Reconstructing Scientific Ignorance: Ignorance Claims in Science and Journalism." *Science Communication.* 15.2 (December 1, 1993): 186–210.

Switzer, John. "Unofficial Summary of the Rush Limbaugh Show for Wednesday, May 22 1996." University of Washington. *Sokal & Social Text.* jwalsh.net. 24 Nov 1997. 8 Dec. 2007 <http://jwalsh.net/projects/sokal/articles/rlimbaugh.html>.

Symonds, William C. "America's Failure in Science Education." *Business Week Online.* 16 March 2004. 8 Dec. 2007 <http://www.businessweek.com/technology/content/mar2004/tc20040316_0601_tc166.htm>.

"Tanks of World War II." *OnWar.com.* 8 Dec 2007 <http://www.onwar.com/tanks/index.htm>.

Tanner, William E., and J. Dean Bishop. *Rhetoric and Change.* Arlington, TX: Liberal Arts Press, 1985.

Taylor, Mark C. *The Moment of Complexity: Emerging Network Culture.* Chicago: U of Chicago P, 2001.

Tenner, Edward. *Our Own Devices: How Technology Remakes Humanity.* New York: Vintage Books, 2004.

—. *Why Things Bite Back: Technology and the Revenge of Unintended Consequences.* New York: Knopf, 1996.

The Third Culture. *Edge.* n.d. 26 May 2005 <http://www.edge.org/3rd_culture/bios/smolin.html>.

Tietge, David J. *Flash Effect: Science and the Rhetorical Origins of Cold War America.* Athens: Ohio UP, 2002.

"Transport of Earth, Lunar, and Asteroidal Materials." *Permanent.* 2002. 7 Dec 2007 <http://www.permanent.com/t-overvw.htm>.

Truscello, Michael. "The Clothing of the American Mind: The Construction of Scientific Ethos in the Science Wars." *Rhetoric Review.* 20.3/4 (2001): 329–50.

Twitchell, James B. "'But First, a Word from Our Sponsor': Advertising and the Carnivalization of Culture." *Dumbing Down: Essays on the Strip Mining of American Culture.* Ed. Katherine Washburn and John F. Thornton. New York: W.W. Norton and Co., 1996. 197–208.

—. *Twenty Ads That Shook the World.* Three Rivers, MI: Three Rivers Press, 2001.

Ulam, Stanislaw M. *Science, Computers, and People: From the Tree of Mathematics.* Ed. Mark C. Reynolds and Gian-Carlo Rota. Boston: Birkhauser, 1986.

Ungar, Sheldon. "Knowledge, Ignorance, and Popular Culture: Climate Change Versus the Ozone Hole." *Public Understanding of Science* 9 (2000): 297–312.

US Department of Education. "Scientifically Based Evluation Methods; Notice." *Federal Register.* 70.15 (January 25, 2005): 1–5.

Van Tassel, David D., and Michael G. Hall, eds. *Science and Society in the United States.* Homewood, IL: Dorsey Press, 1966.

Vorderer, Peter, Cristoph Klimmt, and Ute Ritterfeld. "Enjoyment: At the Heart of Media Entertainment." *Communication Theory* 2004 14(4):388–408.

Waddell, Craig. "The Role of Pathos in the Decision-Making Process: A Study in the Rhetoric of Science Policy." *Landmark Essays on the Rhetoric of Science.* Mahwah, NJ: Hermagoras Press, 1997.

Warnick, Barbara. "Online Ethos: Source Credibility in an 'Authorless' Environment." *The American Behavioral Scientist* 48.2 (Oct. 2005): 256–65.

Washburn, Katharine, and John F. Thornton, eds. *Dumbing Down: Essays on the Strip Mining of American Culture.* New York: Norton, 1996.

Weingart, Peter, et al. "Risks of Communication: Discourse on Climate Change in Science, Politics, and the Mass Media." 9 (2000): 261–83.

Weingart, Peter. "Science and the Media." *Research Policy* 27 (1998): 869–79.

Werner, Vivian. *Scientist Versus Society: Six Profiles.* New York: Hawthorn Books, Inc., 1975.

Wertheim, Margaret. "Science and Religion." *Omni* 17.1 (Oct. 01, 1994): 36+.

Wiebe, Donald. "Is Science Really an Implicit Religion?" *Studies in Religion* 18.2 (1989): 171+.

Williams, James D. *Preparing to Teach Writing: Research, Theory, and Practice.* 3rd ed. Mahwah, NJ: Lawrence Erlbaum Associates, Publishers, 2003.

Wilson, Edward O. *Consilience: The Unity of Knowledge.* New York: Vintage, 1998.

—. "Religion, Science, and the Transformation of Knowledge." *Sophia* 32.2 (1993): 36+.

Willems, Jaap. "The Biologist as a Source of Information to the Press." *Bulletin of Science, Technology, and Society* 15.1 (1995): 21+.

Williams, Kipper. "Pile 'Em High." Cartoon. Online Image. http://pinker.wjh.harvard.edu/about/silly/should_i_buy_this_book.gif.

Williamson, Judith. *Decoding Advertisements: Ideas in Progress.* London: Marion Boyars Publishers, 1994.

Winsten, Jay A. "Science and the Media: The Boundaries of Truth." *Health Affairs.* 4.1 (Spring 1985). 9 Sept. 2007 <http://content.healthaffairs.org/cgi/reprint/4/1/5>.

Wise, Jeff. "The Daring Visionaries of Crackpot Aviation." *Popular Science.* Jan 2005. 7 Dec 2007 <http://www.popsci.com/popsci/aviation/article/0,20967,1006774,00.html>.

Wittgenstein, Ludwig. *Philosophical Investigations,* 3rd ed. Trans. G. E. M. Anscombe. New York: Macmillan, 1968.

Wright, Robert. "The Accidental Creationist: Why Stephen Jay Gould is Bad for Evolution." *Nonzero: The Logic of Human Destiny.* 2000. 5 Dec 2007 <http://www.nonzero.org/newyorker.htm>.

Young, Richard E., et al. *Rhetoric: Discovery and Change.* New York: Harcourt, 1970.

Zehr, Stephen C. "Public Representation of Scientific Uncertainty about Global Climate Change." *Public Understanding of Science.* 9.2 (2000): 85-103.

Index

About the Author

David Tietge is Associate Professor of English at Monmouth University, where he teaches courses in rhetorical theory, the rhetoric of science, composition pedagogy, literature, and writing. He has published on scientific rhetoric in The Journal of Technical Writing and Communication and The Journal of Advanced Composition. His earlier book, *Flash Effect: Science and the Rhetorical Origins of Cold War America* (2002, Ohio University Press), examines the role of science on the ideology of American society during the early Cold War era.

Printed in the United States
209046BV00001B/1/P

9 781602 350700